Kishodai Tokyo Kanku

The Climate of Japan

Kishodai Tokyo Kanku

The Climate of Japan

ISBN/EAN: 9783337164089

Printed in Europe, USA, Canada, Australia, Japan

Cover: Foto ©berggeist007 / pixelio.de

More available books at **www.hansebooks.com**

THE CLIMATE

OF

JAPAN.

CENTRAL METEOROLOGICAL OBSERVATORY

OF JAPAN,

TOKIO.

1893.

PREFACE.

The origin of meteorological observations in this country is not very remote, but it is at the beginning of *Meiji*, i. e. about twenty years ago. Though, before that time, some foreigners had taken meteorological observations at Nagasaki, Hakodate, Yokohama, and some other parts, yet the methods of observations and also their instruments were very imperfect. Almost all of their records are scattered away and at present we have no means of collecting them, even their fragments. The first establishment of meteorological stations by our Government is in 1872, at Hakodate in Hokkaido. After the lapse of three years, in June, 1875, another station was opened in Tokio, which is the Central Meteorological Observatory of the present day. With this establishment, the foundation of the meteorological observations in this country was first laid down, and since that time there was a continual developement from year to year. In 1883, we had twenty stations, and at present, the number is increased to fourtyseven. Besides these, temperature stations, and raingauge stations arose successively in several places, and now we have 366 temperature stations, and 272 raingauge stations distributed over the Empire.

All stations are provided with mercurial barometer, wet and dry bulb thermometers, maximum thermometer, minimum thermometer, windvane, Robinson's anemometer, raingauge, and atmometer (evaporation gauge); and in some stations, solar radiation vacuum thermometer, terrestrial radiation thermometer, ozonometer, sunshine recorder, etc. are added to the above mentioned. All the instruments used in our stations, and temperature and raingauge stations, were once tested at well known foreign observatories, and they were again compared to the standard instruments in our Central Observatory, and after minute examinations, the instrumental errors were found for every one of them. The observations, too, are taken following strictly to the Regulations ordered by the Central Observatory, and moreover observers are generally those who were trained in the Central Observatory. Thus there is no doubt that the results of observations in all stations may be equally trusted.

Before 1886, the number of times of observations taken in a day was not numerous, it was generally three times, and at most did not exceed eight times. But from 1886, hourly observations were taken in Tokio; and six observations,—i. e. 2, 6, 10 o'clock a. m. and 2, 6, 10 o'clock p. m.—in other stations. From 1888, in the nine stations Hakodate, Sapporo, Nemuro, Nagano, Osaka, Wakayama, Hiroshima, Matsuyama and

Kumamoto, the hourly observations were taken. Adding the newly established stations, Nagoya and Tadotsu, to the above mentioned nine stations, we have now eleven stations where hourly observations are taken.

Though the observations do not extend for a long time, yet in about one half of our nearly fifty stations, they have been continued for about ten years.

Now I have entrusted the task of collecting the results hitherto obtained, and publishing the Climatology of Japan, to Mr. K. Nakamura, Chief of the Service of Statistics of the Observatory. I sincerely hope, that by this compilation, the general character of our climate may be made known to the public. For minute details of observations in every year, see our monthly and annual reports issued, from time to time, from the Observatory and other stations.

Tokio, January, 1893.

K. Kobayashi

Director of the Central Meteorological Observatory of Japan.

CHAPTER VI.

PRECIPITATION.

CHAPTER VII.

CONCLUSION.

LIST OF DIAGRAMS AND MAPS.

THE CLIMATE OF JAPAN.

INTRODUCTION.

The Empire of Japan lies to the east of the Continent of Asia, and extends from the lat. 24° 6′ N (Hajokan Island, Liukiu) to the 50° 56′ N (Alaid Island in Shimshugori, Chishima) and from the long. 122° 45′ E (Yonakuni Island, Liukiu) to the 156° 32′ E (Shimshu Island in Shimshugori, Chishima). It consists of a chain of islands stretching in the northeasterly direction from the neighbourhood of Formosa to Kamtchatka. The difference of latitudes of the southern and the northern extremities is about 27° and that of longitudes of the eastern and the western extremities is about 34°. The area of the whole Empire is about 382,400 square kilometres.

Of the numerous islands which form this Empire, there are four principal islands, Kiushu, Shikoku, Honshu, and Hokkaido. Of these four, the largest and the most important is the island of Honshu. It has an area of about 58.7 per cent. of that of the whole Empire. The island of Kiushu has an area of 10.5 per cent. of the whole, Shikoku 4.7 per cent., Hokkaido 20.4 per cent., Chishima Islands 4.2 per cent. and Liukiu Islands 0.6 per cent. The area of the other minor islands, summed up, is only 0.9 per cent. Our country is very mountainous. High mountains generally extend their feet out to the very sea coast, and hence there are only small tracts of level lands. Indeed we have scarcely a place properly to be called a plain, except the plain of Kwanto.

There are two main mountain systems traversing through this country: the Chinese system and the Saghalien system. The former runs from west toward east, the latter from northeast to southwest, and they meet together at the central Nippon. They form, so to speak, the back bone of the Island, and are the boundaries of the two slopes, one toward the Sea of Japan, and the other toward the Pacific. These two systems are not generally very high; in the western part, the height scarcely exceeds 1,000 metres and in the northeastern part it is about 1,500 metres. But in the central Nippon, where these two systems meet, we have many high peaks and jagged precipices of considerable heights, and most of them are 2,500—3,000 metres high.

Besides the above mentioned two main systems, there are several other mountain ranges. In the northern part of Nippon, there is the Kitakami range which runs along the coast of the Pacific in southerly direction parallel to the main system. In the central Nippon, there are two mountain ranges branched out from the main systems, and both of them run across the island; the first starts from the eastern part of the province of Shinano, and running southwards reaches to the southern extremity of the peninsula of Izu, while the second starts from the western part of Shinano, and running southwesterly, meets with another range, which comes along the southern coast of the province of Kii in the direction of from east to west. In short, the province of Shinano, the centre of Nippon, is a plateau surrounded completely by high mountains on all sides.

The mountain ranges in Shikoku run generally from east to west, so that the island is divided into the two slopes, one toward *Setouchi* or the Inland Sea, and the other toward the Pacific.

In Kiushu, the main mountain range runs from south to north dividing the island into two parts, the eastern and the western. In Hokkaido, there are three main mountain systems, (1) that extending southwards along the coast of the Sea of Japan, (2) that traversing through the central part in almost south-and-north direction, and (3) and lastly that starting from the northeastern extremity of the island, and running eastwards meet with the (2) at the central part.

Of the currents flowing through the seas around us, the foremost is *Kuroshio.* It first appears at the eastern side of Luzon and Formosa, then flows in the northern direction, and divides itself into two currents at the south of the Liukiu Islands. The main current turns toward northeast, and passes by the southern coasts of Kiushu, Shikoku and Honshu, and then bends slightly toward east at about 38° N. Since this current comes from the tropical region, it is a warm current, 4° or 5° C warmer than the adjacent waters. The branch current keeps its former direction, i. e. flows northward, and touches the western coast of Kiushu, and then passing through the Strait of Corea, it enters into the Sea of Japan. Here it is called the Tsushima Current. It runs in the northeastern direction along the northwestern coast of Nippon, and then passes through the Soya Strait (or La Perouse Strait as it is sometimes called) at the extreme north end of Hokkaido. Thus entering into the Sea of Okhotsk, it seems to reach the eastern coast of the Saghalien Island.

There are two cold currents near us. They are Liman current and *Oyashio.* The former first appears in the Sea of Okhotsk, passes down toward south through the Gulf of Tartary between the Saghalien Island and Siberia. Then flowing along the northern shores of the Sea of Japan, it passes through the Strait of Corea, and enters into the Yellow Sea in China. The latter current, i. e. *Oyashio* comes from Kamtchatka, runs in the southwestern direction along Chishima, and then touches the eastern and the southern coast of Hokkaido and ultimately reaches to the eastern coast of Nippon.

Thus our Empire extends over 27° of latitude, and is everywhere very mountainous, plains being almost absent. Owing to these irregularities of the configuration, it is manifest that our climate also can, by no means, be uniform, it ought to be very different in different parts of the Empire.

CHAPTER I.

AIR TEMPERATURE.

The thermometers, used in all the meteorological stations in Japan, are tested at the Central Meteorological Observatory, and the observations are taken when they are suspended within Stevenson's double louvre-boarded box. The observations of temperature at different stations, therefore, may be considered to have the same weight. Should there be any difference of weight to be attached to the results given in the present paper, it is due wholly to the difference of periods, over which observations extend.

A. DIURNAL VARIATION OF AIR TEMPERATURE.

The meteorological stations, where hourly observations are taken through day and night, are eleven in number. They are Tokio, Kumamoto, Matsuyama, Hiroshima, Ozaka, Wakayama, Nagano, Nagoya, Hakodate, Sapporo, and Nemuro. The mean hourly temperature for each month and year are given in the following tables.*

KUMAMOTO.

Hour \ Month	Jan.	Feb.	Mar.	Apr.	May	June	July	Aug.	Sept.	Oct.	Nov.	Dec.	Year	Wint.	Spr.	Sum.	Aut.
1 am	0,34	4,50	6,95	12,64	14,86	18,71	23,69	23,50	22,08	14,21	8,00	5,55	12,92	3,46	11,48	21,97	11,76
2 am	0,08	4,24	6,66	12,26	14,38	18,28	23,33	23,17	21,69	13,84	7,58	5,29	12,57	3,20	11,10	21,59	11,37
3 am	0,02	3,95	6,34	11,89	13,85	18,01	23,15	22,89	21,46	13,57	7,36	5,01	12,29	2,99	10,69	21,35	14,13
4 am	0,00	3,62	6,17	11,61	13,52	17,74	22,94	22,61	21,15	13,33	7,05	4,85	12,05	2,82	10,43	21,11	13,81
5 am	99,97	3,34	5,97	11,40	13,21	17,52	22,78	22,39	20,93	13,04	6,78	4,60	11,83	2,64	10,19	20,90	13,58
6 am	99,93	3,16	5,81	11,26	13,41	18,04	22,88	22,35	20,80	12,80	6,51	4,46	11,78	2,52	10,16	21,09	13,37
7 am	0,01	3,06	6,08	12,31	15,41	20,04	24,38	24,10	21,99	13,30	6,50	4,40	12,63	2,49	11,27	22,84	13,93
8 am	0,45	4,21	7,91	14,56	17,78	22,03	25,85	26,10	24,52	16,08	8,44	4,92	14,40	3,19	13,42	24,66	16,35
9 am	2,22	6,48	10,10	16,52	19,95	23,41	27,31	27,92	26,45	18,93	11,87	7,16	16,53	5,29	15,54	26,21	19,08
10 am	4,43	8,65	11,75	17,87	21,42	24,58	28,10	29,11	27,82	21,14	14,86	9,63	18,28	7,57	17,01	27,26	21,27
11 am	5,91	10,01	12,94	18,76	22,36	25,50	28,81	29,87	28,66	22,36	16,74	11,77	19,47	9,23	18,02	28,06	22,59
Noon	6,97	10,95	13,91	19,44	22,99	26,06	29,48	30,34	29,11	23,11	17,74	12,80	20,24	10,24	18,78	28,63	23,32
1 pm	7,46	11,39	14,36	19,73	23,47	26,54	29,85	30,83	29,65	23,58	18,31	13,29	20,68	10,71	19,19	29,07	23,75
2 pm	7,64	11,58	14,45	20,04	23,83	26,72	30,18	31,33	29,52	23,83	18,66	13,70	20,96	10,97	19,44	29,41	24,00
3 pm	7,46	11,68	14,58	20,17	23,99	26,72	30,30	31,20	29,45	23,64	18,43	13,65	20,94	10,93	19,58	29,41	23,84
4 pm	7,11	11,40	14,45	19,70	23,61	26,42	30,05	30,87	28,75	23,00	17,71	12,94	20,50	10,48	19,25	29,11	23,15
5 pm	6,05	10,76	13,68	19,00	23,00	25,64	29,43	29,89	27,49	21,50	15,82	11,39	19,47	9,40	18,56	28,32	21,60
6 pm	4,43	9,22	12,28	17,70	21,65	24,67	28,46	28,68	26,02	19,31	13,60	9,82	17,99	7,82	17,21	27,27	19,64
7 pm	3,36	7,95	10,77	16,35	19,81	23,37	27,07	27,06	24,82	17,79	12,14	8,73	16,60	6,68	15,64	25,83	18,25
8 pm	2,66	7,09	9,81	15,34	18,39	22,05	25,93	26,00	24,01	16,85	11,12	7,93	15,60	5,89	14,51	24,66	17,34
9 pm	2,05	6,54	9,20	14,58	17,61	21,33	25,41	25,33	23,50	16,13	10,24	7,21	14,93	5,27	13,80	24,02	16,62
10 pm	1,40	5,84	8,35	13,92	16,75	20,48	24,88	24,71	23,06	15,47	9,36	6,59	14,23	4,61	13,01	23,36	15,96
11 pm	0,97	5,39	7,81	13,49	16,09	19,90	24,35	24,24	22,72	14,89	8,68	6,08	13,72	4,15	12,46	22,83	15,43
M. N.	0,44	4,96	7,36	13,05	15,57	19,81	24,02	23,81	22,30	14,42	8,18	5,74	13,26	3,71	11,99	22,38	14,97
Mean	2,97	7,08	9,90	15,56	18,62	22,21	26,36	26,00	24,90	17,75	11,74	8,23	15,99	6,09	14,69	25,06	18,13

MATSUYAMA.

1 am	2,25	4,41	5,95	10,78	13,49	17,92	22,81	22,98	21,26	13,96	8,80	6,32	12,58	4,34	10,07	21,24	14,67
2 am	1,99	4,23	5,69	10,46	13,18	17,57	22,63	22,79	21,07	13,80	8,57	6,21	12,30	4,14	9,78	21,00	14,48
3 am	2,03	4,12	5,57	10,32	12,76	17,35	22,45	22,61	20,90	13,51	8,38	6,33	12,20	4,16	9,55	20,80	14,25
4 am	2,04	3,88	5,44	10,13	12,35	17,16	22,22	22,39	20,80	13,28	8,41	6,40	12,04	4,11	9,31	20,59	14,16
5 am	1,93	3,73	5,33	9,94	12,30	17,09	22,17	22,21	20,65	13,03	8,26	6,23	11,90	3,96	9,19	20,49	13,98
6 am	2,04	3,47	5,33	9,83	12,73	18,01	22,49	22,24	20,61	12,84	8,21	6,23	12,00	3,91	9,30	20,91	13,89
7 am	1,97	3,36	5,93	11,27	15,54	20,16	24,05	24,30	21,73	13,87	8,32	6,12	13,05	3,82	10,91	22,84	14,64
8 am	2,77	4,81	7,80	13,40	17,88	21,70	25,33	26,22	23,54	16,80	11,07	6,85	14,85	4,81	13,08	24,42	17,14

* To avoid the minus sign, the degrees below the freezing point are shown by adding 100; thus 99° for −1°; 95° for −5°, etc.

MATSUYAMA.

Month / Hour	Jan.	Feb.	Mar.	Apr.	May	June	July	Aug.	Sept.	Oct.	Nov.	Dec.	Year	Wint.	Spr.	Sum.	Aut.
9 am	4.16	6.69	9.42	14.76	19.23	22.50	25.97	27.23	24.91	19.01	13.96	8.97	16.43	6.71	14.47	25.23	19.30
10 am	6.04	8.00	10.65	15.54	20.16	23.40	26.61	27.77	25.71	20.31	15.08	10.81	17.56	8.28	15.45	25.93	20.57
11 am	6.94	8.85	11.36	16.36	20.66	23.88	27.22	28.20	26.29	20.92	16.20	11.74	18.22	9.18	16.13	26.43	21.13
Noon	7.30	9.37	11.58	16.95	21.23	24.37	27.53	28.71	26.65	21.28	16.63	12.06	18.64	9.61	16.59	26.87	21.52
1 pm	7.59	9.64	11.81	17.24	21.36	24.95	28.00	29.09	27.03	21.59	16.97	12.51	18.98	9.91	16.80	27.35	21.86
2 pm	7.49	9.61	12.06	17.35	21.50	25.14	28.11	29.15	27.00	21.58	17.24	12.56	19.07	9.89	16.97	27.47	21.94
3 pm	7.49	9.74	12.16	17.34	21.44	25.12	28.23	29.26	26.83	21.44	17.07	12.33	19.01	9.85	16.98	27.53	21.78
4 pm	7.01	9.54	11.91	17.11	21.02	24.62	28.11	29.07	26.30	20.84	16.32	11.86	18.64	9.47	16.68	27.27	21.15
5 pm	6.15	8.75	11.39	16.22	20.62	24.14	27.50	28.37	25.57	19.32	14.22	10.50	17.73	8.47	16.08	26.67	19.70
6 pm	4.77	7.53	10.22	15.28	19.43	23.23	26.58	27.16	24.23	16.84	12.27	8.85	16.36	7.05	14.98	25.66	17.78
7 pm	4.08	6.44	8.76	13.74	17.46	22.14	25.69	26.01	23.09	15.89	11.63	8.13	15.25	6.22	13.32	24.61	16.87
8 pm	3.53	6.05	8.09	12.79	16.11	20.79	24.55	25.07	22.48	15.28	10.94	7.74	14.45	5.77	12.33	23.16	16.23
9 pm	3.08	5.49	7.56	12.32	15.40	20.03	24.10	24.61	22.15	14.95	10.46	7.27	13.95	5.28	11.76	22.91	15.85
10 pm	2.74	5.11	7.02	11.69	14.70	19.49	23.62	24.04	21.76	14.44	9.90	6.97	13.45	4.94	11.14	22.38	15.37
11 pm	2.48	4.87	6.61	11.29	14.42	18.92	23.32	23.71	21.44	14.02	9.41	6.71	13.10	4.69	10.77	21.98	14.96
M. N.	2.21	4.57	6.25	11.09	13.96	18.55	23.07	23.37	21.21	13.74	8.97	6.45	12.79	4.41	10.43	21.66	14.64
Mean	4.19	6.34	8.50	13.47	17.04	21.18	25.10	25.69	23.47	16.77	11.99	8.59	15.19	6.57	13.00	23.99	17.41

HIROSHIMA.

1 am	1.61	3.49	6.06	10.92	14.29	19.26	23.11	24.57	21.31	14.33	8.44	5.00	12.70	3.37	10.42	22.31	14.69
2 am	1.44	3.23	5.74	10.57	13.81	18.90	22.85	24.22	21.03	13.97	8.16	4.78	12.39	3.15	10.04	21.99	14.39
3 am	1.28	2.99	5.46	10.27	13.39	18.59	22.65	23.91	20.78	13.62	7.90	4.63	12.12	2.97	9.71	21.72	14.10
4 am	1.12	2.81	5.23	10.00	13.04	18.34	22.45	23.66	20.56	13.34	7.63	4.52	11.89	2.82	9.42	21.48	13.84
5 am	1.04	2.67	5.03	9.76	12.72	18.12	22.25	23.45	20.40	13.05	7.43	4.42	11.69	2.71	9.17	21.27	13.63
6 am	0.99	2.54	4.86	9.69	12.70	18.20	22.26	23.36	20.27	12.87	7.25	4.32	11.61	2.62	9.08	21.27	13.46
7 am	0.88	2.47	4.99	10.18	13.64	18.97	22.80	23.97	20.59	12.96	7.18	4.21	11.90	2.52	9.60	21.91	13.58
8 am	1.12	3.04	6.21	11.69	15.52	20.37	23.82	25.42	21.71	14.36	8.19	4.58	13.00	2.91	11.14	23.20	14.75
9 am	2.39	4.43	7.82	13.36	17.61	21.98	24.69	27.11	23.05	16.03	10.08	6.02	14.57	4.28	12.93	24.69	16.39
10 am	3.93	6.12	9.54	14.80	19.12	22.96	25.98	28.39	24.47	18.14	12.22	7.59	16.11	5.88	14.49	25.78	18.28
11 am	5.21	7.57	10.84	15.82	20.13	23.86	26.65	29.28	25.69	20.02	14.07	8.95	17.34	7.21	15.60	26.60	19.93
Noon	6.19	8.47	11.72	16.50	20.83	24.51	27.14	30.02	26.52	21.17	15.51	10.36	18.24	8.34	16.35	27.22	21.07
1 pm	6.60	8.93	12.37	16.89	21.30	24.91	27.62	30.54	26.97	21.78	16.29	11.13	18.78	8.89	16.85	27.69	21.68
2 pm	6.98	9.24	12.68	16.97	21.47	25.11	27.78	30.73	27.08	21.89	16.59	11.46	18.99	9.22	17.04	27.87	21.85
3 pm	6.89	9.20	12.70	16.90	21.64	25.10	27.78	30.57	27.08	21.69	16.42	11.36	18.94	9.15	17.08	27.82	21.73
4 pm	6.41	8.79	12.34	16.45	21.43	24.67	27.56	30.29	26.50	21.12	15.58	10.63	18.50	8.61	16.74	27.51	21.07
5 pm	5.53	7.95	11.63	16.06	20.84	24.16	27.11	29.48	25.85	19.88	14.10	9.49	17.67	7.66	16.18	26.92	19.94
6 pm	4.58	6.99	10.56	15.13	19.99	23.41	26.43	28.58	24.81	18.41	12.63	8.35	16.66	6.64	15.23	26.14	18.63
7 pm	3.85	6.11	9.51	14.16	18.68	22.58	25.70	27.59	23.82	17.39	11.66	7.53	15.72	5.84	14.12	25.29	17.62
8 pm	3.28	5.50	8.82	13.41	17.65	21.81	25.03	26.81	23.18	16.59	10.88	6.90	14.99	5.23	13.29	24.55	16.88
9 pm	2.93	5.06	8.33	12.95	17.05	21.39	24.67	26.44	22.77	16.08	10.37	6.52	14.55	4.84	12.78	24.17	16.41
10 pm	2.43	4.54	7.59	12.30	16.17	20.77	24.18	25.82	22.16	15.38	9.56	5.92	13.90	4.30	12.02	23.59	15.70
11 pm	2.08	4.13	7.12	11.78	15.53	20.27	23.83	25.34	21.71	14.90	8.98	5.54	13.43	3.92	11.18	23.15	15.20
M. N.	1.76	3.81	6.71	11.38	14.98	19.84	23.53	24.94	21.39	14.56	8.63	5.31	13.07	3.63	11.02	22.77	14.86
Mean	3.35	5.43	8.49	13.26	19.23	21.59	24.92	26.85	23.31	16.81	11.07	7.06	14.95	5.28	12.99	24.45	17.06

OZAKA.

1 am	1.97	3.35	6.19	11.37	14.84	19.32	23.84	24.95	21.34	13.99	9.43	5.42	13.00	3.58	10.80	22.70	14.92
2 am	1.76	3.11	5.98	11.01	14.46	19.05	23.66	24.66	21.05	13.87	9.07	5.22	12.74	3.36	10.49	22.46	14.66
3 am	1.58	2.93	5.76	10.81	14.08	18.79	23.48	24.41	20.82	13.46	8.80	5.09	12.50	3.20	10.22	22.23	14.36
4 am	1.42	2.75	5.47	10.53	13.77	18.49	23.31	24.20	20.60	13.28	8.58	4.99	12.28	3.05	9.92	22.00	14.15
5 am	1.23	2.64	5.27	10.28	13.54	18.32	23.15	24.05	20.45	13.06	8.42	4.79	12.10	2.89	9.70	21.84	13.98
6 am	1.07	2.50	5.15	10.28	14.02	18.80	23.42	24.18	20.43	12.88	8.26	4.78	12.15	2.78	9.82	22.13	13.86
7 am	1.14	2.48	5.63	11.34	15.49	20.13	24.38	25.40	21.27	13.46	8.41	4.71	12.82	2.78	10.82	23.30	14.38
8 am	1.81	3.48	7.06	12.90	17.37	21.52	25.50	26.95	23.09	15.30	9.83	5.45	14.18	3.56	12.14	24.66	16.07
9 am	3.28	4.92	8.94	14.12	18.77	22.73	26.41	28.09	24.77	17.52	11.91	7.06	15.73	5.09	14.04	25.74	18.07

OZAKA.

Month / Hour	Jan.	Feb.	Mar.	Apr.	May	June	July	Aug.	Sept.	Oct.	Nov.	Dec.	Year	Wint.	Spr.	Sum.	Aut.
10 am	4,76	6,39	10,24	15,60	19,92	23,66	27.17	28,92	25,87	19,09	13,88	8,74	17,02	6,63	15,25	26,58	19,60
11 am	5,93	7,56	11,08	16,34	20,00	24.18	27,68	29,56	26,63	20,15	15,03	9,85	17,88	7,78	16,01	27,14	20,66
Noon	6,63	8,22	11,85	16,92	21,13	24.71	28,09	30,08	27,04	20,91	15,81	10,61	18,50	8,49	16,63	27,63	21,25
1 pm	6,96	8,58	12,32	17,22	21,45	24.87	28,36	30,32	27,32	21,13	16,22	11,14	18,84	8,89	17,00	27,85	21,62
2 pm	7,09	8,75	12,47	17,30	21,61	24,92	28,49	30,37	27,31	21,42	16,31	11,31	18,96	9,05	17,13	27,93	21,68
3 pm	6,94	8,69	12,21	17,17	21,32	24,85	28,37	30,33	26,97	20,98	15,90	10,99	18,73	8,87	16,90	27,85	21,28
4 pm	6,36	8,10	11,95	16,80	20,90	24,42	28,06	29,91	26,41	20,40	15,08	10,29	18,22	8,25	16,55	27,46	20,63
5 pm	5,47	7,26	11,21	16,04	20,33	23,94	27,54	29,11	25,49	19,11	13,83	9,24	17,38	7,32	15,86	26,87	19,18
6 pm	4,67	6,32	10,03	15,03	19,40	23,26	26,89	28,18	24,36	17,64	12,76	8,38	16,11	6,46	14,82	26,11	18,25
7 pm	4,03	5,67	9,12	14,04	18,19	22,22	25,96	27,12	23,49	16,69	12,00	7,73	15,52	5,81	13,78	25,10	17,39
8 pm	3,55	5,15	8,51	13,36	17,46	21,47	25,41	26,58	22,92	15,96	11,31	7,18	14,90	5,29	13,11	24,19	16,73
9 pm	3,19	4,74	8,01	12,93	16,94	21,04	25,08	26,21	22,45	15,39	10,81	6,77	14,46	4,90	12,63	24,11	16,22
10 pm	2,77	4,35	7,56	12,45	16,39	20,56	24,73	25,80	22,01	14,92	10,32	6,44	14,02	4,51	12,13	23,70	15,76
11 pm	2.48	3,93	7,13	12,03	15,95	20,16	24,47	25,42	21,66	14,53	9,86	6,03	13,61	4,15	11,70	23,35	15,35
M. N.	2,17	3,66	6,71	11,71	15,51	19,82	24,23	25,00	21,37	14,07	9,54	5,69	13,29	3,84	11,31	23,02	14,99
Mean	3,68	5,23	8,58	13,67	17,64	21,71	25,73	27,07	23,55	16,64	11,72	7.41	15,22	5,44	13,30	24,84	17,30

WAKAYAMA.

1 am	2,88	3,55	6,98	12,01	14,91	19,54	24,10	24,48	21,19	14,39	10,14	6,31	13,38	4,25	11,32	22,70	15,21
2 am	2,80	3,40	6,72	11,72	14,56	19,19	23,91	24,19	20,92	14,17	9,84	6,16	13,13	4,12	11,00	22,43	14,98
3 am	2,62	3,19	6,11	11,42	14,21	18,86	23,70	23,99	20,69	13,90	9,62	6,02	12,89	3,91	10,68	22,18	14,74
4 am	2,52	3,00	6,18	11,18	13,90	18,61	23,49	23,80	20,45	13,71	9,36	5,86	12,67	3,79	10,42	21,97	14,51
5 am	2,39	2,90	5,92	10,96	13,75	18,47	23,35	23,56	20,25	13,54	9,20	5,78	12,51	3,69	10,21	21,79	14,33
6 am	2,29	2,84	5,70	10,90	14,18	18,97	23,65	23,76	20,20	13,29	9,07	5,68	12,54	3,60	10,26	22,13	14,19
7 am	2,19	2,78	6,25	11,89	15,67	20,26	24,57	25,09	21,15	13,83	9,10	5,62	13,20	3,53	11,27	23,31	14,69
8 am	2,78	3,56	7,68	13,18	17,47	21,64	25,58	26,57	22,64	15,43	10,41	6,30	14,41	4,21	12,78	24,60	16,16
9 am	4,09	5,00	9,21	14,51	18,79	22,67	26,35	27,85	24,17	17,16	12,30	7,64	15,81	5,58	14,17	25,62	17,88
10 am	5,46	6,22	10,35	15,19	19,65	23,32	26,98	28,54	25,24	18,73	14,05	9,14	16,93	6,91	15,17	26,28	19,31
11 am	6,52	7,17	11,08	16,20	20,29	23,96	27,38	29,12	25,94	19,70	15,07	10,26	17,72	7,98	15,86	26,82	20,21
Noon	7,07	7,75	11,79	16,73	20,75	24,36	27,82	29,56	26,85	20,30	15,70	11,03	18,27	8,61	16,42	27,25	20,78
1 pm	7,42	8,06	12,15	17,16	21,08	24,16	28,01	29,66	26,53	20,71	16,10	11,37	18,56	8,95	16,80	27,38	21,12
2 pm	7,53	8,17	12,48	17,33	21,25	24,52	28,03	29,82	26,60	20,91	16,23	11,54	18,70	9,08	17,02	27,46	21,25
3 pm	7,41	8,20	12,56	17,27	21,21	24,48	28,01	29,66	26,48	20,81	16,05	11,38	18,63	9,00	17,01	27,38	21,11
4 pm	6,99	7,88	12,31	16,93	20,82	24,18	27,77	29,29	26,16	20,39	15,55	10,91	18,27	8,59	16,70	27,08	20,70
5 pm	6,23	7,36	11,69	16,40	20,29	23,66	27,23	28,67	25,50	19,41	14,43	10,04	17,58	7,88	16,13	26,52	19,79
6 pm	5,49	6,50	10,84	15,56	19,47	23,13	26,61	27,88	24,18	18,17	13,17	9,15	16,73	7,05	15,29	25,88	18,70
7 pm	4,93	5,79	9,94	14,74	18,40	22,24	25,82	26,92	23,61	17,17	12,68	8,47	15,89	6,40	14,36	24,99	17,83
8 pm	4,47	5,25	9,32	14,17	17,68	21,59	25,31	26,35	23,03	16,44	12,07	8,09	15,31	5,94	13,72	24,42	17,18
9 pm	4,15	4,82	8,82	13,70	17,19	21,17	25,11	25,98	22,60	15,95	11,53	7,71	14,89	5,56	13,24	24,09	16,69
10 pm	3,81	4,41	8,35	13,18	16,55	20,71	24,87	25,48	22,10	15,42	11,05	7,32	14,41	5,19	12,69	23,69	16,19
11 pm	3,43	4,11	7,90	12,77	16,00	20,34	24,60	25,49	21,70	14,93	10,57	6,91	14,03	4,82	12,22	23,34	15,73
M. N.	3,13	3,77	7,49	12,42	15,57	19,97	24,30	24,77	21,28	14,57	10,29	6,63	13,69	4,51	11,83	23,04	15,38
Mean	4,52	5,24	9,09	14,08	17,65	21,68	25,69	26,67	23,30	16,80	12,24	8,14	15,43	5,97	13,61	24,68	17,15

NAGANO.

1 am	96,97	98,93	1,94	8,02	10,89	16,52	20,32	22,06	18,34	10,16	4,87	2,33	9,27	99,41	6,95	19,63	11,12
2 am	96,80	98,76	1,80	7,85	10,50	16,17	20,05	21,86	18,05	9,76	4,48	1,89	8,99	99,15	6,72	19,36	10,76
3 am	96,51	98,53	1,64	7,57	10,12	15,82	19,81	21,61	17,81	9,50	4,25	1,71	8,74	98,92	6,41	19,08	10,52
4 am	96,16	98,29	1,44	7,25	9,82	15,56	19,56	21,39	17,11	9,18	4,11	1,47	8,47	98,64	6,17	18,84	10,24
5 am	95,97	98,23	1,22	6,92	9,51	15,47	19,34	21,25	17,20	8,95	3,91	1,38	8,28	98,53	5,89	18,69	10,02
6 am	95,75	97,95	1,03	7,17	10,19	16,40	19,85	21,42	17,25	8,76	3,78	1,06	8,38	98,25	6,13	19,22	9,93
7 am	95,66	97,91	1,72	8,14	11,47	17,57	20,75	22,63	18,31	9,18	3,82	1,09	9,01	98,22	7,14	20,32	10,54
8 am	96,47	99,05	3,01	9,43	12,85	18,96	21,84	23,93	19,64	10,83	4,98	1,73	10,23	99,08	8,41	21,58	11,82
9 am	97,61	0,26	4,38	10,76	14,29	20,37	22,85	25,09	20,85	12,50	6,41	2,96	11,53	0,28	9,79	22,77	13,26
10 am	98,52	1,30	5,87	12,00	15,62	21,73	23,89	26,34	22,05	13,98	7,84	3,94	12,76	1,25	11,16	23,99	14,62

NAGANO.

Month / Hour	Jan.	Feb.	Mar.	Apr.	May	June	July	Aug.	Sept.	Oct.	Nov.	Dec.	Year	Wint.	Spr.	Sum.	Aut.
11 am	99,40	2,30	6,92	13,21	16,88	22,64	24,96	27,45	23,15	15,12	9,00	1,91	13,83	2,20	12,34	25,02	15,76
Noon	0,11	3,15	7,73	14,20	17,76	23,48	25,71	28,29	23,93	16,17	9,80	5,89	14,69	3,06	13,23	25,83	16,63
1 pm	0,82	3,64	8,30	14,76	18,10	24,08	26,26	28,70	24,32	16,76	10,52	6,57	15,26	3,68	13,82	26,35	17,20
2 pm	1,04	3,87	8,44	15,05	18,63	24,06	26,47	28,58	24,40	16,81	10,69	6,80	15,41	3,90	14,04	26,37	17,30
3 pm	0,85	3,75	8,11	14,62	18,32	23,95	26,12	28,09	23,73	16,34	10,35	6,71	15,03	3,77	13,68	25,85	16,81
4 pm	0,16	3,03	7,47	13,91	17,39	22,73	25,31	27,06	22,75	15,18	9,34	5,87	14,18	3,02	12,92	25,03	15,76
5 pm	99,16	1,89	6,43	12,87	16,46	21,84	24,52	26,07	21,32	13,72	8,10	4,76	13,09	1,94	11,92	24,14	14,38
6 pm	98,71	1,16	5,19	11,64	15,44	20,83	23,76	25,10	20,29	12,93	7,31	4,18	12,22	1,36	10,76	23,23	13,51
7 pm	98,38	0,61	4,31	10,64	14,31	19,81	22,78	24,25	19,64	12,24	6,66	3,91	11,46	0,97	9,75	22,28	12,85
8 pm	98,10	0,29	3,85	10,15	13,38	18,85	22,16	23,62	19,20	11,66	6,33	3,60	10,93	0,66	9,13	21,54	12,40
9 pm	97,91	0,11	3,49	9,71	12,79	18,24	21,65	23,29	18,85	11,25	5,97	3,39	10,56	0,47	8,66	21,06	12,02
10 pm	97,60	99,74	3,06	9,13	12,20	17,78	21,29	22,84	18,55	10,88	5,59	3,16	10,15	0,17	8,13	20,64	11,67
11 pm	97,35	99,44	2,73	8,86	11,77	17,38	20,90	22,59	18,39	10,65	5,24	2,87	9,85	99,89	7,79	20,29	11,43
M. N.	97,06	99,16	2,50	8,56	11,36	17,03	20,67	22,35	18,30	10,27	4,88	2,58	9,56	99,60	7,47	20,02	11,15
Mean	98,05	0,47	4,28	10,51	13,76	19,45	22,53	24,40	20,15	12,21	6,59	3,53	11,33	0,68	9,52	22,13	12,98

TOKIO.

1 am	0,85	1,55	5,23	10,68	14,09	18,60	22,33	23,74	20,80	14,12	8,55	3,14	11,97	1,85	10,00	21,56	14,19
2 am	0,54	1,25	4,88	10,36	13,76	18,35	22,13	23,44	20,58	13,81	8,29	2,81	11,68	1,53	9,67	21,31	14,23
3 am	0,25	1,04	4,57	10,03	13,45	18,06	21,96	23,19	20,40	13,57	8,06	2,56	11,43	1,28	9,35	21,07	14,01
4 am	0,06	0,78	4,26	9,75	13,18	17,85	21,82	22,97	20,21	13,38	7,75	2,45	11,20	1,10	9,06	20,88	13,78
5 am	99,86	0,56	3,95	9,49	13,02	17,84	21,76	22,77	20,06	13,17	7,49	2,19	11,01	0,87	8,82	20,79	13,57
6 am	99,61	0,31	3,83	9,61	13,66	18,46	22,32	23,23	20,13	13,02	7,29	1,92	11,12	0,61	9,03	21,34	13,48
7 am	99,46	0,40	4,49	10,56	14,68	19,42	23,26	24,27	20,91	13,65	7,57	1,80	11,71	0,55	9,91	22,32	14,04
8 am	0,58	1,74	5,95	11,78	15,95	20,48	24,34	25,58	22,03	14,93	9,04	3,26	12,97	1,86	11,23	23,47	15,33
9 am	2,46	3,31	7,41	13,01	17,12	21,48	25,29	26,72	23,12	16,31	10,72	5,42	14,37	3,73	12,51	24,50	16,73
10 am	4,08	4,67	8,56	14,08	18,02	22,22	26,02	27,67	24,02	17,50	12,16	7,24	15,52	5,33	13,55	25,30	17,89
11 am	5,37	5,79	9,61	14,85	18,64	22,80	26,48	28,27	24,62	18,49	13,23	8,62	16,40	6,59	14,37	25,85	18,78
Noon	6,23	6,70	10,38	15,50	19,09	23,21	26,97	28,77	25,07	19,18	14,06	9,57	17,06	7,50	14,99	26,22	19,44
1 pm	6,96	7,33	10,86	15,85	19,42	23,42	27,26	29,04	25,31	19,51	14,64	10,21	17,48	8,17	15,37	26,57	19,83
2 pm	7,26	7,63	11,05	16,01	19,50	23,41	27,31	29,10	25,38	19,64	14,85	10,44	17,63	8,44	15,52	26,61	19,96
3 pm	7,12	7,61	11,00	15,89	19,36	23,20	27,08	28,91	25,16	19,49	14,57	10,17	17,46	8,30	15,42	26,40	19,74
4 pm	6,31	7,18	10,53	15,35	19,02	22,81	26,70	28,38	24,59	18,80	13,58	9,15	16,87	7,56	14,97	25,96	18,99
5 pm	5,04	6,25	9,72	14,70	18,38	22,22	26,10	27,64	23,83	17,69	12,17	7,58	15,94	6,29	14,27	25,32	17,90
6 pm	4,01	5,14	8,80	13,78	17,49	21,46	25,28	26,61	22,94	16,83	11,40	6,62	15,03	5,26	13,36	24,45	17,06
7 pm	3,32	4,47	8,10	13,04	16,64	20,67	24,43	25,73	22,46	16,20	10,78	5,88	14,31	4,56	12,59	23,61	16,48
8 pm	2,80	3,89	7,50	12,61	16,08	20,20	23,89	25,26	22,08	15,73	10,18	5,26	13,79	3,98	12,06	23,12	16,00
9 pm	2,27	3,36	7,01	12,17	15,69	19,84	23,51	24,80	21,75	15,28	9,73	4,78	13,36	3,47	11,62	22,75	15,59
10 pm	1,79	2,86	6,62	11,75	15,34	19,53	23,18	24,57	21,39	14,90	9,34	4,34	12,97	3,00	11,24	22,43	15,21
11 pm	1,48	2,39	6,23	11,40	14,99	19,26	22,93	24,27	21,09	14,53	9,00	3,95	12,62	2,59	10,87	22,15	14,87
M. N.	1,07	1,98	5,83	11,12	14,64	18,98	22,66	23,98	20,83	14,21	8,59	3,49	12,28	2,18	10,53	21,87	14,54
Mean	2,86	3,67	7,35	12,64	16,30	20,57	24,38	25,79	22,45	16,00	10,54	5,54	14,01	4,02	12,10	23,58	16,33

HAKODATE.

1 am	94,56	96,67	1,11	4,10	8,12	12,20	16,61	18,78	15,82	7,86	3,94	0,50	6,71	97,24	4,54	15,86	9,21
2 am	94,53	96,82	0,94	4,29	7,99	11,97	16,48	18,67	15,48	7,80	3,81	0,38	6,60	97,24	4,41	15,71	9,05
3 am	94,49	96,67	0,90	4,06	7,67	11,89	16,32	18,57	15,39	7,61	3,66	0,35	6,47	97,17	4,21	15,59	8,90
4 am	94,37	96,47	0,61	3,86	7,25	11,80	16,25	18,39	15,26	7,48	3,35	0,32	6,28	97,05	3,91	15,48	8,70
5 am	94,44	96,57	0,40	3,84	7,61	12,11	16,41	18,33	15,18	7,35	3,34	0,20	6,31	97,07	3,95	15,62	8,62
6 am	94,28	96,39	0,43	4,34	8,85	12,99	17,12	19,03	15,40	7,24	3,39	0,15	6,63	96,94	4,54	16,38	8,67
7 am	94,28	96,49	1,17	5,58	10,27	13,77	17,95	20,30	16,97	8,75	3,53	0,04	7,12	96,94	5,67	17,34	9,75
8 am	94,64	97,44	2,28	6,89	11,68	14,49	18,57	21,28	18,41	11,18	4,88	0,35	8,51	97,48	6,95	18,11	11,49
9 am	95,08	98,66	3,25	7,84	12,54	15,15	19,19	22,18	19,45	12,84	5,86	1,29	9,49	98,54	7,88	18,84	12,72
10 am	96,47	99,60	4,05	8,40	13,19	15,65	19,76	22,84	20,22	14,04	6,88	2,04	10,26	99,37	8,55	19,42	13,71
11 am	97,10	0,27	4,51	8,87	13,70	16,06	20,25	23,17	20,66	14,58	7,45	2,64	10,77	0,00	9,03	19,83	14,23

HAKODATE.

Month / Hour	Jan.	Feb.	Mar.	Apr.	May	June	July	Aug.	Sept.	Oct.	Nov.	Dec.	Year	Wint.	Spr.	Sum.	Aut.
Noon	97.39	0.63	4.78	9.20	13.94	16.27	20.58	23.56	20.87	14.93	7.69	2.91	11.06	0.31	9.31	20.14	11.50
1 pm	97.53	0.80	4.94	9.31	14.24	16.52	20.78	23.73	20.86	15.12	7.59	2.94	11.20	0.42	9.51	20.34	14.52
2 pm	97.45	0.59	4.93	9.37	14.24	16.65	20.90	23.73	20.75	14.92	7.45	2.78	11.14	0.27	9.51	20.43	14.37
3 pm	97.28	0.36	4.58	9.08	13.86	16.46	20.69	23.46	20.42	14.52	7.14	2.63	10.87	0.09	9.17	20.20	14.03
4 pm	96.73	99.84	4.14	8.63	13.38	16.08	20.21	23.00	20.08	13.79	6.39	2.21	10.37	99.59	8.72	19.76	13.42
5 pm	96.26	99.20	3.43	7.99	12.71	15.54	19.60	22.38	19.35	12.61	5.65	1.71	9.70	99.06	8.01	19.17	12.54
6 pm	95.87	98.69	2.78	7.40	11.98	14.97	19.07	21.81	18.49	11.42	5.09	1.39	9.08	98.65	7.39	18.62	11.67
7 pm	95.54	98.33	2.40	6.66	11.03	14.21	18.38	20.88	17.60	10.64	4.95	1.25	8.49	98.37	6.70	17.82	11.05
8 pm	95.19	97.84	2.12	6.05	10.13	13.67	17.81	20.22	17.24	9.89	4.65	1.01	8.01	98.02	6.20	17.24	10.60
9 pm	94.88	97.29	1.86	5.76	9.94	13.24	17.50	19.89	16.84	9.01	4.31	0.94	7.62	97.70	5.85	16.88	10.05
10 pm	94.63	97.17	1.63	5.34	9.33	13.06	17.17	19.52	16.40	8.32	4.21	0.79	7.30	97.53	5.43	16.58	9.64
11 pm	94.57	96.99	1.60	5.19	9.01	12.66	17.01	19.24	16.13	8.14	3.86	0.65	7.08	97.40	5.27	16.30	9.38
M. N.	94.57	97.14	1.38	4.96	8.55	12.39	16.83	19.00	15.89	7.92	3.73	0.56	6.91	97.41	4.96	16.07	9.18
Mean	95.53	98.29	2.51	6.55	10.90	14.16	18.39	20.92	17.88	10.75	5.12	1.25	8.51	98.33	6.65	17.82	11.25

SAPPORO.

1 am	91.09	93.37	98.67	2.78	7.47	11.86	16.05	17.71	14.35	5.80	1.46	97.66	4.86	94.04	2.97	15.21	7.20
2 am	90.90	93.20	98.44	2.53	7.06	11.62	15.82	17.53	14.01	5.38	1.20	97.62	4.61	93.91	2.68	14.99	6.86
3 am	90.89	93.13	98.09	2.53	6.75	11.57	15.68	17.38	13.86	5.25	0.97	97.45	4.46	93.82	2.46	14.88	6.69
4 am	90.79	93.07	97.86	2.33	6.54	11.18	15.48	17.24	13.60	5.26	0.92	97.38	4.33	93.75	2.24	14.73	6.59
5 am	90.61	92.85	97.80	2.22	6.88	11.88	15.78	17.25	13.38	5.25	0.79	97.41	4.34	93.63	2.30	14.97	6.47
6 am	90.36	92.92	97.70	3.05	8.47	13.20	16.99	18.31	13.92	5.35	0.74	97.38	4.87	93.55	3.07	16.17	6.67
7 am	90.21	93.09	98.60	4.52	10.39	14.68	18.52	20.10	15.66	6.84	1.07	97.51	5.93	93.60	4.50	17.77	7.86
8 am	91.12	94.12	0.16	5.99	11.94	15.95	19.87	21.30	17.39	9.11	2.35	98.02	7.27	94.12	6.02	19.04	9.62
9 am	92.67	95.67	1.56	7.23	13.29	17.01	20.77	22.93	18.80	11.08	3.83	99.03	8.66	95.79	7.36	20.24	11.24
10 am	94.02	97.06	2.57	7.96	14.34	18.02	22.13	23.90	19.75	12.32	4.92	99.96	9.75	97.01	8.29	21.35	12.33
11 am	94.92	97.98	3.08	8.44	15.07	18.65	22.79	24.76	20.42	13.19	5.67	0.53	10.46	97.81	8.86	22.07	13.09
Noon	95.41	98.40	3.40	8.70	15.37	19.08	23.08	25.22	20.74	13.41	5.97	0.93	10.81	98.25	9.16	22.46	13.37
1 pm	95.51	98.57	3.41	8.67	15.12	19.04	23.19	25.38	20.59	13.39	5.92	1.00	10.84	98.36	9.17	22.54	13.30
2 pm	95.56	98.55	3.33	8.70	15.14	18.86	23.00	24.99	20.33	13.20	5.59	0.85	10.67	98.32	9.06	22.28	13.04
3 pm	95.27	98.31	3.01	8.14	14.51	18.21	22.39	24.36	19.86	12.58	4.98	0.10	10.20	97.99	8.66	21.65	12.47
4 pm	94.41	97.75	2.56	8.08	13.59	17.54	21.76	23.67	19.20	11.66	3.96	99.56	9.48	97.24	8.08	20.99	11.61
5 pm	93.50	96.95	1.83	7.34	12.66	16.81	20.90	22.96	18.38	10.06	3.04	98.86	8.61	96.44	7.28	20.22	10.49
6 pm	92.84	96.23	0.87	6.33	11.61	16.02	19.94	21.75	17.05	8.79	2.45	98.52	7.70	95.86	6.27	19.24	9.43
7 pm	92.35	95.61	0.32	5.39	10.32	14.82	18.74	20.53	16.30	8.07	2.02	98.25	6.89	95.40	5.34	18.03	8.80
8 pm	91.97	95.27	99.81	4.79	9.61	14.00	17.88	19.73	15.75	7.56	1.86	97.95	6.35	95.06	4.75	17.20	8.39
9 pm	91.81	95.02	99.55	4.42	9.19	13.38	17.31	19.22	15.35	7.27	1.74	97.93	6.02	94.92	4.39	16.65	8.12
10 pm	91.66	94.52	99.23	4.06	8.76	12.91	16.95	18.67	15.11	6.79	1.42	97.69	5.65	94.62	4.02	16.47	7.77
11 pm	91.52	94.37	99.07	3.67	8.22	12.63	16.66	18.36	14.69	6.35	1.40	97.71	5.39	94.53	3.65	15.88	7.48
M. N.	91.38	93.99	99.05	3.23	7.92	12.27	16.36	18.03	14.42	5.96	1.37	97.75	5.14	94.37	3.40	15.55	7.25
Mean	92.53	95.42	0.42	5.17	10.86	15.06	19.09	20.90	16.79	8.75	2.73	98.64	7.22	95.53	5.58	18.35	9.42

NEMURO.

1 am	93.27	93.69	97.80	2.04	5.43	8.32	12.85	16.77	15.03	8.91	3.61	99.24	4.74	95.40	1.76	12.65	9.18
2 am	93.31	93.46	97.72	1.96	5.30	8.28	12.80	16.74	14.99	8.74	3.53	99.09	4.66	95.29	1.66	12.61	9.09
3 am	93.27	93.22	97.61	1.98	5.26	8.17	12.72	16.63	15.01	8.62	3.34	99.06	4.57	95.18	1.62	12.51	8.90
4 am	93.18	93.04	97.41	1.75	5.25	8.26	12.68	16.59	14.92	8.52	3.32	98.91	4.48	95.03	1.48	12.51	8.92
5 am	93.03	92.98	97.18	1.91	5.75	8.76	13.12	16.80	14.79	8.42	3.21	98.90	4.57	94.97	1.61	12.89	8.81
6 am	92.95	92.97	97.43	2.62	6.56	9.40	13.78	17.11	15.31	8.53	3.12	98.85	4.91	94.92	2.20	13.53	8.99
7 am	92.97	93.34	98.08	3.59	7.11	10.16	14.51	18.11	16.28	9.51	3.59	98.84	5.54	95.05	3.04	14.26	9.80
8 am	93.55	94.18	98.94	4.17	8.20	10.87	15.26	18.88	17.02	10.64	4.57	99.61	6.35	95.78	3.87	15.00	10.74
9 am	94.42	95.27	96.67	5.32	8.93	11.41	16.02	19.68	17.69	11.34	5.31	0.46	7.12	96.72	4.64	15.70	11.45
10 am	94.96	95.86	0.29	5.91	9.54	11.88	16.65	20.29	18.24	12.03	5.75	1.11	7.71	97.31	5.25	16.27	12.01
11 am	95.33	96.40	0.69	6.17	9.08	12.26	17.08	20.56	18.63	12.38	6.16	1.58	8.07	97.75	5.51	16.63	12.39
Noon	95.56	96.68	0.77	6.35	9.74	12.18	17.08	20.80	18.69	12.54	6.39	1.83	8.24	98.02	5.62	16.79	12.54

NEMURO.

Hour \ Month	Jan.	Feb.	Mar.	Apr.	May	June	July	Aug.	Sept.	Oct.	Nov.	Dec.	Year	Wint.	Spr.	Sum.	Aut.
1 pm	95,71	96,71	0,77	6,01	9,91	12,11	17,05	20,66	18,66	12,57	6,31	1,85	8,22	98,09	5,56	16,71	12,51
2 pm	95,62	96,71	0,55	5,87	9,70	12,28	16,97	20,30	18,30	12,10	5,92	1,66	8,02	98,00	5,37	16,52	12,21
3 pm	95,33	96,15	0,35	5,11	9,14	11,73	16,54	19,91	17,79	11,80	5,28	1,00	7,57	97,62	4,98	16,06	11,62
4 pm	94,67	95,97	99,88	4,76	8,39	11,16	15,96	19,35	17,08	10,96	4,62	0,32	6,93	96,99	4,34	15,19	10,89
5 pm	94,32	95,39	99,22	4,01	7,56	10,30	15,22	18,58	16,22	10,22	4,25	0,08	6,28	96,60	3,61	14,70	10,23
6 pm	94,22	95,06	98,83	3,25	6,71	9,61	14,32	17,60	15,58	9,90	3,98	0,07	5,76	96,45	2,94	13,84	9,82
7 pm	94,20	95,04	98,61	2,83	6,29	9,01	13,71	17,19	15,28	9,71	3,96	99,95	5,18	96,40	2,59	13,31	9,65
8 pm	94,02	94,95	98,56	2,69	6,05	8,85	13,40	17,00	15,19	9,57	3,79	99,86	5,33	96,28	2,13	13,08	9,52
9 pm	93,89	94,85	98,45	2,65	5,98	8,70	13,36	16,92	15,05	9,40	3,72	99,75	5,22	96,16	2,35	12,99	9,39
10 pm	93,65	94,56	98,23	2,38	5,72	8,50	13,20	16,88	14,89	9,30	3,56	99,69	5,04	95,97	2,11	12,86	9,25
11 pm	93,45	94,32	98,07	2,29	5,55	8,14	13,07	16,85	14,94	9,09	3,51	99,53	4,93	95,77	2,00	12,79	9,18
M. N.	93,40	93,99	98,03	2,21	5,63	8,50	13,05	16,84	14,98	8,98	3,40	99,44	4,87	95,61	1,96	12,79	9,12
Mean	94,09	94,80	98,88	3,69	7,24	9,99	14,60	18,23	16,27	10,17	4,34	0,03	6,03	96,31	3,27	14,27	10,26

The numbers given in the above tables are obtained by observations extending over three to six years. The numb r of years observed is not sufficiently great, so that there may be great departures from truth in the absolute values. But we are sure that the manner of diurnal variation is fairly given by the above tables and that there will be no great difference to the result which would be found by observations continued for many years. To show the legitimacy of this supposition, we give the differences of the hourly mean temperatures in the following table, which is calculated from the observations continued for six years in Tokio.

Difference of temperature from preceding hours

Yearly mean at Tokio.

Year \ Hour	1 am	2 am	3 am	4 am	5 am	6 am	7 am	8 am	9 am	10 am	11 am	Noon	1 pm	2 pm	3 pm	4 pm	5 pm	6 pm	7 pm	8 pm	9 pm	10 pm	11 pm	M.N.
1886	0,34	0,35	−0,28	−0,26	0,21	+0,17	+0,50	+1,21	+1,56	+1,21	+0,95	+0,62	+0,51	+0,13	−0,14	−0,34	0,96	−0,93	0,78	−0,55	−0,39	−0,36	0,34	−0,33
1887	0,34	−0,33	−0,25	−0,24	0,19	+0,12	+0,60	+1,26	+1,17	+1,13	+0,91	+0,72	+0,31	+0,08	−0,18	−0,57	−0,92	−0,90	−0,70	−0,51	−0,11	0,37	−0,36	−0,36
1888	−0,35	−0,29	−0,30	−0,27	0,20	+0,11	+0,63	+1,29	+1,15	+1,21	+0,90	+0,65	+0,11	+0,13	−0,16	−0,58	−0,91	−0,91	−0,72	−0,51	−0,15	−0,41	−0,35	−0,35
1889	0,27	−0,25	−0,21	−0,21	0,21	+0,06	+0,51	+1,18	+1,31	+1,11	+0,86	+0,68	+0,38	+0,18	−0,23	−0,52	−0,87	−0,87	0,69	0,51	−0,14	−0,36	0,33	−0,37
1890	−0,24	−0,21	−0,21	−0,22	0,15	+0,11	+0,55	+1,18	+1,19	+1,03	+0,80	+0,66	+0,37	+0,11	−0,11	−0,53	−0,85	−0,83	0,64	0,45	0,42	−0,38	−0,35	−0,32
1891	−0,39	−0,30	0,28	−0,20	0,20	+0,11	+0,66	+1,10	+1,53	+1,21	+0,89	+0,61	+0,56	+0,25	−0,18	0,70	−1,01	1,01	0,82	−0,58	−0,47	−0,42	−0,36	−0,33
1892	−0,31	−0,29	−0,26	−0,23	0,19	+0,14	+0,59	+1,26	+1,10	+1,17	+0,88	+0,66	+0,42	+0,15	−0,17	0,79	0,93	0,91	0,72	−0,52	−0,43	−0,30	−0,35	−0,31
Mean	−0,32	0,21	−0,25	0,23	0,19	+0,11	+0,58	+1,25	+1,42	+1,16	+0,88	+0,66	+0,43	+0,15	0,17	0,59	0,92	0,91	−0,72	−0,52	−0,43	−0,34	−0,35	−0,31

This table shows us that the difference of temperature at any time of observation from that of the next succeeding is nearly constant in every year. Therefore we may say that the diurnal variations of temperature in Tokio throughout the whole year, have been pretty well determined. In order to find how for the degree of accuracy is attained, let us form the deviations of the mean temperatures of each ye r from their means of the six years. They are:—

Deviations from the mean of seven years observations in $^{1}/_{100}$ C^{o}.

Year \ Hour	1 am	2 am	3 am	4 am	5 am	6 am	7 am	8 am	9 am	10 am	11 am	Noon	1 pm	2 pm	3 pm	4 pm	5 pm	6 pm	7 pm	8 pm	9 pm	10 pm	11 pm	M.N.
1886	−2	−6	−2	−3	−2	+6	−8	−4	+14	+5	+7	−4	+8	−2	+3	−5	−1	−2	−6	−8	+4	+3	+1	+1
1887	−2	−1	+1	+2	0	+1	−2	+1	+5	−3	+3	+6	−9	−7	−1	+2	0	+1	+2	+1	−1	+2	1	2
1888	−3	0	−1	−4	−1	0	−5	+1	+3	+8	+2	−1	−2	−2	+1	+1	+1	0	0	2	−2	−5	0	−1
1889	−5	+1	+5	+2	−2	−5	−7	7	−8	−2	−2	+2	−5	+3	−6	+7	+5	+4	+3	+1	+2	+3	+2	3
1890	+8	+5	+2	+1	+4	0	−3	−7	−23	−13	−8	0	−6	−4	+3	+6	+7	+8	+8	+7	+1	+1	0	+2
1891	−7	−1	−2	+3	−1	0	+8	+15	+11	+5	0	−2	+13	+10	−1	−11	−9	−10	−10	6	4	−8	−1	+1
1892	+1	0	0	0	0	0	+1	+1	−2	+1	0	0	−1	0	0	0	0	−1	0	0	0	0	0	0
$\sqrt{\frac{\Sigma\Delta\Delta}{6\times5}}$	±1,9	±1,5	±1,1	±1,0	±0,8	±1,2	±2,3	±2,9	±4,7	±2,7	±1,8	±1.2	±3,0	±2,1	±1,2	±2,1	±2,0	±2,1	±2,3	±1,6	±1,0	±1,2	±0,4	±0,7

Thus in the mean of the six years, we see that the mean diurnal variation of temperature of the whole year in Tokio is found true to $\frac{1}{100}$ C°. Even if we take the month of February, in which temperature difference from year to year is most great, we can say without any gross mistakes that the diurnal variation deduced from the mean of the six years' observations is true to $\frac{5}{100}$ C°; the probable error in February is in $\frac{1}{100}$ C°.

Hour	1 am	2 am	3 am	4 am	5 am	6 am	7 am	8 am	9 am	10 am	11 am	Noon	1 pm	2 pm	3 pm	4 pm	5 pm	6 pm	7 pm	8 pm	9 pm	10 pm	11 pm	M.N.
	±7	±6	±3	±3	±3	±5	±2	±5	±7	±6	±5	±3	±5	±5	±3	±4	±2	±3	±3	±4	±4	±2	±3	±3

From the above stated, we see that the diurnal variation of temperature in Tokio is shown sufficiently accurate by the mean of the six years' observations, and there is no doubt that the hourly observations continued for only three or four years can show the diurnal variation with the accuracy of one tenth of a degree.

The diurnal variation of temperature presents one maximum and one minimum (See Plate I). The times of their occurrences are given in the following table. (The times given in this table are local times, and not the standard times of our Empire, which is generally the case in the rest of the present volume).

Time of the occurrence of

	Minimum.					Maximum.				
	Win.	Spr.	Sum. AM	Aut.	Year	Wint.	Spr.	Sum. PM	Aut.	Year
	h m	h m	h m	h m	h m	h m	h m	h m	h m	h m
Kumamoto	6.52	5.25	5.19	6.07	5.33	2.13	2.43	2.13	1.58	2.13
Matsuyama	6.51	5.36	5.06	6.11	5.36	1.51	2.41	2.41	1.36	2.11
Hiroshima	7.08	5.50	5.32	6.32	5.56	2.15	2.32	2.20	2.20	2.10
Ozaka	6.42	5.22	5.17	6.08	5.47	2.02	1.52	2.02	1.42	2.02
Wakayama	7.09	5.12	5.15	6.18	5.50	2.20	2.30	2.10	2.00	2.18
Nagano	6.58	5.39	5.21	6.17	5.47	2.23	2.13	1.43	1.49	2.13
Tokio	6.59	5.31	5.19	6.07	5.50	2.25	2.24	1.51	2.14	2.19
Hakodate	6.53	4.43	4.24	5.53	5.08	1.23	1.53	2.23	1.08	1.43
Sapporo	6.55	4.50	4.10	5.45	5.10	1.55	0.55	1.10	0.25	1.10
Nemuro	6.42	4.57	4.12	6.10	5.12	1.42	0.42	0.52	1.12	1.12

It is not easy to determine accurately the time when the temperature attains its maximum or minimum value. For, when it approaches the maximum or the minimum value, its variation is so slow that for some thirty or fourty minutes, it remains in nearly the same value, and there is no appreciable difference to be seen. Hence if there be any small error incurred in the observation, and the form of the curve be slightly modified, then the time of the maximum or the minimum temperature will deviate considerably from truth. The above table shows only the general feature, and can be said in no way to be very accurate. If, however, we group our stations into two parts, and take the mean of each of these parts separately, we shall be able to obtain more trustworthy values. Now the Observatory of Tokio, and the stations of Kumamoto, Matsuyama, Hiroshima, Ozaka, Wakayama, Nagano and Nagoya lie on the southern part of Japan. Group them into one and call it the "southern part". The other three stations, Hakodate, Sapporo and Nemuro, lie on the northern extremity of our country. Let them be grouped into "the northern part". The times of the maximum and the minimum temperatures are:—

	Minimum.					Maximum.				
	Wint.	Spr.	Sum. AM	Aut.	Year	Wint.	Spr.	Sum. PM	Aut.	Year
	h m	h m	h m	h m	h m	h m	h m	h m	h m	h m
Southern Part	6.56	5.34	5.18	6.16	5.46	2.13	2.16	2.09	1.57	2.12
Northern Part	6.50	4.50	4.32	5.56	5.10	1.40	1.10	1.28	0.55	1.22

Thus the time of the minimum temperature depends on the season, that is to say, on the length of the day. It is latest in winter and earliest in summer, and generally just preceeds the sunrise. On the other hand, the time of the maximum temperature does not depend at all on the length of the day. It is nearly constant throughout the year, that is, about two hours after the culmination of the sun. The times of both the maximum and the minimum temperatures of the northern part preceeds those of the southern part by about half an hour.

Since there is one maximum of temperature in a day, there must be two times when the temperature is equal to the mean value of that day. They are given in the following table.

	Forenoon.					Afternoon.				
	Wint. h m	Spr. h m	Sum. h m	Aut. h m	Year h m	Wint. h m	Spr. h m	Sum. h m	Aut. h m	Year h m
Kumamoto	9.03	8.19	7.58	8.21	8.25	7.25	7.33	7.18	6.50	7.19
Matsuyama	8.39	7.49	7.31	7.58	8.06	6.39	7.09	7.21	6.11	6.53
Hiroshima	9.30	8.55	8.47	9.12	9.05	7.45	8.20	8.08	7.35	7.56
Ozaka	9.17	8.39	8.12	8.35	8.38	7.47	7.11	7.22	7.10	7.32
Wakayama	9.20	8.35	8.03	8.45	8.42	7.48	8.15	7.25	7.33	7.53
Nagano	9.38	9.01	8.41	9.03	9.03	8.07	7.33	7.23	6.58	7.28
Tokio	9.29	9.01	8.25	9.01	9.04	8.19	8.19	7.19	7.37	7.55
Hakodate	9.09	8.08	7.56	8.13	8.23	7.38	7.59	7.23	7.01	7.20
Sapporo	9.13	8.10	7.50	8.18	8.25	7.10	7.07	7.10	6.30	7.04
Nemuro	9.12	8.02	7.48	8.10	8.18	8.42	6.12	6.12	8.00	6.10
Southern part	9.25	8.36	8.14	8.42	8.43	7.11	7.50	7.28	7.08	7.34
Northern part	9.11	8.07	7.51	8.14	8.22	7.50	7.06	6.55	7.10	6.50

Thus in forenoon, the temperature at some instant lying between half past eight and half past nine is equal to the mean value, while in afternoon this occurs at about half past seven. Here, as in the cases of the maximum and the minimum temperatures, the northern part preceeds the southern. We see that the period during which the temperature remains higher than the mean value is a little shorter than that of lower temperature.

The amplitude of the diurnal variation depends on the locality and also on the season. The smallest amplitude is 2.°8 C while the greatest is 12.°3 C. With regard to different localities, we see that it gets smaller as we go from the southwestern part of this country to the northeastern. With regard to different months, we see that the maximum amplitude is in October and May, and the minimum in July. However, Tokio is an exception to this rule. The maximum amplitude in Tokio is in December, as the weather is most clear there in this month. For the detail, see the next table.

MEAN DAILY AMPLITUDE OF TEMPERATURE.

Locality.	Jan.	Feb.	Mar.	April	May	June	July	Aug.	Sept.	Oct.	Nov.	Dec.	Year
	°	°	°	°	°	°	°	°	°	°	°	°	°
Kumamoto	7.7	8.6	8.8	8.9	11.0	9.2	7.5	9.0	8.7	11.0	12.2	9.3	9.2
Matsuyama	5.7	6.4	6.8	7.5	9.2	8.0	6.1	7.0	6.4	8.7	9.0	6.4	7.2
Hiroshima	6.1	6.7	7.8	7.3	8.9	7.0	5.5	7.4	6.8	9.0	9.4	7.2	7.4
Ozaka	6.0	6.3	7.3	7.0	8.1	6.6	5.3	6.3	6.7	8.5	8.0	6.6	6.8
Wakayama	5.3	5.4	6.9	6.4	7.5	6.0	4.7	6.3	6.4	7.6	7.2	5.9	6.2
Nagano	5.4	6.0	7.4	8.1	9.1	8.6	7.1	7.4	7.2	8.0	6.9	5.7	7.1
Tokio	7.8	7.4	7.3	6.4	6.5	5.8	5.5	6.1	5.4	6.8	7.6	8.5	6.6
Hakodate	3.2	4.4	4.5	5.5	7.0	4.8	4.6	5.4	5.7	7.9	4.3	2.9	4.9
Sapporo	5.3	5.7	5.7	6.5	8.9	7.6	7.7	8.1	7.4	8.2	5.2	3.6	6.5
Nemuro	2.8	3.7	3.6	4.6	4.7	4.3	4.4	4.2	3.9	4.1	3.3	3.0	3.8

The close examination of this table shows us that the amplitude is small along the coast, but large in inland regions.

If we consider amplitudes of different seasons, the greatest is in autumn and the smallest in winter. Tokio, however, is an exception to this for the same reason as mentioned before. There the maximum is in winter, and the minimum in summer. Again if we consider amplitudes in different localities, we see that in the southern part, the maximum is in spring and autumn, and the minimum in winter and summer; while in the northern part, maximum is in summer and the minimum in winter.

	Winter.	Spring.	Summer.	Autumn.
Kumamoto	8.5	9.4	8.5	10.6
Matsuyama	6.1	7.8	7.0	8.0
Hiroshima	6.7	8.0	6.6	8.4
Ozaka	6.3	7.4	6.1	7.8
Wakayama	5.5	6.8	5.7	7.1
Nagano	5.7	8.1	7.7	7.4
Tokio	7.9	6.7	5.8	6.5
Hakodate	3.5	5.6	4.9	5.9
Sapporo	4.8	6.9	7.8	6.9
Nemuro	3.2	4.1	4.3	3.7
Southern part	6.6	7.7	6.8	8.0
Northern part	3.8	5.5	5.7	5.5

Since, we have mentioned before, the period during which the temperature remains higher than the mean value of the day is shorter than that of the lower temperature, it is clear that the excess of the higher temperature over the mean can not be equal to the defect of the lower temperature. The higher temperature must differ from the mean farther than the lower. This fact is clearly shown in the following table.

{Maximum-Mean}—{Mean-Minimum}.

	Winter.	Spring.	Summer.	Autumn.	Year.
Kumamoto	1.28	0.36	0.19	1.11	0.76
Matsuyama	0.99	0.17	0.04	1.01	0.59
Hiroshima	1.18	0.18	0.24	1.19	0.70
Ozaka	0.95	0.23	0.09	0.94	0.61
Wakayama	0.67	0.01	0.11	0.54	0.35
Nagano	0.76	0.89	0.80	1.27	1.03
Tokio	0.93	0.13	0.23	0.79	0.63
Hakodate	0.70	0.12	0.27	0.64	0.46
Sapporo	0.85	0.25	0.57	1.00	0.73
Nemuro	0.37	0.56	0.76	0.83	0.66

When observations are taken a few times a day, not hourly, some corrections must be applied to the direct result of observations in order to deduce the values more approximate to the truth. These temperature corrections suitable to our country are found to be as follows:—

Corrections to be applied to the mean of

	Jan.	Feb.	Mar.	April	May	June	July	Aug.	Sept.	Oct.	Nov.	Dec.	Year
$\frac{2^a+6^a+10^a+2^p+6^p+10^p}{6}$	−0.02	+0.03	+0.04	+0.04	+0.00	+0.00	+0.01	+0.02	+0.04	+0.05	−0.01	0.02	+0.02
$\frac{6^a+2^p+10^p}{3}$	−0.06	−0.03	+0.18	+0.18	−0.10	+0.10	+0.11	+0.16	+0.15	+0.15	+0.05	−0.03	+0.10
$\frac{\text{Maximum}+\text{Minimum}}{2}$	0.36	−0.30	−0.06	−0.17	−0.00	−0.27	0.33	0.06	−0.36	−0.51	−0.51	0.31	−0.31

This shows us that when six observations are taken in a day, viz. at 2, 6, and 10 o'clock am and 2, 6, and 10 o'clock pm then the direct result of observations nearly coincides throughout the year with that of hourly observations, and consequently no corrections are required. But if the observations are taken three times a day, viz. at 6 o'clock am and 2 and 10 o'clock pm then though the direct result nearly coincides with that of hourly observations in winter, yet it is lower by 0.°2 C than that of hourly observations during eight months from March to October. The mean temperature deduced from the mean of the readings of maximum and minimum thermometers is always higher than the mean obtained by hourly observations; that is, about 0.°5 C higher toward the end of autumn and the beginning of winter, and about 0. 3 C higher in the average of the whole year. In spring it nearly coincides with the true values.

B. ANNUAL VARIATION OF AIR TEMPERATURE.

As we have shown in the preceding article, six observations a day give the mean temperatures nearly coinciding with those of hourly observations. Consequently, it seems that the monthly mean temperatures of all stations may at once be compared with those found by hourly observations without applying any correction. But here one thing must be noticed, namely, the number of years of observations differ very widely, in some stations temperature observations have been continued over fifteen years, while in others for only two years. Now, the monthly mean temperatures do not remain constant every year, but change considerably year to year. The consequence is that they can not be directly compared without removing those errors arising from the difference of periods of observation.

It is obvious that when the period of observations is short, we can not obtain accurate values. The number of years requisite to obtain a determinate accuracy of the result differs for every country. Thus to find the mean temperature of winter with the accuracy of 0.°1 C, about 400 years are required for Vienna and 800 years for West Siberia; while for the mean temperature of summer, the same accuracy can be obtained with only 100 years for both places. On the other hand, for the mean temperature of Batavia, 5 years are already sufficient to obtain the same accuracy (*Handbuch der Klimatologie von Prof. Dr. Hann*) Now we have calculated, from the observations of sixteen years in Tokio, the number of years for giving the monthly mean temperature with errors less than 0.°1 C. They are:—

REQUIRED NUMBER OF YEARS.*

Jan.	Feb.	March	April	May	June	July	Aug.	Sept.	Oct.	Nov.	Dec.	Year
70	180	140	70	100	100	100	65	130	50	90	210	30

Thus the required number of years differs very much for each month. The mean temperature of October can be found true to 0.°1 C by observations of about fifty years, while for December more than two hundred years are required to obtain the same degree of accuracy. The mean temperature of the year can be found equally accurately by observations of thirty years. In the case of the four seasons, about fifty years are sufficient for summer and autumn, while from sixty to eighty years are required for winter and spring for the same accuracy of 0.°1 C, as shown in the next table:—

* This is calculated in the following manner. Let Δ be the difference between the mean of sixteen year's observations and the mean monthly temperature of each year. Then the mean error of observation E of each year is

$$E = \pm \sqrt{\frac{\Sigma\Delta\Delta}{n}}$$

Then the required number of years x is obtained by

$$0.1 = \frac{E}{\sqrt{x-1}}$$

Season.	Winter.	Spring.	Summer.	Autumn.
Number of Years.	80	62	43	45

The actual observations hitherto made in our country fall quite short of the above numbers. Even the longest is only sixteen or seventeen years, and the shortest is not longer than a year, so that the result will be far from being true. But if corrections are applied to the observations of those stations, where the observed period is short, so as to give the mean of ten years, by comparing them with those of neighbouring stations, then we are sure that the monthly mean will be true to 0.°4 C and the annual mean to 0.°2 C. (In the tables given hereafter, the aforesaid corrections for giving the mean of ten years are applied to short period observations.)

The temperature varies by the height of the station. Now, in our country, the observations hitherto made, to which we can refer for the information about this change of temperature with the height, are very small in number. If we enumerate temperature observations undertaken in high localities, we have (1) meteorological observations on Mt. Gozaishodake in the province of Ise, by Mr. H. Masato in 1888 (the 1st, September—the 3rd, October), (2) on Mt. Hoben in the province of Nagato by Mr. H. Tonno in 1889 (the 1st, August—the 31st, October), (3) on Mt. Fuji by Messrs. K. Nakamura, and H. Kondo, in the same year (the 30th, July—the 7th, Sept.), and lastly (4) on Mt. Ontake in the province of Shinano by Messrs. H. Masato, and H. Kondo in 1891 (the 1st August—the 11th, September). These observations are made on heights varying from 800 metres to 3,700 metres above the sea level, but the observed periods are not quite long enough, and they are all in mid-summer. For this reason, though the temperature gradient at July and August, due to different heights, is known, yet that in autumn, winter and spring remains quite unknown. According to the above mentioned observations, the temperature gradient during July and August is about 0.°6 C per 100 metres. Now this nearly coincides with the result obtained by Wild by the observations in Russia. We have therefore made direct use of his result given in "*Temperaturverhältnis des Russischen Reichs von Wild*" by which we have reduced the mean of the observed values of all stations to the sea-level. One thing must be noticed here. Almost all of our stations, with the exceptions of Nagano, Yamagata, Kamikawa and few others lie on heights not exceeding sixty or seventy metres above the sea level. Hence the error, if any, arising from using Wild's numbers must be very small.

The next table shows the monthly mean temperatures in our country. The numbers attached to the names of stations are the numbers of years over which observations extend. (For those stations, where the observed years are less than 10, the aforesaid corrections are applied, and for all stations, the corrections for reducing to the sea level are applied by Wild's numbers.)

MEAN AIR TEMPERATURE.

Locality.		Jan.	Feb.	March	April	May	June	July	Aug.	Sept.	Oct.	Nov.	Dec.	Year
Naha	(2)	16.5	15.5	16.5	20.0	22.6	21.4	27.7	27.6	25.5	21.1	19.8	16.1	21.4
Kagoshima	(9)	6.8	7.3	10.9	16.0	18.8	22.4	26.2	26.6	21.3	19.1	13.4	8.8	16.7
Miyazaki	(9)	6.7	7.1	10.7	16.0	19.0	22.4	25.9	26.5	23.8	18.7	13.0	8.6	16.5
Kochi	(10)	5.3	6.2	9.7	15.2	18.4	21.7	25.2	26.0	23.6	18.5	12.4	7.4	15.8
Wakayama	(13)	4.3	4.7	7.9	13.6	17.5	21.6	25.9	26.9	23.5	17.1	11.6	7.1	15.2
Oita	(5)	5.4	5.3	8.0	13.1	16.7	21.0	25.0	25.8	22.9	17.8	11.9	7.6	15.0
Yamaguchi	(4)	2.7	3.6	6.7	12.2	16.0	20.6	24.4	25.7	22.2	15.7	9.6	5.4	13.7
Hiroshima	(13)	3.5	4.4	7.3	12.8	17.1	21.1	25.4	26.8	23.0	16.9	10.6	5.7	14.5
Matsuyama	(2)	4.0	4.4	7.3	12.9	16.4	21.1	25.4	25.9	22.3	16.8	11.2	6.8	14.5
Okayama	(3)	3.0	3.5	7.0	12.6	16.8	20.9	25.7	26.7	23.1	16.6	10.5	5.5	14.3
Ozaka	(10)	3.5	3.7	7.3	13.0	17.1	21.5	25.8	26.9	23.3	17.2	11.0	6.2	14.7
Kioto	(12)	2.5	3.1	6.4	12.6	16.7	21.4	25.5	26.6	22.8	16.1	9.7	4.6	14.0
Kumamoto	(2)	3.8	5.0	9.0	14.4	18.2	21.8	26.1	26.7	23.6	17.5	10.9	6.2	15.3
Saga	(2)	4.2	5.0	8.6	14.2	18.0	21.8	25.4	26.4	23.2	17.3	11.4	6.5	15.2

Mean air temperature.

Locality.		Jan.	Feb.	March	April	May	June	July	Aug.	Sept.	Oct.	Nov.	Dec.	Year
Nagasaki	(14)	5,8	6,6	9,6	14,9	18,5	22,0	26,3	27,2	24,1	18,8	12,5	7,7	16,1
Fukuoka	(2)	4,7	4,7	7,7	13,0	16,3	20,6	24,9	26,1	22,3	16,3	11,1	6,7	14,5
Isugahara	(6)	4,2	4,8	7,9	12,8	16,8	20,5	24,3	25,9	22,2	17,2	11,2	6,6	14,5
Akamagaseki	(9)	5,2	5,1	8,2	13,0	16,8	20,7	24,6	26,3	23,2	18,0	12,4	7,9	15,1
Sakai	(9)	3,5	3,5	6,9	11,8	15,8	20,3	24,5	26,1	22,2	16,5	11,0	6,5	14,0
Tsu	(4)	3,7	3,7	6,9	12,6	16,8	20,8	24,9	26,2	22,5	16,6	10,9	6,1	14,3
Nagoya	(2)	2,6	3,1	6,8	12,6	16,8	21,3	25,4	26,4	22,7	16,3	10,1	5,5	14,1
Gifu	(11)	2,5	3,4	6,9	12,8	17,0	21,4	25,5	26,6	22,7	16,5	10,3	5,5	14,2
Hamamatsu	(19)	4,5	5,0	8,3	13,7	17,2	21,0	24,7	26,1	23,1	17,6	11,9	7,3	15,0
Numazu	(10)	4,8	5,1	8,5	13,5	16,9	20,8	24,4	25,6	22,7	17,2	11,8	7,6	14,9
Tokio	(17)	2,7	3,6	7,1	12,3	16,6	20,4	24,4	25,6	22,1	15,8	10,0	5,3	13,9
Utsunomiya	(2)	0,3	1,7	5,7	10,8	15,9	19,9	23,8	24,9	21,4	14,7	7,9	3,3	12,6
Choshi	(6)	5,2	5,7	8,8	13,1	16,8	19,9	23,4	25,4	23,2	18,1	12,7	7,8	15,0
Kanazawa	(10)	2,2	2,1	5,4	11,1	15,3	19,8	24,1	25,7	21,8	15,6	9,8	5,4	13,1
Fushiki	(7)	1,6	1,7	5,5	10,5	14,7	19,1	23,6	25,7	21,8	15,8	10,3	5,4	13,0
Nagano	(3)	99,5	0,8	5,1	11,7	16,4	21,2	25,3	26,5	22,2	14,4	7,2	2,6	12,8
Niigata	(11)	1,4	1,5	4,7	10,2	14,7	19,0	23,5	26,0	21,9	15,4	9,5	4,4	12,7
Yamagata	(3)	98,4	98,1	2,9	9,5	14,9	19,2	23,3	24,5	20,4	13,0	6,4	1,8	11,0
Akita	(19)	98,1	98,4	2,3	8,1	12,9	17,5	22,2	23,7	19,5	12,5	6,7	2,0	10,3
Fukushima	(3)	0,3	0,0	4,4	10,4	15,1	18,8	23,1	24,3	20,3	13,1	7,4	2,8	11,7
Ishinomaki	(11)	99,4	0,1	3,7	8,8	13,2	17,2	21,7	23,7	20,3	13,9	7,7	3,0	11,1
Miyako	(9)	98,9	99,5	2,7	7,9	11,9	15,3	20,2	22,3	18,8	12,5	7,2	2,7	9,9
Aomori	(10)	96,9	97,4	0,8	6,7	11,3	15,8	20,7	22,7	18,4	11,9	5,5	0,1	9,0
Hakodate	(19)	97,2	97,9	1,4	6,0	10,3	14,2	19,3	21,2	17,8	11,3	5,1	0,2	8,5
Suttsu	(8)	96,1	97,4	1,1	5,9	10,2	13,8	18,8	21,2	17,4	11,4	4,6	99,9	8,1
Sapporo	(15)	94,1	95,3	99,1	4,9	10,4	14,9	19,7	21,2	16,3	9,3	2,5	97,3	7,1
Kamikawa	(4)	90,2	92,1	96,9	3,5	10,2	15,4	20,7	20,9	15,3	7,4	0,1	95,5	5,7
Soya	(7)	94,2	94,8	98,6	3,8	7,1	10,6	17,0	19,7	16,4	9,8	2,2	97,2	6,0
Abashiri	(3)	93,0	93,0	97,7	4,0	8,4	11,8	18,2	20,4	16,2	9,4	2,6	97,3	6,0
Nemuro	(13)	95,3	95,0	98,3	3,0	6,9	10,1	15,3	18,5	15,9	10,5	4,4	99,5	6,0
Kushiro	(3)	90,6	92,6	97,7	3,3	8,0	12,3	17,4	19,8	15,8	8,1	1,2	95,1	5,1
Erimo	(6)	97,7	97,7	99,8	3,4	7,3	11,7	16,2	18,7	17,4	12,3	5,8	0,9	7,4
Ogasawarajima	(3)	18,0	16,8	18,8	21,2	23,7	27,0	28,1	27,2	27,0	26,0	23,2	20,5	23,1

N. B. The temperature below the freezing point is represented by adding 100 to; thus 95° for −5°, 90° for −10° &c.

The mean temperature is highest, in August, not in July, as may be supposed. In the extreme south of our country July and August have nearly the same mean temperature, but as we go to higher latitudes, the difference of their mean temperatures gets greater,—that is, in the south of Kiushu Island the mean temperature of August is higher than that of July by only 0.°6 C, but in Hokkaido, this difference is more than 2.°5 C. The differences of the mean temperatures of the two months for every one degree of latitude, from 26° N to 45° N are given in the next table.

Latitude.	26°–27°	27°–28°	28°–29°	29°–30°	30°–31°	31°–32°	32°–33°	33°–34°	34°–35°	35°–36°	36°–37°	37°–38°	38°–39°	39°–40°	40°–41°	41°–42°	42°–43°	43°–44°	44°–45°
Temperature of Aug. is higher than that of July by	0	—	—	—	—	0.5	0.7	1.0	1.2	1.3	1.5	1.8	1.6	1.8	2.0	2.2	2.4	2.3	2.4

Thus throughout the country, the mean temperature of August is higher than that of July, and their difference increases proportionally to the increase of latitude. This may be accidental, yet still a remarkable fact!

Generally speaking, the mean temperature is lowest in January, but in certain cases, it is lowest in February If we calculate the difference of the mean temperatures of these two months, we find that, on the average, in January it is lower by 0.°47 C than in February. The difference being such a small quantity quite comparable with errors, from which the mean temperatures of winter months can not be free, on accounts of the period of observations being short, we can not tell beforehand how it will become in future. Hence, it will be safer for us to say that the lowest mean temperature is in January and February.

Let us now examine the manner of variation of the monthly mean temperature. It is minimum, as we have just seen, in January and February. It then increases with the rate of about 4° C per month, and reaches its maximum value in August. The decrease from August to September is not so very quick, being about 3.°5 C. But the transitions from September to October, and from October to November are very quick, the rate of decrease per month being about 6° C. In short, at the increasing stage, the rate is very slow, (though from March to April, the rate is somewhat great, being 5.°3 C) and a long period of seven months is required. But at the decreasing stage, the change is very quick, and consequently the period is short, requiring only five month.

If we calculate the mean temperature at each season, we get:

MEAN AIR TEMPERATURE FOR FOUR SEASONS.

Locality.	Winter	Spring	Summer	Autumn	Locality.	Winter	Spring	Summer	Autumn
Naha	16.1	19.7	26.6	23.1	Numazu	5.8	13.0	23.6	17.2
Kagoshima	7.5	15.2	25.1	19.0	Tokio	3.9	12.0	23.5	16.0
Miyazaki	7.5	15.2	24.9	18.5	Utsunomiya	1.8	10.7	22.9	14.7
Kochi	6.3	14.2	24.1	18.0	Choshi	6.2	12.9	22.9	18.0
Wakayama	5.5	13.0	24.8	17.5	Kanazawa	3.2	10.6	23.2	15.7
Oita	6.1	12.6	23.9	17.5	Fushiki	2.9	10.2	22.8	16.0
Yamaguchi	3.8	14.5	23.4	15.7	Nagano	0.9	11.1	24.3	14.6
Hiroshima	4.5	12.4	24.4	16.8	Niigata	2.5	9.9	22.8	15.6
Matsuyama	5.1	12.2	24.1	16.8	Yamagata	99.4	9.1	22.3	13.3
Okayama	4.0	12.1	24.1	16.8	Akita	99.5	7.8	21.1	12.9
Ozaka	4.5	12.5	24.7	17.2	Fukushima	1.0	10.0	22.1	13.6
Kioto	3.2	11.9	24.5	16.2	Ishinomaki	0.8	8.6	20.9	14.0
Kumamoto	5.0	13.9	24.9	17.3	Miyako	0.3	7.5	19.3	12.8
Saga	5.2	13.6	24.5	17.3	Aomori	98.1	6.3	19.7	11.9
Nagasaki	6.7	14.3	25.2	18.4	Hakodate	98.4	5.9	18.2	11.4
Fukuoka	5.4	12.3	23.9	16.6	Suttsu	97.8	5.7	17.9	11.1
Itsugahara	5.2	12.5	23.6	16.9	Sapporo	95.6	4.8	18.6	9.4
Akamagaseki	6.1	12.7	23.9	17.9	Kamikawa	92.6	3.5	19.0	7.2
Sakai	4.5	11.5	23.6	16.6	Soya	95.1	3.1	15.8	9.5
Tsu	4.5	12.1	24.0	16.7	Abashiri	91.4	3.4	16.8	9.4
Nagoya	3.7	12.1	24.4	16.4	Nemuro	96.6	2.7	14.6	10.3
Gifu	3.8	12.3	24.5	16.5	Kushiro	92.8	3.0	16.5	8.3
Hamamatsu	5.6	13.1	23.9	17.5	Erimo	98.6	3.5	15.5	14.8

N. B. The temperature below the freezing point are represented by adding 100 to: thus 95° for —5°, 90° for —10° &c.

Thus, autumn is far warmer than spring and throughout the country, the difference of temperatures between spring and winter is nearly equal to that between summer and autumn.

The range of mean temperatures between the warmest and the coolest months is given in the following table:—

Locality.	Amplitude.	Locality.	Amplitude.	Locality.	Amplitude.
Naha	12.2	Itsugahara	21.7	Akita	25.6
Kagoshima	19.8	Akamaga-eki	21.0	Fukushima	23.8
Miyazaki	19.8	Sakai	22.6	Ishinomaki	24.2
Kochi	20.6	Tsu	22.4	Miyako	23.4
Wakayama	22.6	Nagoya	23.8	Aomori	25.8
Oita	20.4	Gifu	24.0	Hakodate	24.0
Yamaguchi	22.8	Hamamatsu	21.5	Suttsu	25.1
Hiroshima	23.3	Numazu	20.7	Sapporo	27.1
Matsuyama	21.8	Tokio	22.9	Kamikawa	30.5
Okayama	23.8	Utsunomiya	24.3	Soya	25.5
Ozaka	23.1	Choshi	20.2	Abashiri	27.3
Kioto	24.0	Kanazawa	23.4	Nemuro	23.1
Kumamoto	22.9	Fushiki	24.1	Kushiro	29.1
Saga	22.2	Nagano	26.0	Erimo	20.9
Nagasaki	21.2	Niigata	24.6	Ogasawarajima	11.3
Fukuoka	21.1	Yamagata	25.7	Fusan	22.5

Thus in Bonin Island and Okinawa, which lie on the ocean and their latitudes are low, the temperature range is very small, i. e. 9°—12° C; but it increases rapidly as we proceed toward north, that is to say, into inland. Indeed, the range reaches to about 20° C even in the southern part of Kiushu Island, and in the central part of Hokkaido, it is more than 30°. (See Plate XII).

C. DISTRIBUTION OF TEMPERATURE OVER JAPAN.

In order to make the temperature distribution over this country more intelligible to us, we have drawn isotherms showing the mean temperatures of the four seasons and the year. (See the Plate X). For the drawing of these lines, it is very necessary for us to know meteorological conditions on the opposite continent, i. e. of Corea, Siberia, and the eastern coast of China. For these informations, we made use of Wild's "*Die Temperaturverhältniss des Russischen Reichs*", Fritsche's "*The climate of eastern Asia*", and "*Bulletins mensuels de l'observatoire magnétique et météorologique de Zikawei*", W. Doberck's "*Observations made at the Hongkong observatory*" and lastly Wild's "*Annalen des physikalischen Centralobservatoriums zu St. Petersbourg*". We have felt great difficulty in drawing these isotherms, owing to the absence of the observations on the sea. The only observations on the wide Pacific Ocean are those at Naha and Bonin Island; and the area covered by the Sea of Japan is far greater than that of our country, and unhappily here we have nearly no informations to refer for. Therefore, we are obliged to draw the isotherms on the sea by considering carefully the general course of isotherms on land, assuming the uniformity of the temperature distributions over the sea, which is generally the case, as there are no local disturbances on the sea as on the land. We are sure that we have made no sensible mistakes here by making this assumption.

The following conclusions will be drawn by the close examinations of the isothermal maps.

(1) The air temperature generally decreases as we proceed toward north, though the influence, which the distribution of land and sea produces upon it, is more remarkable than that of the change of latitude.

(2) During winter, all isotherms on the land turn their convexity toward southwest; while, on the contrary, the isotherms on the sea, especially on the Sea of Japan, turn their convexity toward northeast. Thus, during winter, the land is colder than the sea, and it is manifest that the Tsushima current in the Sea of Japan warms our northwestern coast. One of the remarkable features of the isotherms during winter is that they are very close together, so that even on the Ocean, temperature

variation is about 1° C for every 10 kilometres, and in Hokkaido and Siberia it is about 5° C for every 10 kilometres.

(3) Just contrary is the case in summer. The isotherms of the land are convex toward north, and those of the sea toward south. Or in other words, the land is warmer than the sea. The cold Liman current flows through the Sea of Japan southward, and the cold current Oyashio flows along our eastern coast and comes down so far south as to Choshi. Thus those places on our eastern coast are very much colder than those places, on the same latitude, facing the Sea of Japan.

(4) In spring, there is no great difference of temperature between land and sea. In southwestern part, isotherms are almost parallel to the parallels of latitudes. As we proceed eastward they bend slightly to the north, and from about 38° N, they begin to turn to the south, and still farther north this inclination is more and more increased, so that in Hokkaido, they run nearly from NNW to SSE. Thus as in summer, the eastern coast is colder than the western.

(5) In autumn, again there is a difference of temperature between land and sea:—it is lower on land than on sea. The isotherms have close resemblance to those of winter, except that the space rate of the change of the temperature is slower than in winter.

(6) In the mean of the year, we find nearly no difference between land and sea in southern provinces, though inland regions are rather colder than the sea coast. But in those places lying north of 37° N, the difference is somewhat considerable, being most remarkable in Hokkaido, and also in Hokkaido, the western coast is very much warmer than the eastern.

We see that the temperature range over our country, arising whether from the difference of latitude or from the distribution of land and sea, is greatest in winter, least in summer, and spring and autumn lie intermediate between them. It reaches 23.°8 C in winter, while it is only 11°.5 C in summer; in the yearly mean it is 16.°4 C. The next table shows the detail.

	Winter.		Spring.		Summer.		Autumn.		Year.	
	Temp.	Locality	Temp.	Locality	Temp.	Locality	Temp.	Locality	Temp.	Locality
Maximum	16°.1	Naha	19°.7	Naha	26°.6	Naha	23°.1	Naha	21°.4	Naha
Minimum	−7°.7	Kamikawa	2°.6	Nemuro	15°.1	Erimo	7°.2	Kamikawa	5°.0	Kushiro
Difference	23°.8		17°.1		11°.5		15°.9		16°.4	

D. TEMPERATURE ANOMALY.

The question which naturally arises in the course of investigating the climate of our country is "Is the temperature of this country higher or lower than the normal temperature of the latitude?" Now this question is very difficult to answer satisfactorily. As is well known, Dove constructed the isotherms over the surface of the globe, and by their means calculated the normal temperatures of every ten degrees of north latitude, viz. those on the equator, 10° N, 20° N and so on (See *Handbuch der klimatologie von Dr. Hann; Etude sur la distribution des températures à la surface du globe par Mr. Teisserenc de Bort.*) After him, owing to the great advancement of meteorological investigations, the method of temperature observations has undergone great improvement, and the number of stations is also greatly increased. The isotherms over the globe have been drawn very accurately by the investigations of Teisserenc de Bort, Hann, Wild, and others. But the calculations of the normal temperatures present many great difficulties, and indeed none has ever tried this after Dove. So we must be content with the numbers obtained by Dove, in order to know anything about temperature anomaly of this country. It must be remarked here that as Dove had obtained the normal temperatures by means of imperfect isotherms, there must be some incomplete points; it is allowable for us, however, to believe that the number found by a authority as Dove, is not very far from truth. We may use his results for finding the general feature of the temperature anomaly.

We calculated temperature anomalies at all our stations by Dove's numbers. They are given in the following table.

TEMPERATURE ANOMALY.

Locality.	Latitude N.		January.	July.	Year.
Naha	26°	13′	— 1.1°	+ 0.9°	— 1.6°
Ogasawara Jima	26	42	+ 0.8	+ 1.3	+ 0.3
Kagoshima	31	35	— 6.7	+ 0.7	— 3.2
Miyazaki	31	56	— 6.3	+ 0.6	— 3.2
Nagasaki	32	44	— 6.6	+ 1.2	— 3.0
Kumamoto	32	48	— 8.5	+ 1.0	— 3.7
Saga	33	12	— 7.7	+ 0.5	3.5
Oita	33	13	— 6.5	+ 0.1	— 3.7
Kochi	33	33	— 6.3	+ 0.4	— 2.7
Fukuoka	33	35	— 6.8	+ 0.1	— 3.9
Matsuyama	33	50	— 7.3	+ 0.6	— 3.8
Akamagaseki	33	58	— 5.9	— 0.1	— 3.0
Yamaguchi	34	11	— 7.8	+ 0.1	— 1.1
Itsugahara	34	12	— 6.7	— 0.3	— 3.5
Wakayama	34	14	— 6.6	+ 1.3	— 2.8
Hiroshima	34	23	— 7.2	+ 0.8	— 3.3
Okayama	34	40	— 7.4	+ 1.2	— 3.3
Ozaka	34	42	— 6.9	+ 1.3	— 2.9
Tsu	34	43	— 6.7	+ 0.4	— 3.3
Hamamatsu	34	43	— 5.9	+ 0.2	— 2.6
Kioto	35	01	— 7.6	+ 1.1	— 3.4
Numazu	35	06	— 5.2	+ 0.0	— 2.4
Nagoya	35	10	— 7.3	+ 1.1	— 3.2
Gifu	35	27	— 7.1	1.3	— 2.7
Sakai	35	33	— 6.0	+ 0.3	— 3.0
Tokio	35	41	— 6.6	+ 0.2	— 3.0
Choshi	35	44	— 4.1	— 0.8	— 1.9
Kanazawa	36	33	— 6.3	+ 0.2	— 3.2
Utsunomiya	36	34	— 8.1	— 0.1	— 3.6
Nagano	36	40	— 8.8	+ 1.5	— 3.3
Fushiki	36	47	— 6.6	— 0.2	— 3.1
Fukushima	37	45	— 7.0	— 0.3	— 3.7
Niigata	37	55	— 5.7	+ 0.3	— 2.6
Yamagata	38	14	— 8.4	+ 0.2	— 4.0
Ishinomaki	38	26	— 6.9	— 1.3	— 3.8
Miyako	39	38	— 6.2	— 2.4	— 4.0
Akita	39	42	— 6.9	— 0.3	— 3.5
Aomori	40	51	— 6.7	— 1.3	— 4.0
Hakodate	41	46	— 5.2	— 2.3	— 3.8
Erimo	41	55	— 4.4	— 5.2	— 4.8
Suttsu	42	48	— 5.1	— 2.2	— 3.5
Sapporo	43	03	— 6.7	— 1.1	— 4.2
Nemuro	43	20	— 5.3	— 3.2	— 5.2
Kushiro	43	23	— 9.9	— 3.2	— 6.0
Kamikawa	43	45	— 9.9	+ 0.2	— 5.1
Abashiri	44	02	— 6.8	— 2.1	— 4.6
Soya	45	31	— 3.9	— 2.5	— 3.6

Locality.	Latitude N.	January.	July.	Year.
Fusan	35° 06′	— 5°.2	0°.1	— 1°.8
Chemulpo	37 29	— 9.4	1.3	— 3.1
Söul	37 35	—11.0	2.9	— 2.8
Yuensan	39 10	— 7.7	1.2	— 2.1
Wladiwostock	43 07	—14.9	— 0.8	— 6.7
Korssakowkij	46 39	— 8.9	— 4.4	— 5.8
St. Olga	43 46	—13.1	— 0.2	— 6.2
Kussnai	48 00	— 6.7	— 2.5	— 4.7

We thus see that:—

(1) In January, the temperature is lower throughout the country than the normal temperatures. This is especially the case in islands. At those places projecting into the sea, as Choshi, Erimo, Soya etc., it is lower than the normal temperature by about 4° C. In inland districts, such as Nagano, Utsunomiya, Yamagata etc., the anomaly is more than 8° C. In Central Hokkaido, as in Kamikawa, Shibecha, it is still greater, being about 10°C.

(2) In July, our temperature is slightly higher than the normal temperature in the southern part about 0.°5 along the coast and more than 1°C in inland. In the northern part, it is generally lower than the normal temperature. Along the eastern coast, the anomaly is rather great, at Miyako, Erimo, Nemuro etc., their temperatures are about 3° C lower than the normal. At the central part of Hokkaido, the temperature is almost equal to the normal temperature.

(3) If we compare our yearly mean temperature with the normal temperature, it is about 3° C lower in the average of the whole country, there being no considerable difference between land and sea. As we go northward, however, the anomaly increases and at the central Hokkaido it reaches the value greater than 5° C.

Thus our temperature is lower in winter and higher in summer than the normal. If we consider the whole year, the cold prevails over the warmth. (See Plate XIII.)

E. MAXIMUM AND MINIMUM TEMPERATURES AND THEIR RANGE.

With regard to the maximum and the minimum values in the regular diurnal variation of our air temperature, we have already spoken of in the article A. We shall now give the monthly means of the maximum and the minimum temperatures and also their ranges shown by self recording instruments.

MEAN DAILY MAXIMUM OF AIR TEMPERATURE.

Locality.	Jan.	Feb.	Mar.	Apr.	May	June	July	Aug.	Sept.	Oct.	Nov.	Dec.	Year	Wint.	Spr.	Sum.	Aut.
Naha	19.6	19.8	20.2	23.5	26.3	27.4	31.5	31.3	30.1	27.9	24.4	21.8	25.3	20.4	23.3	30.1	27.5
Kagoshima	11.9	12.2	16.1	20.7	23.8	26.7	30.5	31.2	29.1	24.6	19.1	14.4	21.7	12.8	20.2	29.5	24.3
Miyazaki	12.7	12.7	16.3	21.0	24.1	26.8	30.4	31.1	28.6	24.0	19.3	15.0	21.8	13.5	20.5	29.4	24.0
Kochi	10.4	10.8	14.4	19.2	22.8	25.0	28.2	29.7	27.4	23.0	17.7	12.9	20.1	11.4	18.6	27.6	22.7
Wakayama	8.4	8.7	12.7	18.4	22.4	25.6	29.7	31.4	27.8	22.0	16.2	11.6	19.6	9.6	17.8	28.9	22.0
Oita	9.7	9.9	13.3	17.7	21.2	24.9	28.5	30.0	26.6	21.8	17.8	13.5	19.6	11.0	17.4	27.8	22.1
Yamaguchi	7.9	9.2	13.7	18.1	22.9	25.9	28.7	30.9	27.5	22.2	17.0	12.2	19.7	9.8	18.2	28.5	22.2
Hiroshima	8.0	9.1	12.3	17.5	22.2	25.2	29.4	31.4	27.7	22.4	16.3	10.9	19.4	9.3	17.3	28.7	22.1
Matsuyama	8.4	10.8	13.3	18.6	22.7	26.3	29.4	30.2	28.1	22.5	18.0	13.6	20.2	10.9	18.2	28.6	22.9
Okayama	6.8	8.3	13.3	17.6	24.0	25.2	28.6	29.7	28.4	22.6	17.0	11.1	19.4	8.7	18.3	27.8	22.7
Ozaka	8.0	8.4	12.4	18.3	22.3	26.2	30.1	31.8	28.4	22.7	16.4	11.2	19.7	9.2	17.7	29.4	22.5
Kioto	7.4	7.9	11.9	18.1	22.3	26.0	30.0	31.6	27.8	22.0	15.9	10.4	19.3	8.6	17.4	29.2	21.9
Kumamoto	8.6	12.8	15.8	21.2	24.9	27.7	31.3	32.1	30.7	24.7	19.5	14.8	22.0	12.1	20.6	30.4	25.0
Saga	7.8	9.9	11.7	19.4	24.4	27.2	28.8	30.5	29.6	23.4	18.2	13.6	20.6	10.4	19.5	28.8	23.7

Mean daily maximum of air temperature.*

Locality.	Jan.	Feb.	Mar.	Apr.	May	June	July	Aug.	Sept.	Oct.	Nov.	Dec.	Year	Wint.	Spr.	Sum.	Aut.
Nagasaki	9.7	10.7	14.2	19.2	23.0	25.6	29.7	31.1	28.3	23.3	17.2	12.1	20.3	10.8	18.8	28.8	22.9
Fukuoka	9.8	11.5	13.8	18.8	22.7	26.5	29.4	30.6	28.9	22.3	17.8	13.7	20.5	11.7	18.4	28.8	23.0
Itsugahara	8.6	10.0	13.4	17.5	21.7	25.0	27.7	30.2	26.6	22.4	17.1	12.5	19.4	10.4	17.5	27.6	22.0
Akamagaseki	8.1	8.0	11.7	16.4	20.5	23.8	27.8	29.7	26.6	21.6	15.8	10.8	18.4	9.0	16.2	27.1	21.3
Sakai	6.5	6.5	10.9	16.1	20.7	24.4	28.4	30.3	26.0	20.7	15.0	10.1	18.0	7.7	15.9	27.7	20.6
Tsu	9.2	10.4	13.3	18.1	22.4	25.2	28.0	29.8	26.6	21.5	16.6	12.7	19.5	10.8	18.0	27.7	21.6
Nagoya	7.1	9.0	13.7	18.8	25.2	25.6	30.1	31.4	28.8	22.8	16.8	12.5	20.1	9.5	19.2	29.0	22.8
Gifu	7.2	8.3	12.1	17.9	22.3	26.0	29.7	31.4	27.6	22.4	16.0	10.4	19.3	8.6	17.4	29.0	22.0
Hamamatsu	9.3	10.0	13.5	18.4	21.7	25.0	28.5	30.6	27.7	22.7	17.1	12.1	19.7	10.5	17.9	28.0	22.5
Numazu	10.2	10.1	13.5	17.8	21.0	24.5	27.7	29.6	27.1	22.2	17.2	13.2	19.5	11.2	17.4	27.3	22.2
Tokio	8.2	8.6	12.2	17.0	21.1	24.3	28.4	29.7	26.1	20.5	15.6	11.3	18.6	9.4	16.8	27.5	20.7
Utsunomiya	7.1	7.9	13.0	16.2	24.0	23.7	27.9	29.0	27.8	20.5	15.8	12.2	18.7	9.1	17.7	26.9	21.2
Choshi	9.5	10.0	13.6	17.0	20.3	23.6	27.0	29.3	26.5	21.3	17.1	13.5	19.1	11.0	17.0	26.6	21.6
Kanazawa	5.5	5.7	10.1	16.2	20.4	24.0	28.0	30.4	26.3	20.7	14.3	9.3	17.6	6.8	15.6	27.5	20.4
Fushiki	4.4	5.6	10.3	14.4	19.3	22.2	26.8	28.9	25.1	19.6	14.4	9.8	16.7	6.6	14.7	26.0	19.7
Nagano	2.3	5.1	10.6	16.5	21.7	25.3	27.9	29.8	26.5	18.9	12.3	8.0	17.1	5.1	16.3	27.7	19.2
Niigata	3.8	4.2	8.1	14.3	19.3	22.6	27.2	30.0	25.7	19.5	13.1	7.6	16.3	5.2	14.0	26.6	19.4
Yamagata	1.1	3.2	9.0	15.6	21.9	23.8	26.7	28.4	25.4	17.9	11.2	6.8	15.9	3.7	15.5	26.3	18.2
Akita	1.2	2.1	6.4	12.8	17.7	21.9	26.3	28.5	21.5	18.1	11.1	5.4	14.7	2.9	12.3	25.6	17.9
Fukushima	4.7	6.4	11.3	17.2	22.2	24.7	27.2	29.1	25.9	19.2	13.2	9.0	17.5	6.7	16.9	27.0	19.4
Ishinomaki	3.0	3.9	8.8	13.2	18.0	20.4	24.7	26.8	23.8	18.1	12.8	8.0	15.1	5.0	13.3	24.0	18.3
Miyako	4.5	5.1	8.6	14.0	17.7	19.9	24.7	27.4	23.7	18.8	13.6	8.5	15.5	6.0	13.4	24.0	18.7
Aomori	0.1	1.3	5.0	11.9	16.5	20.0	24.5	27.1	23.1	17.3	9.6	3.6	13.3	1.7	11.1	23.9	16.7
Hakodate	0.2	1.4	4.9	10.5	14.8	18.0	22.9	25.5	22.4	16.4	8.9	3.5	12.4	1.7	10.1	22.1	15.9
Suttsu	98.5	0.3	4.0	9.5	14.1	17.3	22.3	24.5	20.5	14.7	7.6	2.4	11.3	0.4	9.2	21.4	14.3
Sapporo	98.3	99.7	3.4	10.0	16.2	20.6	21.9	26.3	21.6	15.0	6.9	1.4	12.0	99.8	9.9	23.9	14.5
Kamikawa	94.2	97.6	3.4	8.9	17.6	21.5	26.0	25.9	20.3	13.6	6.4	1.4	11.4	97.7	10.0	24.5	13.4
Soya	94.5	96.0	0.9	6.2	9.5	12.0	18.9	21.7	18.5	12.3	5.0	99.7	7.9	96.7	5.5	17.5	11.9
Abashiri	95.2	97.3	4.1	9.9	15.3	15.8	20.9	24.4	20.4	13.7	7.0	2.3	10.5	98.3	9.8	20.4	13.7
Nemuro	98.5	98.5	1.7	6.7	11.0	13.9	19.2	22.1	19.0	13.8	7.5	2.5	9.5	99.8	6.5	18.4	13.4
Kushiro	96.7	99.6	4.7	10.5	15.8	17.8	22.1	23.6	21.3	14.4	8.2	3.1	11.5	99.8	10.3	21.2	14.6
Erimo	98.7	99.4	2.1	5.7	9.5	12.3	17.4	20.5	18.7	11.1	8.7	4.2	9.3	0.8	5.8	16.7	13.8

Mean daily minimum of air temperature.*

Locality.	Jan.	Feb.	Mar.	Apr.	May	June	July	Aug.	Sept.	Oct.	Nov.	Dec.	Year	Wint.	Spr.	Sum.	Aut.
Naha	12.2	14.0	14.3	16.8	19.9	21.9	24.8	24.9	24.0	20.9	17.0	15.0	18.8	13.7	17.0	23.9	20.6
Kagoshima	2.1	2.7	5.9	11.8	14.3	18.7	22.8	23.2	20.7	15.4	8.3	3.7	12.5	2.8	10.7	21.6	14.8
Miyazaki	1.4	2.0	5.5	11.4	14.2	18.7	22.4	22.9	20.1	14.1	7.4	2.8	11.9	2.1	10.4	21.3	13.9
Kochi	0.5	1.5	4.7	10.9	13.9	18.1	21.9	22.4	20.0	14.2	7.6	2.3	11.5	1.4	9.8	20.8	13.9
Wakayama	0.5	1.0	3.2	8.9	12.6	17.8	22.5	23.0	19.5	13.0	7.3	2.8	11.0	1.4	8.2	21.1	13.3
Oita	1.8	1.9	4.0	9.1	12.2	17.1	21.8	22.6	19.2	13.3	7.8	4.0	11.2	2.6	8.4	20.5	13.4
Yamaguchi	98.6	0.1	2.7	7.5	10.7	16.3	21.3	21.9	18.5	10.3	4.1	1.7	9.5	0.1	7.0	19.8	11.0
Hiroshima	99.6	0.5	2.6	8.0	12.3	17.4	22.0	22.9	19.0	12.5	6.1	1.5	10.4	0.5	7.6	20.8	12.5
Matsuyama	99.6	1.7	3.5	7.9	11.1	16.2	21.3	21.6	19.5	11.4	6.2	3.2	10.3	1.5	7.5	19.7	12.4
Okayama	98.4	0.4	3.3	6.7	11.1	16.2	22.0	22.4	21.2	11.3	6.0	0.5	10.0	99.8	7.0	20.2	12.8
Ozaka	99.3	99.5	2.5	8.2	12.2	17.5	22.2	22.9	19.2	12.4	6.2	1.4	10.3	0.1	7.6	20.9	12.6
Kioto	97.4	97.9	0.2	6.3	10.2	16.0	20.7	21.6	17.8	10.4	4.1	99.1	8.5	98.1	5.6	19.4	10.8
Kumamoto	97.6	1.8	4.3	10.0	12.1	16.8	22.0	21.7	19.8	11.4	4.8	2.1	10.4	0.5	8.8	20.2	12.0
Saga	99.8	2.5	4.1	8.9	12.4	16.7	22.0	22.6	20.6	12.9	7.5	4.1	11.2	2.2	8.5	20.1	13.7
Nagasaki	1.7	2.3	4.6	10.1	13.6	18.1	22.8	23.3	20.0	14.2	8.1	3.6	11.9	2.5	9.4	21.4	14.1
Fukuoka	99.9	2.2	3.5	8.1	10.7	16.0	21.0	21.7	19.2	10.5	6.0	3.6	10.2	1.9	7.4	19.6	11.9
Itsugahara	99.8	0.6	3.5	8.4	11.9	16.4	21.2	22.2	18.2	11.9	6.3	2.1	10.2	0.9	7.9	19.9	12.1
Akamagaseki	2.2	2.3	5.0	9.5	13.3	17.5	21.9	23.4	20.2	14.8	9.5	5.0	12.0	3.2	9.3	20.9	14.8
Sakai	0.5	0.4	2.9	7.4	11.0	16.7	21.4	22.7	18.7	12.2	7.1	3.0	10.3	1.3	7.1	20.3	12.7
Tsu	99.6	1.4	4.0	8.6	12.8	17.1	21.5	22.5	19.6	12.0	6.6	3.0	10.8	1.3	8.5	20.4	32.1

* To avoid the minus sign, the degrees below 0° are shown by adding 100. Thus 99° for −1°, 95° for −5° etc.

Mean daily minimum of air temperature.

Locality.	Jan.	Feb.	Mar.	Apr.	May	June	July	Aug.	Sept.	Oct.	Nov.	Dec.	Year	Wint.	Spr.	Sum.	Aut.
Nagoya	97.8	99.0	3.1	7.6	12.1	16.6	22.2	22.8	20.8	11.9	5.5	2.6	10.2	99.8	7.6	20.5	12.7
Gifu	98.1	98.6	1.6	7.6	11.5	16.9	21.7	22.5	18.6	11.5	5.3	1.0	9.6	99.2	6.9	20.4	11.8
Hamamatsu	0.6	0.9	3.8	9.3	12.8	17.4	21.6	22.4	19.6	13.6	7.7	3.2	11.1	1.6	8.6	20.5	13.6
Numazu	99.5	0.1	3.4	9.0	12.3	17.1	21.4	21.9	19.0	12.8	6.8	2.0	10.4	0.5	8.2	20.1	12.9
Tokio	97.8	99.0	1.7	7.3	11.7	16.6	20.9	21.7	18.4	11.5	5.0	0.1	9.3	99.0	6.9	19.7	11.6
Utsunomiya	93.4	96.6	1.3	4.6	10.2	15.5	20.2	20.4	19.2	9.5	2.6	99.1	7.7	96.4	5.4	18.7	10.4
Choshi	1.1	2.1	5.9	10.2	13.6	17.0	20.9	22.9	20.6	15.0	9.6	4.3	11.9	2.5	9.9	20.3	15.1
Kanazawa	99.0	98.8	1.2	6.2	10.3	15.8	20.1	21.5	17.8	11.4	5.8	1.8	9.2	99.9	5.9	19.2	11.7
Fushiki	99.1	99.2	2.5	6.9	10.9	16.2	21.1	22.1	18.0	11.5	6.7	2.7	9.7	0.3	6.8	19.8	12.1
Nagano	93.8	95.6	99.5	4.3	8.0	14.2	18.9	19.7	16.4	6.8	1.6	98.4	6.4	95.9	3.9	17.6	8.3
Niigata	98.5	98.5	1.0	6.1	10.1	15.5	20.7	22.3	18.4	11.5	6.1	1.4	9.2	99.5	5.8	19.5	12.0
Yamagata	94.1	94.8	99.7	3.8	8.5	14.1	18.1	19.3	15.8	6.9	1.7	98.9	6.3	95.9	4.0	17.2	8.1
Akita	94.5	94.1	98.3	3.5	8.1	13.3	18.5	19.5	15.3	7.5	2.6	98.6	6.1	95.7	3.3	17.1	8.5
Fukushima	95.9	96.6	1.3	5.3	8.5	14.4	18.7	20.1	17.0	7.3	2.2	99.9	7.3	97.5	5.0	17.7	8.8
Ishinomaki	95.4	95.8	0.2	4.7	9.0	13.5	18.3	20.4	16.5	8.9	3.6	99.4	7.1	96.9	4.6	17.4	9.7
Miyako	93.9	91.3	97.8	2.5	6.5	11.4	16.1	18.3	14.6	7.3	2.0	97.8	5.3	95.3	2.3	15.4	8.0
Aomori	93.2	93.0	96.6	2.1	6.8	12.5	17.6	19.2	14.4	7.3	1.7	96.6	5.1	91.3	1.8	16.4	7.8
Hakodate	92.9	93.5	97.4	1.4	5.7	10.3	15.6	17.7	13.3	6.0	0.6	96.2	4.2	94.2	1.5	14.5	6.6
Suttsu	93.0	94.4	97.9	2.0	6.2	10.4	15.7	18.0	14.2	7.1	1.2	96.7	4.8	94.7	2.0	14.7	7.6
Sapporo	88.4	89.5	94.1	99.7	5.1	10.0	15.2	16.6	11.5	4.1	98.3	92.6	2.1	90.2	99.6	13.9	4.6
Kamikawa	79.3	82.6	89.0	97.2	2.5	8.2	13.2	13.9	9.3	0.8	94.9	90.2	98.4	84.0	96.2	11.8	1.7
Soya	90.9	91.8	96.3	1.3	3.8	6.9	13.7	16.5	13.3	6.5	0.3	95.7	3.1	92.8	0.5	12.4	7.7
Abashiri	86.7	88.2	91.6	0.2	5.1	8.5	13.9	16.0	12.6	4.6	98.5	91.1	1.9	89.7	0.0	12.8	5.2
Nemuro	91.2	93.5	93.8	98.5	2.2	5.7	11.0	14.6	12.1	6.5	0.7	95.7	1.9	92.5	98.2	10.4	6.1
Kushiro	81.0	85.3	93.1	98.3	3.2	8.3	12.6	16.2	12.7	2.4	95.6	90.3	99.9	85.5	98.3	12.4	3.6
Erimo	93.5	91.4	97.3	0.7	4.0	7.7	12.6	15.7	11.5	9.2	3.3	98.3	4.3	95.4	0.7	12.0	9.0

Mean daily range of air temperature.

Naha	7.4	5.8	5.9	6.7	6.4	5.6	6.7	6.4	6.1	7.0	7.4	6.8	6.5	6.7	6.3	6.2	6.9
Kagoshima	9.8	9.5	10.2	8.9	9.5	8.0	7.7	8.0	8.4	9.2	10.8	10.7	9.2	10.0	9.5	7.9	9.5
Miyazaki	11.3	10.7	10.8	9.6	9.9	8.1	8.0	8.2	8.5	9.9	11.9	12.2	9.8	11.4	10.1	8.1	10.1
Kochi	9.9	9.3	9.7	8.3	8.4	6.9	6.3	7.3	7.4	8.8	10.1	10.6	8.6	10.0	8.8	6.8	8.8
Wakayama	7.9	7.7	9.5	9.5	9.8	7.8	7.2	8.4	8.3	9.0	8.9	8.8	8.6	8.2	9.6	7.8	8.7
Oita	7.9	8.0	9.3	8.6	9.0	7.8	6.7	7.4	7.4	8.5	10.0	9.5	8.4	8.4	9.0	7.3	8.7
Yamaguchi	9.3	9.1	11.0	10.6	12.2	9.6	7.4	9.0	9.0	11.9	12.9	10.5	10.2	9.7	11.2	8.7	11.2
Hiroshima	8.4	8.6	9.7	9.5	9.9	7.8	7.4	8.5	8.7	9.9	10.2	9.4	9.0	8.8	9.7	7.9	9.6
Matsuyama	8.8	9.1	9.8	10.7	11.6	10.1	8.1	8.6	8.6	11.1	11.8	10.4	9.9	9.4	10.7	8.9	10.5
Okayama	8.1	7.9	10.0	10.9	12.9	9.0	6.6	7.3	7.2	11.3	11.0	10.6	9.4	8.9	11.3	7.6	9.9
Ozaka	8.7	8.9	9.9	10.1	10.1	8.7	7.9	8.9	9.2	10.3	10.2	9.8	9.3	9.1	10.1	8.5	9.9
Kioto	10.0	10.0	11.7	11.8	12.1	10.0	9.3	10.0	10.0	11.6	11.8	11.3	10.8	10.5	11.8	9.8	11.1
Kumamoto	11.0	11.0	11.5	11.2	12.8	10.9	9.3	10.4	10.9	13.3	14.7	12.7	11.6	11.6	11.8	10.2	13.0
Saga	8.0	7.4	10.6	10.5	12.0	10.5	6.8	7.9	9.0	10.5	10.7	9.2	9.4	8.2	11.0	8.4	10.0
Nagasaki	8.0	8.4	9.6	9.1	9.4	7.5	6.9	7.8	8.3	9.1	9.1	8.5	8.4	8.3	9.4	7.4	8.8
Fukuoka	9.9	9.3	10.3	10.7	12.0	10.5	8.4	8.9	9.7	11.8	11.8	10.1	10.3	9.8	11.0	9.2	11.1
Itsugahara	8.8	9.4	9.9	9.1	9.8	8.6	6.5	8.0	8.1	10.5	10.8	10.1	9.2	9.5	9.6	7.7	9.9
Akamagaseki	5.9	5.7	6.7	6.9	7.2	6.3	5.9	6.3	6.4	6.8	6.3	5.8	6.4	5.8	6.9	6.2	6.5
Sakai	6.0	6.1	8.0	8.7	9.7	7.7	7.0	7.6	7.3	8.5	7.9	7.1	7.7	6.4	8.8	7.4	9.9
Tsu	9.6	9.0	9.3	9.8	9.6	8.1	6.5	7.3	7.0	9.5	10.0	9.7	8.7	9.5	9.5	7.3	8.8
Nagoya	9.3	10.0	10.6	11.2	13.1	9.0	7.9	8.6	8.0	10.9	11.3	9.9	10.0	9.7	11.6	8.5	10.1
Gifu	9.1	9.7	10.5	10.3	10.8	9.1	8.0	8.9	9.0	10.9	10.7	9.4	9.7	9.4	10.5	8.6	10.2
Hamamatsu	8.7	9.1	9.7	9.1	8.9	7.6	6.9	8.2	8.1	9.1	9.4	8.9	8.6	8.9	9.3	7.5	8.9
Numazu	10.7	10.0	10.1	8.8	8.7	7.4	6.3	7.7	8.1	9.4	10.4	11.2	9.1	10.7	9.2	7.2	9.3
Tokio	10.4	9.6	10.5	9.7	9.4	7.7	7.5	8.0	7.7	9.0	10.6	11.2	9.3	10.4	9.9	7.8	9.1
Utsunomiya	13.7	11.3	11.7	11.6	13.8	8.2	7.7	8.6	8.6	11.0	12.7	13.1	11.0	12.7	12.3	8.2	10.8

Mean daily range of air temperature.

Locality.	Jan.	Feb.	Mar.	Apr.	May	June	July	Aug.	Sept.	Oct.	Nov.	Dec.	Year	Wint.	Spr.	Sum.	Aut.
Choshi	8.4	7.9	7.7	6.8	6.7	6.6	6.1	6.4	5.9	6.3	7.5	9.2	7.2	8.5	7.1	6.3	6.5
Kanazawa	6.5	6.9	8.9	10.0	10.1	8.2	7.6	8.9	8.5	9.3	8.5	7.5	8.4	6.9	9.7	8.3	8.7
Fushiki	5.3	6.1	7.8	7.5	8.4	6.0	5.7	6.8	7.1	8.1	7.7	7.1	7.0	6.3	7.9	6.2	7.6
Nagano	8.5	9.5	11.1	12.2	13.7	11.1	9.0	10.1	10.1	12.1	10.7	9.6	10.7	9.2	12.4	10.1	10.9
Niigata	5.3	5.7	7.4	8.2	8.9	7.1	6.5	7.7	7.3	8.0	7.0	6.2	7.1	5.7	8.2	7.1	7.4
Yamagata	7.0	8.4	9.3	11.8	13.1	9.7	8.6	9.1	9.6	11.0	9.5	7.9	9.6	7.8	11.5	9.1	10.1
Akita	6.7	8.0	8.1	9.3	9.6	8.6	7.8	9.0	9.2	10.6	8.5	6.8	8.6	7.2	9.0	8.5	9.4
Fukushima	8.8	9.8	10.0	11.9	13.7	10.3	8.5	9.0	8.9	11.9	11.0	9.1	10.2	9.2	11.9	9.3	10.6
Ishinomaki	7.6	8.1	8.6	8.5	9.0	6.9	6.4	6.4	7.3	9.5	9.2	8.6	8.0	8.1	8.7	6.6	8.6
Miyako	10.6	10.8	10.8	11.5	14.2	8.5	8.3	9.1	9.1	11.5	11.6	10.7	10.2	10.7	11.1	8.6	10.7
Aomori	6.9	8.3	8.4	9.8	9.7	7.5	6.9	7.9	8.7	10.1	7.9	7.0	8.2	7.4	9.3	7.5	8.9
Hakodate	7.3	7.9	7.5	9.1	9.1	7.7	7.3	7.8	9.1	10.4	8.3	7.3	8.2	7.5	8.6	7.6	9.3
Suttsu	5.5	5.9	6.1	7.5	7.9	6.9	6.6	6.5	6.3	7.3	6.4	5.7	6.5	5.7	7.2	6.7	6.7
Sapporo	9.9	10.2	9.3	10.3	11.1	10.6	9.7	9.7	10.1	10.9	8.6	8.8	9.9	9.6	10.3	10.0	9.9
Kamikawa	14.9	15.0	14.4	11.7	15.1	13.3	12.8	12.0	11.0	12.8	11.5	11.2	13.0	18.7	13.8	12.7	11.7
Soya	3.6	4.2	4.6	4.9	5.7	5.1	5.2	5.2	5.2	5.8	1.6	4.0	4.6	3.9	5.0	5.1	4.2
Abashiri	8.5	9.1	9.5	9.7	10.2	7.3	7.0	8.4	7.8	9.1	8.5	8.2	8.6	8.6	9.8	7.6	8.5
Nemuro	7.3	8.0	7.9	8.2	8.8	8.2	8.2	7.5	6.9	7.3	6.8	6.8	7.6	7.3	8.3	8.0	7.0
Kushiro	15.7	11.3	11.3	12.2	12.6	9.5	9.5	7.4	8.6	12.0	12.6	12.8	11.6	14.3	12.0	8.8	11.0
Erimo	5.2	5.0	4.8	6.0	5.5	4.6	4.8	4.8	4.2	4.9	5.4	5.9	5.0	5.4	5.1	4.7	4.8

These tables show that the highest of the daily maximum temperatures is in August, and the lowest of the minimum in January. The greatest range lies in spring and autumn, whilst the smallest range is in July. The region along the Pacific, from Numazu to Choshi, has the greatest range in winter, for there the weather is generally very clear in this season. On the other hand, it is cloudy along the Sea of Japan in winter, and consequently we have then the smallest range.

The annual ranges of the mean maximum and the mean minimum temperature, i. e. the differences between the lowest of the mean monthly minimum temperature and the highest of the mean monthly maximum, are given in the following table.

Annual range of mean maximum & minimum temperature.

Locality.	Range.	Locality.	Range.	Locality.	Range.
Naha	19.3	Itsugahara	30.4	Akita	34.4
Kagoshima	29.1	Akamagaseki	27.5	Fukushima	33.2
Miyazaki	29.7	Sakai	29.8	Ishinomaki	31.4
Kochi	29.2	Tsu	30.2	Miyako	33.5
Wakayama	30.9	Nagoya	33.6	Aomori	34.1
Oita	28.2	Gifu	33.3	Hakodate	32.6
Yamaguchi	32.2	Hamamatsu	30.0	Suttsu	31.5
Hiroshima	31.8	Numazu	30.1	Sapporo	37.9
Matsuyama	30.6	Tokio	31.9	Kamikawa	46.7
Okayama	31.3	Utsunomiya	35.6	Soya	30.8
Ozaka	32.5	Choshi	28.2	Abashiri	37.7
Kioto	34.2	Kanazawa	31.6	Nemuro	31.6
Kumamoto	34.5	Fushiki	29.8	Kushiro	42.6
Saga	30.7	Nagano	36.0	Erimo	27.0
Nagasaki	29.4	Niigata	31.5		
Fukuoka	30.7	Yamagata	34.3		

The ranges given above, ought to be far greater than those given in the article B. Hence even in Naha, an island in the Southern Pacific, this annual range is 19.° 3. In Nippon it is about 30° and in the central part of Hokkaido it is more than 45.°

Instead of taking the monthly mean, if we take the actual maximum and the minimum values ever attained since the first date of our observations, we have the extreme maximum value of 37.°5 at Hiroshima and the extreme minimum of —36.°7 at Kamikawa. For details, see the following table.

Extremes of air temperature.*

Locality.	Maximum	Day	Month	Year	Minimum	Day	Month	Year	Range
Naha	34.4	3	VII	1890	7.4	21	I	1891	27.0
Kagoshima	34.8	15	IX	1884	93.9	2	II	1886	40.9
Miyazaki	36.3	13	VII	1886	93.9	11;22	II;XII	1883;84	42.4
Kochi	34.8	3	VIII	1889	94.1	18	I	1891	40.7
Wakayama	36.6	25	VII	1883	94.6	9	II	1883	42.0
Oita	34.6	20	VIII	1888	94.8	15	I	1888	39.8
Yamaguchi	34.4	17	VIII	1888	92.3	3	II	1888	42.1
Hiroshima	37.5	19	VIII	1886	91.6	31	XII	1883	45.9
Matsuyama	31.1	24	VI	1891	91.7	16	II	1891	39.4
Okayama	33.3	1	IX	1891	94.9	16	I	1891	38.4
Ozaka	35.8	17	VII	1886	92.9	25	I	1891	42.9
Kioto	36.2	25	VII	1883	88.1	16	I	1891	48.1
Kumamoto	35.6	2	VIII	1891	93.1	25	I	1891	42.5
Saga	34.7	1;15	VIII	1891	95.7	26	I	1891	39.0
Nagasaki	35.7	25	VII	1887	95.1	13	I	1886	40.6
Fukuoka	36.4	13	VII	1890	95.4	29	I	1891	41.0
Itsugahara	34.0	5	VIII	1889	92.2	19	II	1891	41.8
Akamagaseki	35.8	12	VIII	1883	95.6	18	I	1883	40.2
Sakai	37.2	3	VIII	1886	91.8	5	II	1886	45.4
Tsu	35.2	2	VIII	1889	94.3	14	I	1890	40.9
Nagoya	35.8	11;19	VII;VIII	1890	93.3	12	II	1891	42.5
Gifu	36.9	16	VII	1886	88.3	1	I	1884	48.6
Hamamatsu	36.8	14	VII	1886	94.8	19	II	1885	42.0
Numazu	34.1	5	VIII	1886	91.3	19	II	1885	42.8
Tokio	36.6	14	VII	1886	90.8	13	I	1876	45.8
Utsunomiya	33.8	13	VII	1891	89.7	14	II	1891	44.1
Choshi	33.6	24	VIII	1889	95.1	3	II	1888	38.5
Kanazawa	36.8	2	VIII	1886	92.6	14	I	1885	44.2
Fushiki	34.9	12;24	VII;VIII	1888;91	91.3	20	II	1891	43.6
Nagano	31.6	11	IX	1891	81.1	12	II	1891	50.5
Niigata	36.1	5	VIII	1888	90.6	29	I	1891	45.5
Yamagata	31.0	22	VI	1890	80.0	29	I	1891	51.0
Akita	35.0	6	VIII	1885	75.1	5	II	1888	59.9
Fukushima	35.6	22	VI	1890	81.5	4	II	1891	54.1
Ishinomaki	33.7	10	VIII	1888	86.9	4	II	1891	46.8
Miyako	36.2	3;16	VIII	1886	84.6	21	I	1885	51.6
Aomori	34.0	1	VIII	1882	81.0	4,20	II;XII	1882;91	53.0
Hakodate	33.6	29	VII	1876	78.3	29	I	1891	55.3
Suttsu	32.0	26	VII	1887	85.6	12	I	1886	46.4
Sapporo	34.1	15	VIII	1883	74.4	18	II	1885	59.7
Kamikawa	34.9	29	VII	1888	63.3	29	I	1891	71.6
Soya	29.4	5	IX	1890	82.7	29	I	1889	46.7
Abashiri	33.0	23	VIII	1889	78.5	19	I	1890	54.5
Nemuro	31.9	25	VII	1883	78.9	18	III	1885	53.0
Kushiro	30.6	18	VII	1891	72.6	12	I	1891	58.0
Erimo	29.6	4	IX	1890	85.7	28	I	1891	43.9

* To avoid the minus sign, the degrees below 0° are shown by adding 100. Thus 99° for —1°, 95° for —5° etc.

Thus the maximum range of extremes,—viz. the greatest difference between the extremes of the maximum and the minimum temperatures, which we had since the beginning of our observations is generally 40° in the coast land of the southern provinces, and 45° in inland. In the northern provinces, we have still greater range of more than 50° even along the coast. In the central part of Hokkaido, it has the great value of 70° C. These great ranges are due to the frequent occurrences of very low temperatures.

The range of the mean maximum temperature and the mean minimum during the course of a day is 10° C on an average. If we take the actual maximum and the minimum ever attained, but not their mean values, then we have sometimes a great range of temperature. The next table shows the maximum diurnal ranges of temperature ever observed at several stations since their establishments.

ABSOLUTE MAXIMUM OF DAILY RANGE.

Locality.	Range.	Date.	Locality.	Range.	Date.
Naha	13.8	17th Jan. 1891	Tokio	22.1	17th March 1882
Kagoshima	18.1	9th Jan. 1888	Utsunomiya	21.4	21st May 1891
Miyazaki	20.7	22nd Jan. 1883	Choshi	17.1	2nd Jan. 1887
Kochi	17.2	6th Nov. 1891	Kanazawa	23.6	13th April 1883
Wakayama	19.1	21st May 1891	Fushiki	20.6	30th March 1891
Oita	18.9	15th Dec. 1891	Nagano	24.1	14th May 1891
Hiroshima	18.1	7th Sept. 1888	Niigata	18.2	26th April 1885
Matsuyama	19.3	15th Dec. 1891	Yamagata	23.2	15th March 1891
Okayama	19.4	13th May 1891	Akita	21.9	5th Feb. 1888
Ozaka	18.4	10th April 1889	Fukushima	25.4	15th May 1891
Kioto	21.0	22nd May 1891	Ishinomaki	17.6	8th March 1889
Kumamoto	21.2	15th Dec. 1891	Miyako	23.5	27th April 1888
Saga	19.6	14th Nov. 1891	Aomori	22.0	6th Feb. 1890
Nagasaki	18.2	16th Jan. 1888	Hakodate	22.0	8th Dec. 1891
Fukuoka	22.1	30th March 1891	Suttsu	18.7	11th May 1890
Itsugahara	19.5	12th May 1891	Sapporo	24.6	15th May 1887
Akamagaseki	14.7	8th Jan. 1886	Kamikawa	30.2	6th March 1889
Sakai	23.1	15th May 1887	Soya	19.1	29th Oct. 1889
Tsu	17.4	13th May 1891	Abashiri	22.6	6th Feb. 1890
Nagoya	18.9	21st May 1891	Nemuro	22.6	7th June 1885
Gifu	20.2	22nd March 1888	Kushiro	26.9	31st Jan. 1890
Hamamatsu	18.0	21st May 1891	Erimo	15.3	13th Dec. 1889
Numazu	19.8	9th Feb. 1890			

Even in Naha, a solitary island in the Southern Pacific, the maximum range of temperature in a day is about 14.°C. On the main island Nippon, it is about 20° C and in the central part of Hokkaido it exceeds 30° C. The maximum range of temperature in a day is nearly equal to and sometimes exceeds the annual range of mean temperatures—viz. the difference of the mean temperatures of the hottest and the coldest months.

F. VARIABILITY OF TEMPERATURE.

By what we have said in the preceding articles, the general variations of temperature in this country will be clearly understood, but there remains one more important question "What is the manner of variation? Is it quick or slow?" In order to answer this, we have calculated the variability of temperature or the differences of mean temperatures of consecutive days, as first proposed by Hann, the Director of the Austrian Meteorological Observatory, and also their means irrespective of their signs are tabulated as follows:—

Mean variability of temperature.

Locality.	Jan.	Feb.	Mar.	Apr.	May	June	July	Aug.	Sept.	Oct.	Nov.	Dec.	Year	Wint.	Spr.	Sum.	Aut.
Naha	1.19	1.71	1.82	1.17	1.66	0.79	0.61	0.64	0.68	0.71	1.11	1.62	1.17	1.61	1.55	0.68	0.83
Kagoshima	2.05	1.71	2.01	1.77	1.30	1.10	0.78	0.75	0.95	1.35	1.75	2.18	1.48	1.98	1.69	0.88	1.35
Miyazaki	2.08	1.79	1.99	2.01	1.35	1.29	0.90	0.78	1.01	1.52	1.65	2.16	1.54	2.01	1.78	0.99	1.39
Kochi	1.76	1.60	1.74	1.65	1.25	0.99	0.74	0.66	0.97	1.33	1.55	1.81	1.34	1.72	1.55	0.80	1.28
Wakayama	1.61	1.67	1.88	2.03	1.15	1.22	0.88	0.75	1.23	1.40	1.54	1.75	1.45	1.68	1.79	0.95	1.39
Oita	1.59	1.55	1.94	1.91	1.38	1.40	1.10	0.74	1.19	1.33	1.46	2.02	1.47	1.72	1.74	1.08	1.33
Hiroshima	1.50	1.42	1.62	1.72	1.37	1.17	0.86	0.83	1.19	1.39	1.42	1.66	1.35	1.53	1.57	0.95	1.33
Matsuyama	1.63	1.72	1.66	1.98	1.33	1.03	1.02	0.64	1.10	1.19	1.54	2.21	1.42	1.85	1.66	0.90	1.28
Okayama	1.31	1.71	1.93	2.16	1.52	1.25	1.08	0.74	1.29	1.43	1.35	1.89	1.47	1.64	1.87	1.02	1.36
Ozaka	1.42	1.45	1.61	1.86	1.39	1.15	0.86	0.73	1.17	1.28	1.11	1.62	1.33	1.50	1.63	0.91	1.30
Kioto	1.56	1.60	1.85	2.01	1.53	1.36	0.96	0.81	1.32	1.47	1.69	1.70	1.49	1.62	1.81	1.04	1.49
Kumamoto	1.77	1.94	1.88	1.83	1.51	0.98	0.87	0.76	1.21	1.24	1.67	2.54	1.52	2.08	1.74	0.87	1.37
Saga	1.57	1.77	1.66	1.41	1.31	0.98	1.00	0.81	1.24	1.14	1.25	2.05	1.35	1.80	1.47	0.93	1.21
Nagasaki	2.03	1.67	1.66	1.86	1.29	1.09	0.84	0.78	1.06	1.27	1.52	2.18	1.41	1.96	1.60	0.90	1.28
Fukuoka	1.69	1.71	1.59	2.07	1.41	1.28	1.16	0.62	1.31	1.28	1.43	2.08	1.47	1.83	1.69	1.02	1.34
Itsugahara	2.09	1.69	1.91	1.79	1.39	1.11	0.98	0.73	1.02	1.28	1.66	2.47	1.51	2.08	1.70	0.94	1.32
Akamagaseki	1.66	1.36	1.55	1.49	1.12	1.13	0.81	0.65	0.91	1.10	1.29	1.66	1.23	1.56	1.39	0.86	1.10
Sakai	1.47	1.24	1.77	2.11	1.45	1.31	0.99	0.88	1.14	1.22	1.49	1.62	1.39	1.44	1.78	1.06	1.28
Tsu	1.67	1.98	1.70	2.01	1.48	1.14	0.96	0.81	0.96	1.41	1.06	1.69	1.46	1.78	1.73	0.97	1.34
Nagoya	1.30	1.75	1.75	1.73	1.38	1.31	1.00	0.81	0.96	1.37	1.66	1.76	1.40	1.60	1.62	1.05	1.33
Gifu	1.50	1.43	1.74	1.87	1.52	1.30	0.99	0.80	1.11	1.42	1.43	1.48	1.38	1.47	1.71	1.03	1.33
Hamamatsu	1.55	1.46	1.66	1.67	1.37	1.18	0.85	0.82	1.06	1.22	1.31	1.63	1.32	1.55	1.57	0.95	1.21
Numazu	2.02	1.79	1.91	2.00	1.42	1.07	0.85	0.80	1.09	1.50	1.65	2.04	1.51	1.95	1.79	0.91	1.41
Tokio	1.42	1.62	1.82	2.22	1.72	1.32	1.09	0.90	1.38	1.46	1.57	1.74	1.52	1.59	1.92	1.10	1.47
Utsunomiya	1.20	1.68	2.09	1.93	1.71	1.18	1.21	0.82	1.15	1.83	1.90	2.31	1.58	1.73	1.91	1.07	1.63
Choshi	1.82	1.68	1.96	1.85	1.31	0.97	0.81	0.67	0.97	1.23	1.54	2.01	1.10	1.84	1.72	0.82	1.25
Kanazawa	1.74	1.48	1.96	2.37	1.85	1.49	1.12	1.02	1.31	1.40	1.74	2.04	1.63	1.75	2.06	1.21	1.48
Fushiki	1.55	1.57	1.83	2.09	1.61	1.21	1.14	0.91	1.11	1.26	1.59	1.91	1.48	1.68	1.84	1.10	1.32
Nagano	1.28	2.03	2.37	2.91	1.98	1.55	1.29	1.16	1.49	1.59	1.84	2.13	1.80	1.81	2.42	1.33	1.61
Niigata	1.38	1.30	1.62	2.11	1.95	1.42	1.06	1.07	1.35	1.40	1.60	1.63	1.49	1.41	1.89	1.18	1.45
Yamagata	1.66	2.10	1.69	2.52	2.06	1.37	1.35	0.98	1.45	1.67	1.92	2.29	1.75	2.02	2.09	1.23	1.68
Akita	1.90	1.83	1.62	2.09	2.01	1.53	1.06	1.22	1.43	1.83	1.90	1.91	1.70	1.88	1.92	1.27	1.72
Fukushima	1.66	2.21	1.91	2.47	1.90	1.50	1.37	1.04	1.82	1.88	2.44	2.32	1.88	2.06	2.09	1.30	2.05
Ishinomaki	1.79	1.61	1.77	1.91	1.64	1.30	1.24	1.08	1.42	1.72	1.95	1.94	1.61	1.78	1.77	1.21	1.70
Miyako	2.03	1.63	1.95	2.70	2.39	1.83	1.60	1.43	1.59	1.79	2.09	2.19	1.93	1.95	2.35	1.62	1.82
Aomori	1.98	1.83	1.65	1.71	1.64	1.40	1.18	1.13	1.31	1.71	2.41	2.03	1.64	1.95	1.67	1.24	1.71
Hakodate	2.40	2.00	1.85	1.54	1.46	1.33	1.17	1.26	1.50	2.18	2.44	2.56	1.81	2.32	1.62	1.25	2.04
Suttsu	1.96	1.70	1.65	1.64	1.72	1.39	1.18	1.01	1.16	1.79	2.20	2.47	1.66	2.01	1.67	1.19	1.72
Sapporo	2.32	2.26	1.96	1.72	1.81	1.56	1.25	1.22	1.53	2.02	2.18	2.43	1.85	2.34	1.83	1.34	1.91
Kamikawa	3.16	3.76	2.66	2.05	1.92	1.48	1.36	1.36	1.45	1.94	2.51	2.72	2.20	3.21	2.21	1.40	1.97
Soya	1.54	1.69	2.20	1.63	1.97	1.74	1.62	1.15	1.31	1.71	2.11	2.07	1.73	1.77	1.98	1.50	1.71
Abashiri	2.23	2.76	2.14	2.39	2.81	2.19	2.05	1.43	1.51	1.99	2.23	2.41	2.18	2.47	2.45	1.89	1.91
Nemuro	1.95	2.11	1.83	1.57	1.84	1.67	1.78	1.45	1.41	1.72	2.01	2.20	1.80	2.09	1.75	1.63	1.72
Kushiro	3.22	2.80	2.00	1.87	2.27	1.72	1.68	0.98	1.46	2.29	2.69	3.37	2.20	3.13	2.05	1.46	2.15
Erimo	1.77	1.26	1.13	1.09	1.13	0.98	1.01	0.92	1.12	1 50	2.11	2.32	1.36	1.78	1.12	0.98	1.58

Frequency of the variation.

Locality.	< 2°	2°—4°	4°—6°	6°—8°	8°—10°	10°—12°	12°—14°	14°—16°	16°—18°
Naha	81.2	15.8	2.9	0.1	0.1				
Kagoshima	72.7	21.8	4.2	1.0	0.3	..			
Miyazaki	70.5	23.6	4.9	0.8	0.2	0.1			
Kochi	76.2	19.8	3.7	0.4	0.0	0.0			
Wakayama	73.4	21.7	4.1	0.8	0.1	..			
Oita	72.8	22.8	3.9	0.3	0.2				

FREQUENCY OF THE VARIATION.

Locality.	< 2°	2°—4°	4°—6°	6°—8°	8°—10°	10°—12°	12°—14°	14°—16°	16°—18°
Hiroshima	76.2	20.5	3.1	0.2	0.1	0.0			
Matsuyama	74.2	21.0	3.6	1.1	0.1				
Okayama	72.7	22.5	4.4	0.4					
Ozaka	76.2	20.3	3.2	0.3	0.1				
Kioto	71.7	23.1	4.5	0.6	0.1	0.0			
Kumamoto	69.5	25.2	4.3	0.8	0.3				
Saga	76.9	19.3	3.6	0.2	0.1				
Nagasaki	74.1	20.0	4.5	1.0	0.4	0.0			
Fukuoka	71.5	23.6	4.3	0.4	0.3				
Itsugahara	70.9	22.8	5.2	1.0	0.2	0.1			
Akamagaseki	79.5	17.8	2.5	0.2	0.1				
Sakai	75.3	19.9	4.0	0.8	0.1	.			
Tsu	74.1	20.8	4.1	0.3	0.2	0.1			
Nagoya	75.2	20.7	3.4	0.7	.				
Gifu	74.5	21.9	3.1	0.3	0.0				
Hamamatsu	76.8	20.2	3.0	0.2	0.0				
Numazu	72.3	21.5	5.0	1.2	0.1	0.0			
Tokio	72.1	21.9	4.8	1.0	0.2	0.1		0.0	
Utsunomiya	70.3	23.6	5.6	0.4	0.1				
Choshi	74.3	20.5	4.4	0.7	0.1				
Kanazawa	68.6	21.1	5.7	1.1	0.1	0.1			
Fushiki	73.7	20.6	4.3	1.2	0.1	0·1			
Nagano	64.3	26.2	6.3	2.3	0.6	0.2		0.1	
Niigata	71.3	23.5	4.1	0.7	0.2				
Yamagata	66.9	23.5	7.8	1.0	0.6	0.1			
Akita	65.8	26.5	6.6	0.9	0.2				
Fukushima	62.4	28.5	6.1	2.4	0.4		0.1		
Ishinomaki	69.2	21.3	5.3	1.1	0.2	0.0	0.0	..	
Miyako	60.7	27.9	8.7	2.1	0.4	.	.	0.0	
Aomori	68.9	21.3	5.4	1.2	0.2	0.0	0.0		
Hakodate	64.2	26.1	7.8	1.5	0.6	0.0	0.0		
Suttsu	66.9	24.2	5.8	1.5	0.3	.	0.1		
Sapporo	63.1	26.7	7.8	1.9	0.4	0.1	0.1		
Kamikawa	55.9	29.3	8.8	3.8	1.2	0.5	0.1		0.1
Soya	65.0	27.0	6.4	1.1	0.4	0.1	0.1		
Abashiri	55.6	29.0	10.6	3.8	0.6	0.3	0.1		
Nemuro	63.5	27.5	7.2	1.6	0.3	0.0			
Kushiro	55.3	30.4	9.3	3.1	1.6	0.1	0.3		
Erimo	76.7	18.9	3.5	0.8	0.1	.	0.1	..	

We have the greatest rate of the variation of temperature in December over the whole country,—the difference of the temperatures in two consecutive days in about 2° C on an average. The variability is in July, being 0.°9 per day.

As we have said just before, if we take the difference of the mean temperatures of two consecutive days, it is about 2° C even in the month when the variability is greatest. But if, instead of the mean temperatures, we take the difference of the observed temperatures, it has frequently great value. Even in the southern provinces, this often exceeds 10° C and in the central part of Hokkaido, it is no rare occurrences that this difference is 16° C—18° C. However, it is usual to find this difference smaller than 2° C.

6. MISCELLANEOUS ON THE TEMPERATURE.

In the article D, we have shown that our temperature, especially in winter, is very much lower than the normal temperature; hence everywhere in our Empire, with only exceptions of those islands on the Pacific, as Okinawa and Bonin, the air temperature falls very frequently below the freezing point. So for example, in Hokkaido, almost every day during the three winter months, the minimum temperature remains below the freezing point, and this happens for 150 days in a year. Even in the southern parts of Kiushu, lying on the lat. 32° N, the minimum temperature of more than 20 days in a year is below the freezing point. For details, see following table:—

MEAN NUMBER OF DAYS WITH MINIMUM TEMPERATURE <0°.

Locality.	Jan.	Feb.	Mar.	Apr.	May	June	July	Aug.	Sept.	Oct.	Nov.	Dec.	Year	Wint.	Spr.	Sum.	Aut.
Naba	..	..	..								..	..	..	..	..		..
Kagoshima	9.4	7.2	1.2								0.1	3.6	21.5	20.2	1.2		0.1
Miyazaki	12.0	10.0	3.0								1.0	7.0	33.0	29.0	3.0		1.0
Kochi	14.8	10.8	3.0	..							0.3	7.9	36.8	33.5	3.0		0.3
Wakayama	13.8	11.5	4.4	0.2							0.1	4.2	34.2	29.5	4.6		0.1
Oita	10.6	9.2	4.0	0.4		..					0.2	4.8	29.2	24.6	4.4		0.2
Yamaguchi	23.7	17.2	10.2	1.2							3.0	12.3	67.6	53.2	11.4		3.0
Hiroshima	18.8	13.8	6.8	0.2							0.8	8.2	48.6	40.8	7.0		0.8
Matsuyama	20.0	11.0	5.0								1.0	4.0	41.0	35.0	5.0		1.0
Okayama	25.0	14.0	4.0	..							3.0	7.5	53.5	46.5	4.0		3.0
Ozaka	20.2	17.6	6.7	0.3							0.3	8.9	54.0	46.7	7.0		0.3
Kioto	26.0	21.3	16.7	2.5							3.6	21.3	91.4	68.6	19.2		3.6
Kumamoto	26.0	12.5	8.0								2.0	12.0	60.5	50.5	8.0		2.0
Saga	18.0	10.0	4.0								..	2.0	34.0	30.0	4.0		..
Nagasaki	8.3	7.7	1.8	..							..	3.0	20.8	19.0	1.8		..
Fukuoka	15.0	9.5	5.5	0.5							1.0	3.5	35.0	28.0	6.0		1.0
Itsugahara	17.2	14.2	5.4								1.6	8.8	47.2	40.2	5.4		1.6
Akamagaseki	6.8	4.0	0.2	..							..	0.8	11.8	11.6	0.2		..
Sakai	13.0	13.0	5.0	0.0							0.0	3.0	33.0	28.0	5.0		0.0
Tsu	17.0	10.0	4.0								..	5.0	36.0	32.0	4.0		..
Nagoya	27.0	18.0	5.0	..							0.5	9.0	59.5	54.0	5.0		0.5
Gifu	24.4	20.3	11.8	0.7							2.3	12.0	71.5	56.7	12.5		2.3
Hamamatsu	13.3	11.8	2.4	..							..	3.7	31.2	28.8	2.4		..
Numazu	18.3	14.5	4.9	0.3						..	0.9	10.2	49.1	43.0	5.2		0.9
Tokio	25.6	19.2	9.5	0.6						0.1	1.4	15.9	72.3	60.7	10.1		1.5
Utsunomiya	31.0	22.0	10.0	1.0						0.5	9.5	22.0	96.0	75.0	11.0		10.0
Choshi	11.4	8.2	1.6	..								1.6	22.8	21.2	1.6		
Kanazawa	20.0	20.0	10.0	1.0						..	..	6.0	57.0	46.0	11.0		
Fushiki	18.0	13.0	2.0	..						..	..	1.0	34.0	32.0	2.0		..
Nagano	30.7	25.7	19.3	5.3						1.3	10.7	23.7	116.0	79.3	24.7		12.0
Niigata	24.0	21.9	11.3	0.6						.	0.6	7.3	65.7	53.2	11.9		0.6
Yamagata	30.0	26.0	17.5	4.5						0.3	10.3	22.7	111.3	78.7	22.0		10.6
Akita	30.0	27.0	24.0	3.0						..	7.0	23.0	114.0	80.0	27.0	..	7.0
Fukushima	29.5	23.0	12.0	1.5						0.7	7.7	18.3	92.7	70.8	13.5		8.4
Ishinomaki	30.5	25.7	17.0	2.0	..					..	6.2	20.8	102.2	77.0	19.0		6.2
Miyako	30.6	27.2	24.7	7.4	0.3					0.8	10.8	24.7	126.5	82.5	32.4		11.6
Aomori	30.2	27.2	26.2	7.0	0.4					0.2	9.2	26.0	126.4	83.4	33.6		9.4
Hakodate	29.9	25.5	23.4	8.4	0.9					1.7	13.4	25.8	129.0	81.2	32.7		15.1
Suttsu	30.5	27.2	21.5	3.8	0.2					0.5	9.5	25.5	118.7	83.2	25.5		10.0
Sapporo	31.0	27.8	27.3	13.7	2.7				..	5.0	18.5	28.5	154.5	87.3	43.7		23.5
Kamikawa	31.0	28.0	30.0	23.0	8.0				1.0	14.0	25.0	29.0	189.0	88.0	61.0		40.0
Soya	31.0	28.2	26.8	8.4	1.4					0.8	12.0	27.0	135.6	86.2	36.6		12.8
Abashiri	31.0	28.0	26.0	16.0	1.0					4.0	20.0	27.7	153.7	86.7	43.0		24.0
Nemuro	31.0	28.2	29.2	15.0	2.0					0.3	10.7	26.8	143.2	86.0	46.2		11.0
Kushiro	31.0	28.0	27.0	21.5	6.5					10.5	25.5	27.0	177.0	86.0	55.0		36.0
Erimo	30.2	28.0	27.7	10.2	..						6.5	22.0	124.6	80.2	37.9	..	6.5

Not only the minimum temperature, but also the mean temperature falls frequently below the freezing point everywhere in our country. Thus in Hokkaido, the mean temperature is below 0° C almost every day during the three winter months and also for 110 days in a year. Even in the extreme south of our Empire, about once in three years, the mean temperature falls below the freezing point, as shown in the following table in details:—

Mean number of days with mean temperature <0°.

Locality.	Jan.	Feb.	Mar.	April	May	June	July	Aug.	Sept.	Oct.	Nov.	Dec.	Year	Wint.	Spr.	Sum.	Aut.
Naha	..	..	..			..							..	..			
Kagoshima	0.2	0.1							..				0.3	0.3			
Miyazaki	..												..	..	..		
Kochi	0.2	..											0.2	0.2			
Wakayama	0.3	0.4											0.7	0.7	..		..
Oita	0.4	..										.	0.4	0.4			
Yamaguchi	3.7	0.7		..								..	4.4	1.4			
Hiroshima	2.2	1.1										0.4	3.7	3.7			
Matsuyama	1.0	..							..				1.0	1.0			..
Okayama	6.0	2.0											8.0	8.0			
Ozaka	1.6	1.0	..	..								..	2.6	2.6	..		
Kioto	4.5	4.8	0.1			.						1.1	10.5	10.4	0.1		
Kumamoto	5.0	..											5.0	5.0			
Saga	2.0	..			..								2.0	2.0			
Nagasaki	0.5	0.3		..		..							0.8	0.8			
Fukuoka	1.0	..										..	1.0	1.0		..	
Itsugahara	2.4	0.4										0.2	3.0	3.0			
Akamagaseki	1.0	0.3										0.2	1.5	1.5			
Sakai	2.0	1.0		..								0.0	3.0	3.0			
Tsu	..	1.0											1.0	1.0			
Nagoya	1.5	1.5	..									..	3.0	3.0	..		..
Gifu	4.2	2.6	0.1									0.3	7.2	7.1	0.1		
Hamamatsu	0.6	0.4	..		..							0.1	1.1	1.1	..		
Numazu	0.9	0.4	..			..						..	1.3	1.3	..		
Tokio	2.9	1.4	0.1								..	0.2	4.6	4.5	0.1		..
Utsunomiya	19.0	8.0			..	..					0.5	2.5	30.0	29.5			0.5
Choshi	..	..		..	..								..	..	..	..	
Kanazawa	6.0	6.0	..			..							12.0	12.0	..		
Fushiki	7.0	9.0	1.0								..	..	17.0	16.0	1.0		..
Nagano	26.0	15.7	3.0				..				1.0	7.3	53.0	49.0	3.0		1.0
Niigata	8.9	8.8	0.1								..	0.3	18.1	18.0	0.1		..
Yamagata	26.5	16.0	2.5	..							0.7	4.7	50.4	47.2	2.5		0.7
Akita	21.0	20.0	7.0	1.0								8.0	60.0	52.0	8.0		
Fukushima	15.5	8.5	..									2.0	26.0	26.0	..		
Ishinomaki	20.0	14.5	2.2	..				..			..	2.2	38.9	36.7	2.2	..	..
Miyako	20.7	17.2	5.8	0.4				..			0.1	6.4	50.6	41.3	6.2		0.1
Aomori	26.0	22.8	12.9	0.6			..				2.9	15.4	80.6	64.2	13.5		2.9
Hakodate	24.2	21.1	8.4	0.3							2.8	13.8	70.6	59.1	8.7	..	2.8
Suttsu	28.2	23.5	8.8	..						..	4.2	15.7	80.4	67.4	8.8		4.2
Sapporo	30.3	26.5	17.8	1.3				..		0.2	7.2	22.5	105.8	79.3	19.1		7.4
Kamikawa	31.0	28.0	22.0	3.0						..	1.5	25.0	124.0	84.0	25.0		15.0
Soya	30.8	27.6	18.8	3.6					..	0.2	8.0	22.0	111.0	80.4	22.4		8.2
Abashiri	31.0	27.5	18.5	2.5			..				11.5	32.5	123.5	91.0	21.0		11.5
Nemuro	30.3	27.3	20.5	3.8			..			..	3.0	18.0	102.9	75.6	24.3		3.0
Kushiro	31.0	26.5	16.5	1.5						0.5	9.0	22.0	107.0	79.5	18.0		9.5
Erimo	28.0	25.2	16.0	1.0						..	1.5	10.2	81.9	63.4	17.0		1.5

It is very rare in the south of 37° N that the maximum temperature of a day falls below the freezing point. Yet in inland provinces, this happens sometimes. Thus in Hokkaido, the month of January and also February has each more than twenty days with the maximum temperature below the freezing point, and in the whole year, this happens more than sixty days. In its central part, for over ninety days in a year, the maximum temperature falls below 0° C, as shown in the following table:—

MEAN NUMBER OF DAYS WITH MAXIMUM TEMPERATURE <0°.

Locality.	Jan.	Feb.	Mar.	April	May	June	July	Aug.	Sept.	Oct.	Nov.	Dec.	Year	Wint.	Spr.	Sum.	Aut.
Naha							..	..	..		..						
Kagoshima	..								..					..			
Miyazaki	..					..			..		..	..					
Kochi			..						..	..			..	..			
Tokushima					..					..						..	
Wakayama	..	..			..	..		..									
Oita	..			..				..		..		..	..	..			
Yamaguchi	0.5	..					..		..	..	..	..	0.5	0.5			
Hiroshima	..	..	..		..				..		..	..		..			
Matsuyama	..				..				..		..			..			
Okayama					..		..		..	..	..		..				
Ozaka						..		..	..		..	..				..	
Kioto			..		..	..						..	..	..			
Kumamoto									..	..	..			..			
Saga	..	..			..					..	..						
Nagasaki	..				..				..		..	..		..			
Fukuoka	..	..						..		..	..		..	..			
Itsugahara	0.2				.:				..			..	0.2	0.2			
Akamagaseki	..	..	..						..	..			..	..			
Sakai	0.0	0.0	..						..	..			0.0	0.0			
Tsu		..			..				..	..			..				
Nagoya	..	..	..			..						..	..	..			..
Gifu	0.1		..					..	..	..			0.1	0.1	..	..	
Hamamatsu			..	..		..									..		
Numazu	..			..			..			..				..	..		
Tokio	0.1								..				0.1	0.1		..	
Utsunomiya	..					..		..								..	
Choshi	..	..	..					..		..			..	..			
Kanazawa	1.0	..	..					..					1.0	1.0		..	
Fushiki	1·0	1.0	..						..			..	2.0	2.0			
Nagano	2.7	3.0			..				..	..		0.3	6.0	6.0	..		
Niigata	1.3	1.4	..						..		..	..	2.7	2.7			
Yamagata	10.5	4.0		..				..				0.3	14.8	14.8			
Akita	11.0	6.0		..					..		..	2.0	19.0	19.0	..		
Fukushima	0.5	1.0	..	..				..	..			..	1.5	1.5		..	
Ishinomaki	2.7	3.7	..	..						..		0.2	6.6	6.6		..	
Miyako	1.9	0.7	..								..	..	2.6	2.6	..		..
Aomori	15.4	9.7	1.7	..							0.1	5.7	32.6	30.8	1.7	..	0.1
Hakodate	14.2	8.4	1.3	0.1				..			0.4	5.8	30.2	28.4	1.4		0.4
Suttsu	20.7	12.0	2.7							..	1.3	6.8	43.5	39.5	2.7	..	1.3
Sapporo	23.7	13.7	3.3		..			..		..	1.5	9.2	51.4	46.6	3.3	..	1.5
Kamikawa	30.0	21.0	7.0	..							..	12.0	70.0	63.0	7.0	..	..
Soya	29.2	25.6	11.8	1.8						..	5.8	16.0	90.2	70.8	13.6		5.8
Abashiri	28.5	20.0	6.5	..					..		1.7	9.3	66.0	57.8	6.5	..	1.7
Nemuro	26.0	22.2	10.7	0.8							0.2	7.2	67.1	55.4	11.5	..	0.2
Kushiro	26.0	14.5	3.5	..				..		..	0.5	7.5	51.5	47.5	3.5	..	0.5
Erimo	21.0	18.7	5.7	..	..	..	..			..		3.7	49.1	43.4	5.7		

As is mentioned above, in the northern part of our Empire, especially in Hokkaido, it is of common occurrences that the minimum or the mean temperature of the day remains below the freezing point for several days successively. In the southern parts also, this sometimes happens for several days. The longest period, during which the minimum or the mean temperature of the day remained below the freezing point are given in the following table:—

LONGEST PERIOD DURING WHICH MINIMUM TEMPERATURE OR MEAN TEMPERATURE WAS CONSTANTLY BELOW 0°C.

Locality.	Minimum No. of days.	Mean No. of days.	Locality.	Minimum No. of days.	Mean No. of days.
Naha	0	0	Numazu	13	1
Kagoshima	7	3	Tokio	23	4
Miyazaki	8	0	Utsunomiya	37	8
Kochi	8	2	Choshi	8	0
Wakayama	8	0	Kanazawa	24	4
Oita	6	2	Fushiki	14	4
Yamaguchi	27	5	Nagano	46	31
Hiroshima	15	4	Niigata	41	11
Matsuyama	5	0	Yamagata	64	27
Okayama	9	4	Akita	71	22
Ozaka	30	2	Fukushima	32	10
Kioto	25	5	Ishinomaki	88	20
Kumamoto	17	2	Miyako	86	22
Saga	6	2	Aomori	81	34
Nagasaki	7	2	Hakodate	88	41
Fukuoka	4	2	Suttsu	87	43
Itsugahara	26	3	Sapporo	116	64
Akamagaseki	7	2	Kamikawa	148	91
Sakai	11	4	Soya	120	91
Tsu	12	3	Abashiri	105	64
Nagoya	17	3	Nemuro	126	69
Gifu	29	4	Kushiro	114	65
Hamamatsu	7	0	Erimo	112	56

Thus in Kagoshima, which lies on the southern extremity of Kiushu (lat. 32° N), the minimum temperature remained below the freezing point during seven days, and even the daily mean for three days successively. In the inland districts of Hokkaido, for 148 days, the minimum temperature remained constantly below the freezing point, and the daily mean for 91 days.

Owing to the above mentioned fact, there is no place in this country, where there is no snow-fall. Especially in the eastern provinces along the Sea of Japan, and also in Hokkaido, the precipitation in winter months is almost wholly in the form of snow, and the number of snowy days always exceeds sixty or seventy days. In Hokkaido, there are many provinces, where snow falls for more than 100 days in a year. Even in the south of Kiushu, there are generally three or four snow-falls in a year. The numbers of snowy days are given in the following table:—

Mean number of days with snow.

Locality.	Jan.	Feb.	Mar.	April	May	June	July	Aug.	Sept.	Oct.	Nov.	Dec.	Year	Wint.	Spr.	Sum.	Aut.
Naha	..	..										..	..		..		
Kagoshima	2.2	1.4	0.3									0.3	4.2	3.9	0.3		
Miyazaki	0.2	0.4	..								..	0.1	0.7	0.7	..		
Kochi	1.8	1.1	0.3								0.1	0.4	3.7	3.3	0.3		0.1
Wakayama	6.7	6.6	1.5	..							0.1	2.5	17.4	15.8	1.5		0.1
Oita	3.7	2.8	0.8	0.2			..			.	0.8	0.8	9.1	7.3	1.0		0.8
Yamaguchi	11.0	8.2	3.0	..						0.2	1.0	2.2	25.6	21.4	3.0		1.2
Hiroshima	8.3	6.4	2.7	0.2						..	0.3	4.3	22.2	19.0	2.9		0.3
Matsuyama	3.5	2.0	..								1.0	0.5	7.0	6.0			1.0
Okayama	8.0	9.0	.							.	1.0	2.0	20.0	19.0			1.0
Ozaka	5.7	7.1	2.0									1.6	16.4	14.4	2.0		
Kioto	12.5	12.5	6.3	0.4							0.2	5.2	37.1	30.2	6.7		0.2
Kumamoto	8.0	1.5	1.0								0.5	0.5	11.5	10.0	1.0		0.5
Saga	9.0	2.0	2.0								1.0	3.0	17.0	14.0	2.0		1.0
Nagasaki	4.8	2.9	1.2								0.2	3.0	12.1	10.7	1.2		0.2
Fukuoka	7.0	3.0	1.5									1.5	13.0	11.5	1.5		
Itsugahara	4.2	2.6	1.2					..			0.1	2.0	10.4	8.8	1.2		0.1
Akamagaseki	8.0	7.9	2.0	0.3				..			0.1	3.8	22.4	19.7	2.3		0.1
Sakai	16.7	15.3	6.0	0.7							1.0	8.8	48.5	40.8	6.7		1.0
Tsu	8.5	7.5	0.5								0.3	0.7	17.5	16.7	0.5		0.3
Nagoya	7.0	5.0	..								1.0	1.0	14.0	13.0	.		1.0
Gifu	9.7	6.9	2.6	0.3							0.2	4.1	23.8	20.7	2.9		0.2
Hamamatsu	2.3	2.1	0.2	..							0.1	0.9	5.9	5.6	0.2		0.1
Numazu	1.7	2.8	0.1	..							..	0.1	5.0	4.6	0.1		
Tokio	3.7	4.8	2.2	0.1							0.1	1.2	12.1	9.7	2.3		0.1
Utsunomiya	6.0	5.0	..							..	1.0	0.0	12.0	11.0	.		1.0
Choshi	0.6	3.2	0.1	..							..	..	4.2	3.8	0.4		
Kanazawa	19.7	17.2	10.0	0.8							1.1	10.3	59.1	47.2	10.8		1.1
Fushiki	22.0	17.4	9.4	0.4							0.7	7.1	57.3	46.8	9.8		0.7
Nagano	21.7	16.7	7.3	3.3							1.0	10.0	63.0	48.4	10.6		1.0
Niigata	23.5	19.6	11.5	1.1							2.0	14.2	71.9	57.3	12.6		2.0
Yamagata	26.0	17.5	8.0	2.0	..					.	5.3	11.0	69.8	54.5	10.0		5.3
Akita	26.2	21.4	14.7	1.1	0.1					0.2	6.9	18.7	89.6	66.3	16.2		7.1
Fukushima	17.5	10.5	4.0	1.5							2.3	7.7	43.5	35.7	5.5		2.3
Ishinomaki	14.2	11.2	6.2	1.7	..						1.8	7.1	42.5	32.8	7.9		2.8
Miyako	11.0	9.6	8.7	2.9						..	2.8	6.3	41.3	26.9	11.6		1.8
Aomori	27.6	21.9	18.2	3.1						0.5	9.7	21.5	102.5	71.0	21.3		10.2
Hakodate	24.2	19.8	17.2	3.8	..					0.6	9.0	20.0	94.6	64.0	21.0		9.6
Suttsu	27.4	22.9	17.7	4.0	0.1					0.6	12.1	22.7	107.5	73.0	21.8		12.7
Sapporo	24.4	20.6	17.3	5.6	0.6				0.2	1.2	12.2	22.1	104.5	67.1	23.5		13.6
Kamikawa	24.7	21.0	17.7	9.7	0.7					4.0	14.2	25.0	117.0	70.7	28.1		18.2
Soya	25.4	17.2	15.0	6.8	0.8					2.0	12.2	22.4	101.8	65.0	22.6		14.2
Abashiri	20.5	20.0	10.5	6.0	.	..				1.0	9.3	14.7	82.0	55.2	16.5		10.3
Nomuro	15.6	14.5	11.4	5.7	1.1	0.1				0.2	5.9	12.4	70.2	42.5	21.5	..	6.1
Kushiro	14.0	15.0	10.5	5.5						1.0	5.5	10.5	62.0	39.5	16.0		6.5
Erimo	18.0	12.6	9.6	4.0	0.4				..	0.4	5.8	13.6	64.4	44.2	14.0		6.2

The mean dates in which the minimum temperature falls, (earliest in Autumn and latest in Spring) below the freezing point are:—

MEAN DATES OF FIRST AND LAST OCCURRENCE OF MINIMUM TEMPERATURE BELOW 0°.

Locality.	First date.		Last date.		Locality.	First date.		Last date.	
Naha	....		....		Numazu	3	Dec.	25	March
Kagoshima	15	Dec.	8	March	Tokio	28	Nov.	27	„
Miyazaki	5		18		Utsunomiya	3	„	8	April
Kochi	8	„	16	„	Choshi	25	Dec.	5	March
Wakayama	17	„	23	„	Kanazawa	10	„	31	„
Oita	6	„	26	„	Fushiki	15	„	20	„
Yamaguchi	23	Nov.	4	April	Nagano	25	Oct.	16	April
Hiroshima	2	Dec.	26	March	Niigata	8	Dec.	27	March
Matsuyama	28	Nov.	21	„	Yamagata	6	Nov.	12	April
Okayama	11	Dec.	25	„	Akita	7	„	14	„
Ozaka	5	„	23	„	Fukushima	5	Dec.	6	„
Kioto	17	Nov.	11	April	Ishinomaki	18	Nov.	10	„
Kumamoto	7	„	21	March	Miyako	5	„	20	„
Saga	21	Dec.	25	„	Aomori	9	„	20	„
Nagasaki	21	„	13	„	Hakodate	26	Oct.	30	„
Fukuoka	25	Nov.	4	April	Suttsu	8	Nov.	19	„
Itsugahara	2	Dec.	22	March	Sapporo	18	Oct.	7	May
Akamagaseki	26	„	14	Feb.	Kamikawa	3	„	19	„
Sakai	19	„	26	March	Soya	5	„	24	April
Tsu	10	„	20	„	Abashiri	26	„	7	May
Nagoya	8	„	25	„	Nemuro	3	Nov.	13	April
Gifu	27	Nov.	30	„	Kushiro	9	Oct.	19	„
Hamamatsu	16	Dec.	6	„	Erimo	18	Nov.	27	„

Thus the earliest of the first mean date from which the minimum temperature begins to fall below 0°C is the 3rd of October at Kamikawa in the centre of Hokkaido, and the latest is the 25th of December at Choshi. Of the means of the last dates until which the minimum temperature in spring remained below the freezing point, the latest is the 19th of May at Kamikawa and the earliest is the 5th of March at Choshi.

The extremes of the first and the last dates for every station, which we have since their establishment, are as follows:—

EXTREME DATES OF FIRST AND LAST OCCURRENCE OF MINIMUM TEMPERATURE BELOW 0°.

Locality.	First date.		Last date.		Locality.	First date		Last date.	
Naha	....		....		Numazu	15	Nov.	8	April
Kagoshima	29	Nov.	28	March	Tokio	31	Oct.	15	„
Miyazaki	13	„	2	April	Utsunomiya	31	„	8	„
Kochi	26		25	March	Choshi	8	Dec.	30	March
Wakayama	30		4	April	Kanazawa	25	Nov.	17	April
Oita	15		5	„	Fushiki	17	„	25	March
Yamaguchi	13		25	„	Nagano	21	Oct.	20	April
Hiroshima	13		3		Niigata	17	Nov.	9	„
Matsuyama	27	„	25	March	Yamagata	31	Oct.	29	„
Okayama	26	„	25	„	Akita	26		30	
Ozaka	15	„	10	April	Fukushima	26	„	20	
Kioto	7		25	„	Ishinomaki	3	Nov.	25	„
Kumamoto	25	„	25	March	Miyako	17	Oct.	5	May
Saga	10	Dec.	25	„	Aomori	26	„	6	„
Nagasaki	29	Nov.	28	„	Hakodate	11	„	29	
Fukuoka	20		21	April	Suttsu	20	„	1	
Itsugahara	12	„	27	March	Sapporo	8	„	20	„
Akamagaseki	5	Dec.	14	„	Kamikawa	11	Sept.	25	„
Sakai	26	Nov.	10	April	Soya	29	Oct.	9	
Tsu	27	„	25	March	Abashiri	11	„	15	„
Nagoya	27	„	25	„	Nemuro	5		28	June
Gifu	12	„	10	April	Kushiro	8	„	25	May
Hamamatsu	5	Dec.	19	March	Erimo	2	Nov.	24	„

Thus the 11th of September at Kamikawa is the earliest date, which we ever have in this country, and the 28th of June at Nemuro is the latest.

The last two tables show us that the distribution of the period, during which the minimum temperature is below the freezing point, is very complicated and apparently has no direct relation with latitude. It seems to depend mainly on the character of the locality. In short, even in the southern extremity of this country, from the end of October toward the end of April, there is every probability of having the minimum temperature lower than the freezing point.

Though our temperature during winter is far below the normal temperature, yet our mean temperature during summer is nearly equal to the normal, as we have said before. Thus the daily maximum temperature exceeds 30°C for more than 80 days in a year in the neighbourhood of Okinawa Island; over 50 days in Kiushu and the southern Nippon, and about 10 days even in Hokkaido. But we had no instances even in Okinawa that the daily maximum temperature remained constantly over 30°C during whole summer. The mean numbers of days with daily maximum temperature over 30°C are:—

MEAN NUMBER OF DAYS WITH MAXIMUM TEMPERATURE >30°.

Locality.	Jan.	Feb.	Mar.	April.	May	June	July	Aug.	Sept.	Oct.	Nov.	Dec.	Year	Wint.	Spr.	Sum.	Aut.
Naha						6.0	28.5	27.5	17.0	8.5			87.5			62.0	25.5
Kagoshima						2.0	20.1	21.0	11.2	0.1	..		57.4			46.1	11.3
Miyazaki						4.0	19.0	24.0	9.0				56.0			47.0	9.0
Kochi						0.2	6.8	14.1	2.5	..			23.6		..	21.1	2.5
Wakayama					0.1	0.9	15.6	25.2	8.5	..			50.3		0.1	41.7	8.5
Oita					..	1.0	11.6	17.2	3.6				33.4			29.8	3.6
Hiroshima				..		0.1	15.2	24.7	7.8			..	48.1			40.3	7.8
Matsuyama				..	..	3.0	15.0	21.0	7.0	..	..		46.0		..	39.0	7.0
Okayama					1.0	..	8.0	15.0	11.0	..			35.0		1.0	23.0	11.0
Ozaka					0.1	2.0	17.8	26.6	12.4	0.1			59.0		0.1	46.4	12.5
Kioto					0.1	3.0	17.3	24.5	8.4	0.1		..	53.7		0.1	41.8	8.5
Kumamoto					..	7.0	23.5	26.5	19.0	0.5			76.5		..	57.0	19.5
Saga					1.0	5.0	12.0	19.5	16.5		..	..	54.0		1.0	36.5	16.5
Nagasaki						1.3	18.0	25.0	10.3	..			54.6			41.3	10.3
Fukuoka						2.5	17.0	16.5	12.0	..			48.0			36.0	12.0
Itsugabara						0.8	10.6	19.6	3.2				34.2			31.0	3.2
Akamagaseki				..	.	0.2	8.5	14.5	2.7				25.9		..	23.2	2.7
Sakai					0.0	2.0	14.0	20.0	4.0				40.0		0.0	36.0	4.0
Tsu					..	3.0	7.0	15.0	3.0				28.0		..	25.0	3.0
Nagoya					1.0	2.0	16.0	24.5	12.0				55.5		1.0	42.5	12.0
Gifu					0.6	2.3	15.7	24.3	7.6			..	50.5		0.6	42.3	7.6
Hamamatsu				..	0.1	0.6	9.3	20.9	6.4				37.3		0.1	30.8	6.4
Numazu						0.3	4.3	14.8	4.3				23.6		..	19.4	4.2
Tokio					..	0.7	9.9	16.0	4.2				30.8		..	26.6	4.2
Utsunomiya					2.0	..	6.0	9.0	6.5				23.5		2.0	15.0	6.5
Choshi					..	0.2	5.2	13.4	5.1				24.2		..	18.8	5.4
Kanazawa						1.0	12.0	19.0	6.0				38.0			32.0	6.0
Fushiki				..	..	..	7.0	11.0	2.0				20.0		..	18.0	2.0
Nagano	..			0.3	1.7	3.0	10.7	18.0	7.0				40.7		2.0	31.7	7.0
Niigata	..				..	0.9	8.4	16.3	4.4				30.0		..	25.6	4.4
Yamagata		..			1.0	2.5	5.3	10.7	3.0				22.5		1.0	18.5	3.0
Akita		..			..	2.0	5.0	11.0	2.0				20.0		..	18.0	2.0
Fukushima	..				1.3	3.3	8.0	14.0	4.3				30.9		1.3	25.3	4.3
Ishinomaki					..	..	2.0	3.2	0.2				5.1		..	5.2	0.2
Miyako					0.1	0.4	5.3	9.1	2.1			..	17.3		0.1	15.1	2.1
Aomori	..				..	0.2	2.6	5.7	0.5				9.0		..	8.5	0.5
Hakodate		..				..	0.8	1.9	0.1				2.8		..	2.7	0.1
Suttsu		..		..	..	..	0.7	1.5	..				2.2		..	2.2	..
Sapporo		..	..		..	0.2	3.2	5.7	0.3				9.4		..	9.1	0.3
Kamikawa	..	..	..			..	6.0	6.0	..				12.0	..	..	12.0	
Soya					..	..	..	..	..				..		..	..	..
Abashiri				..	..		1.0	4.0	0.7	0.3			6.0	..	..	5.0	1.0
Nemuro	..			..		..	..	..					..		..	..	
Kushiro	..		..			..	1.0	..	..				1.0	..	..	1.0	..

We have never met with instances in all stations, since their establishment, that the daily mean temperature reached 30°. It is also a rare phenomenom that, in our country even in southern part, the minimum temperature reaches or exceeds 25°, as the next table shows:—

Mean number of days with minimum temperature >25°.

Locality.	Jan.	Feb.	Mar.	April	May	June	July	Aug.	Sept.	Oct.	Nov.	Dec.	Year	Wint.	Spr.	Sum.	Aut.
Naha				..		3.0	15.0	15.5	4.5	0.5			38.5			33.5	5.0
Kagoshima						0.1	3.6	3.8	1.4				8.9			7.5	1.4
Miyazaki						..	2.0	1.0	..	..			3.0			3.0	
Kochi						0.1	0.5	0.4	0.2	..			1.2			1.0	0.2
Wakayama				..		0.1	3.1	2.8	0.2				6.2			6.0	0.2
Oita							0.2	0.4	0.2				0.8			0.6	0.2
Yamaguchi							..	0.3					0.3			0.3	
Hiroshima							2.3	2.3					4.6			4.6	
Matsuyama									..				..				
Okayama							..	..	2.0				2.0			..	2.0
Ozaka							1.3	2.8	0.1				4.2			4.1	0.1
Kioto							0.5	0.6	0.1				1.2			1.1	0.1
Kumamoto							0.5	..					0.5			0.5	
Saga				..			2.0	1.0					3.0			3.0	..
Nagasaki							4.5	3.5	0.7				8.7			8.0	0.7
Fukuoka							..	..					..			..	
Itsugahara							1.6	0.8	.	..			2.4			2.4	..
Akamagaseki							0.3	2.0	0.2				2.5			2.3	0.2
Sakai							1.0	2.0					3.0			3.0	
Tsu							..	..	..				..			.	..
Nagoya							0.5	1.5	0.5				2.5			2.0	0.5
Gifu							1.4	1.9	0.4				3.7			3.3	0.4
Hamamatsu							0.6	0.8	0.3				1.7			1.4	0.3
Numazu							0.4	1.8	..				2.2			2.2	
Tokio								0.3	0.2				0.5			0.3	0.2
Utsunomiya									..	..			.				..
Choshi									0.6				0.6				0.6
Kanazawa																	
Fushiki														..	..		
Nagano								..	..							..	..
Niigata							0.4	2.9	0.1				3.4			3.3	0.1
Yamagata																	..
Akita									..				..				..
Fukushima									0.3				0.3				0.3
Ishinomaki																	
Miyako										..				..			
Aomori							..						..		..	..	..
Hakodate							0.1						0.1			0.1	
Suttsu																	..
Sapporo																	
Kamikawa					..												
Soya				..													
Abashiri					..												
Nemuro					..							..					
Kushiro					..												
Erimo	..	..	..		..					..							..

If we now take the maximum temperature over 25° C, it happens very frequently in our country. Thus in the vicinity of Okinawa, it happens in every month of the year without any exception. And for all the nine months from April till October, it happens everywhere in our country, excepting Hokkaido. In August, the day does not pass, in which its maximum temperature is not over 25°C.

The numbers of days having their maximum temperatures over 25° C are given in the following table:—

Mean number of days with maximum temperature >25°.

Locality.	Jan.	Feb.	Mar.	April	May	June	July	Aug.	Sept.	Oct.	Nov.	Dec.	Year	Wint.	Spr.	Sum.	Aut.
Naha	1.0	5.0	2.0	9.0	23.0	27.0	31.0	31.0	30.0	27.0	13.0	3.5	202.5	0.5	34.0	89.0	70.0
Kagoshima				1.4	10.7	21.1	10.6	6.7	17.4	16.0	0.2		84.4		12.4	38.4	33.6
Miyazaki		..		4.0	12.0	23.0	30.0	31.0	28.0	13.0	1.0	..	142.0	..	16.0	81.0	42.0
Kochi				0.4	3.9	15.6	29.4	31.0	26.1	7.8	0.2		114.4		4.3	76.0	34.1
Wakayama		..		0.9	4.9	18.7	29.9	31.0	25.2	4.5	0.2		115.3		5.8	79.6	29.9
Oita				0.8	3.0	12.6	24.0	26.8	20.2	2.0	..		89.4	..	3.8	63.4	22.2
Yamaguchi				0.5	8.5	20.2	27.7	30.5	24.5	4.5	..		116.4	..	9.0	78.4	29.0
Hiroshima		..		0.1	4.6	17.7	28.6	30.7	21.6	7.0	.		113.3		4.7	77.0	31.6
Matsuyama					11.0	21.0	29.0	31.0	25.0	7.0			124.0	..	11.0	81.0	32.0
Okayama		..		..	14.0	18.0	30.0	30.0	25.0	7.0			124.0		11.0	78.0	32.0
Ozaka		..		0.4	6.0	21.0	29.4	30.9	25.8	6.8			120.3		6.4	81.3	32.6
Kioto		..		0.7	6.5	19.0	28.7	30.7	21.1	4.5	..		114.5		7.2	78.4	28.9
Kumamoto				4.5	17.5	26.0	30.5	31.0	28.5	16.5	0.5		155.0	..	22.0	87.5	45.5
Saga		..		..	16.0	23.0	28.0	31.0	27.5	10.5	..	..	136.0	..	16.0	82.0	38.0
Nagasaki				0.5	6.3	21.2	29.7	31.0	27.2	13.0	0.5		129.4		6.8	81.9	40.7
Fukuoka				2.0	7.0	21.5	29.5	30.5	26.0	4.0	..		120.5		9.0	81.5	30.0
Itsugahara				..	5.4	14.8	23.2	30.2	22.4	5.4		..	101.4		5.4	68.2	27.8
Akamagaseki				0.2	2.2	10.7	25.3	30.8	21.8	2.5			93.5		2.4	66.8	24.3
Sakai			0.0	1.0	4.0	12.0	24.0	30.0	16.0	1.0			89.0		5.0	67.0	17.0
Tsu				1.0	5.0	15.0	27.0	30.0	22.0	4.0			104.0		6.0	72.0	26.0
Nagoya				.	17.0	19.0	29.0	30.0	27.0	8.5	..		130.5		17.0	78.0	35.5
Gifu			..	0.8	6.2	20.1	27.9	30.0	23.6	7.4	0.1		116.1		7.0	78.0	31.1
Hamamatsu	..			0.3	3.4	15.3	27.2	30.6	26.3	6.8	0.1		110.0		3.7	73.1	33.2
Numazu				0.3	1.3	12.5	26.5	30.8	23.2	1.1			99.0		1.6	69.8	27.6
Tokio				0.3	3.1	13.1	26.6	29.9	19.5	1.9			94.4		3.4	69.6	21.4
Utsunomiya					10.0	12.0	26.0	23.5	25.0	2.5			99.0		10.0	61.5	27.5
Choshi				..	0.4	9.4	23.4	29.6	19.0	3.2			85.0		0.4	62.1	22.2
Kanazawa				1.0	4.0	12.0	24.0	30.0	18.0	3.0			92.0		5.0	66.0	21.0
Fushiki				..	4.0	6.0	21.0	28.0	11.0	1.0			74.0		4.0	55.0	15.0
Nagano				1.3	7.3	16.7	24.0	29.3	18.7	1.3			98.6		8.6	70.0	20.0
Niigata				0.1	3.1	7.2	21.6	29.8	16.0	1.1			78.9		3.2	58.6	17.1
Yamagata				1.0	7.5	12.0	19.7	29.0	16.3	1.7			87.2		8.5	60.7	18.0
Akita				2.0	4.0	25.0	14.0	10.0	..	..			55.0		2.0	43.0	10.0
Fukushima				2.0	8.7	15.0	21.3	28.3	18.0	1.0			94.3		10.7	64.6	19.0
Ishinomaki				.	.	1.7	14.0	10.0	10.6	0.8			37.1		..	25.7	11.4
Miyako				0.4	2.4	4.8	14.7	21.3	10.2	0.8			54.6		2.8	40.8	11.0
Aomori					0.4	2.9	12.7	12.5	7.5	0.1			46.1		0.4	38.1	7.6
Hakodate				..		0.7	9.9	19.4	4.2				34.2			30.0	4.2
Suttsu				..	..	1.2	7.0	17.2	2.7				28.1		..	25.4	2.7
Sapporo				..	1.0	3.3	15.5	20.7	5.2				45.7		1.0	39.5	5.2
Kamikawa					3.0	7.0	17.0	15.0	2.0	..			44.0	..	3.0	39.0	2.0
Soya					..	0.2	0.6	5.2	0.6				6.6	..	..	6.0	0.6
Abashiri					0.5	2.0	5.0	11.3	4.7				26.5		0.5	21.3	4.7
Nemuro		..			.	0.2	3.8	7.8	2.0				13.8		..	11.8	2.0
Kushiro					0.5	2.5	8.0	11.0	7.5	..	..		32.5		0.5	21.5	7.5

Let us now consider when the daily mean temperature remains over 25° C. Throughout the Empire, excepting Hokkaido, we have the daily mean temperature over 25° C for the four months, from June till September. In Okinawa, this occurs from May till October, but in Hokkaido, it is rare that the daily mean reaches or exceeds 25° C, and if any it happens only in July and August—indeed, in its eastern coast, we have such places where the daily mean never reached 25° C. The details are given in the following table:—

Mean number of days with mean temperature ≥25°.

Locality.	Jan.	Feb.	Mar.	April	May	June	July	Aug.	Sept.	Oct.	Nov.	Dec.	Year	Wint.	Spr.	Sum.	Aut.
Naha					2.0	9.0	30.5	31.0	29.0	10.5			112.0		2.0	70.5	39.5
Kagoshima						4.2	24.7	28.2	14.3	0.1			71.5			57.1	11.4
Miyazaki						4.0	22.0	27.0	11.0				64.0			53.0	11.0
Kochi						1.6	18.0	26.7	9.3				55.6			46.3	9.3
Wakayama						2.0	21.7	28.5	10.5				62.7			52.2	10.5
Oita						1.4	18.0	26.4	5.8				51.6			45.8	5.8
Yamaguchi					..	1.0	14.0	21.7	5.7				42.4			36.7	5.7
Hiroshima						0.6	17.6	27.5	8.5				54.2			45.7	8.5
Matsuyama						2.0	17.0	24.0	7.0				50.0			43.0	7.0
Okayama						..	13.0	20.0	13.0				46.0			33.0	13.0
Ozaka						2.0	20.7	28.0	10.8				61.5			50.7	10.8
Kioto			.			1.8	18.4	26.4	8.5				55.1			46.6	8.5
Kumamoto						3.5	24.5	28.0	14.5				70.5			56.0	14.5
Saga						.	18.0	23.5	14.5				56.0			41.5	14.5
Nagasaki			..			1.8	22.8	29.0	11.0				64.6			53.6	11.0
Fukuoka						2.0	17.5	23.5	9.5				52.5			43.0	9.5
Itsugahara						0.8	15.0	22.8	4.4				43.0			38.6	4.4
Akamagaseki					..	0.5	15.5	25.0	6.8				47.8			41.0	6.8
Sakai					..	2.0	15.0	24.0	7.0				48.0			41.0	7.0
Tsu						3.0	12.0	25.0	5.0	..			45.0			40.0	5.0
Nagoya			..			..	19.5	25.0	13.5				58.0			44.5	13.5
Gifu						1.4	18.3	26.0	8.3				54.0			45.7	8.3
Hamamatsu						0.8	15.2	24.8	6.9				47.7			40.8	6.9
Numazu						1.1	15.1	22.0	6.7				44.9			38.2	6.7
Tokio						1.1	14.7	22.0	5.5	..			43.3			37.8	5.5
Utsunomiya							5.0	5.0	5.0				15.0			10.0	5.0
Choshi						..	10.6	17.2	8.6				36.4			27.8	8.6
Kanazawa						1.0	14.0	21.0	6.0				41.0			35.0	6.0
Fushiki						1.0	12.0	21.0	5.0				39.0			34.0	5.0
Nagano							3.3	10.0	2.7				16.0			13.3	2.7
Niigata			..			0.9	11.2	19.4	6.1				37.6			31.5	6.1
Yamagata						0.5	2.0	7.0	1.7				11.2			9.5	1.7
Akita						.	4.0	9.0	3.0				16.0			13.0	3.0
Fukushima						1.7	7.7	10.4	3.0				22.8			19.8	3.0
Ishinomaki						0.5	3.0	10.7	1.7				15.9			14.2	1.7
Miyako							2.6	6.7	1.0				10.3			9.3	1.0
Aomori							3.3	5.9	0.2				9.4			9.2	0.2
Hakodate							1.4	4.5	0.3				6.2			5.9	0.3
Suttsu							0.7	1.8	..				2.5			2.5	..
Sapporo							1.2	2.7	0.2				4.1			3.9	0.2
Kamikawa							2.0	1.0					3.0			3.0	
Soya							0.2	0.2					0.4			0.4	
Abashiri							..	1.3					1.3			1.3	
Nemuro									..								
Kushiro				..					..	..							

The longest periods during which the daily maximum temperature remained constantly higher than 30° C are as follows:—

LONGEST PERIOD DURING WHICH THE MAXIMUM TEMPERATURE WAS CONSTANTLY OVER 30°.

Locality.	No. of days.	Locality.	No. of days.	Locality.	No. of days.
Naha	27	Itsugahara	21	Akita	13
Kagoshima	24	Akamagaseki	10	Fukushima	9
Miyazaki	21	Sakai	23	Ishinomaki	12
Kochi	15	Tsu	7	Miyako	10
Wakayama	29	Nagoya	15	Aomori	9
Oita	16	Gifu	30	Hakodate	3
Yamaguchi	25	Hamamatsu	34	Suttsu	2
Hiroshima	32	Numazu	20	Sapporo	6
Matsuyama	15	Tokio	21	Kamikawa	8
Okayama	9	Utsunomiya	6	Soya	0
Ozaka	43	Choshi	11	Abashiri	3
Kioto	44	Kanazawa	26	Nemuro	0
Kumamoto	21	Fushiki	10	Kushiro	1
Saga	15	Nagano	17	Erimo	0
Nagasaki	37	Niigata	18		
Fukuoka	16	Yamagata	6		

Thus the long continuation of the hottest season is in no way dependent upon latitudes, but the chief cause is its locality, whether inland or along the coast. For this reason, the longest period during which the maximum temperature remained constantly above 30° C is generally within 30 days in those southern districts as Naha and Kagoshima, while in Kioto this continued for more than 40 days.

The mean first and the mean last days whose maximum temperatures are above 30° C are given in the following table. This table tells us that in the southern Japan, this begins generally from the middle of June, and ends at the last decades of September, and that in the northern Japan it begins from the first decades of July and lasts till the end of August.

MEAN DATES OF FIRST AND LAST OCCURRENCE OF MAXIMUM TEMPERATURE >30°.

Locality.	First date.	Last date.	Locality.	First date.	Last date.
Naha	27 June	20 Oct.	Kumamoto	7 June	30 Sept.
Kagoshima	28 „	24 Sept.	Saga	21 May	29 „
Miyazaki	19 „	20 „	Nagasaki	24 June	20 „
Kochi	9 July	8 „	Fukuoka	12 „	29 „
Wakayama	19 June	14 „	Itsugahara	26 „	7 „
Oita	29 „	8 „	Akamagaseki	12 July	6 „
Yamaguchi	9 „	6 „	Sakai	14 June	10 „
Hiroshima	2 July	13 „	Tsu	23 „	4 „
Matsuyama	13 June	18 „	Nagoya	12 „	20 „
Okayama	22 May	14 „	Gifu	11 „	14 „
Ozaka	13 June	17 „	Hamamatsu	24 „	13 „
Kioto	11 „	12 „	Numazu	7 July	8 „

Locality.	First date.	Last date.	Locality.	First date.	Last date.
Tokio	26 June	11 Sept.	Miyako	29 June	4 Sept.
Utsunomiya	21 May	20 „	Aomori	13 July	29 Aug.
Choshi	17 July	6 „	Hakodate	6 Aug.	13 „
Kanazawa	14 June	20 „	Suttsu	30 July	13 „
Fushiki	27 „	31 Aug.	Sapporo	17 „	12 „
Nagano	9 „	15 Sept.	Kamikawa	10 „	19 „
Niigata	3 July	7 „	Soya	.	—
Yamagata	16 June	6 „	Abashiri	27 July	20 Aug.
Akita	18 July	31 Aug.	Nemuro	—	—
Fukushima	1 June	13 Sept.	Kushiro	—	—
Ishinomaki	16 July	18 Aug.	Erimo	—	—

The extremes of these dates for every station, are

Extreme dates of first and last occurrence of maximum temperature (>30°).

Locality.	First date.	Last date.	Locality.	First date.	Last date.
Naha	24 June	26 Oct.	Numazu	25 June	16 Sept.
Kagoshima	7 „	2 „	Tokio	18 „	24 „
Miyazaki	29 „	11 „	Utsunomiya	21 May	22 „
Kochi	25 „	20 Sept.	Choshi	25 June	25 „
Wakayama	21 May	28 „	Kanazawa	23 May	23 Oct.
Oita	19 June	17 „	Fushiki	30 „	15 Sept.
Yamaguchi	30 May	18 „	Nagano	25 June	28 „
Hiroshima	8 June	22 „	Niigata	12 „	23 „
Matsuyama	13 „	20 „	Yamagata	15 May	15 „
Okayama	22 May	11 „	Akita	22 June	21 „
Ozaka	19 „	1 Oct.	Fukushima	10 May	16 „
Kioto	14 „	1 „	Ishinomaki	10 July	5 „
Kumamoto	3 June	2 „	Miyako	16 May	18 „
Saga	21 May	30 Sept.	Aomori	21 June	15 „
Nagasaki	7 June	30 „	Hakodate	2 Aug.	16 Aug.
Fukuoka	11 „	30 „	Suttsu	23 July	15 „
Itsugahara	6 „	19 „	Sapporo	17 „	6 Sept.
Akamagaseki	23 „	18 „	Kamikawa	10 „	2 „
Sakai	14 May	28 „	Soya	—	—
Tsu	21 „	9 „	Abashiri	17 July	23 Aug.
Nagoya	21 „	22 „	Nemuro	—	—
Gifu	14 „	23 „	Kushiro	18 July	18 July
Hamamatsu	21 „	23 „	Erimo	—	—

The earliest date is the 29th of April, 1890 at Miyazaki and the latest is the 23rd of October, 1887 at Kanazawa.

CHAPTER II.

ATMOSPHERIC PRESSURE.

At all our meteorological stations, the atmospheric pressure is measured by mercurial barometers. The observations made by these instruments have absolutely the same weight, for all these barometers were tested at Kew-observatory, and again compared with the standard barometer at our Central Meteorological Observatory, and thereby the corrections for each of them are calculated.

The barometric readings in the art. A, following, are the readings, to which temperature corrections were applied. For those in the art. B, et seq. the corrections for (1) the temperature, (2) the height above the sea level, and (3) the gravity were applied. In other words, the readings given in the art. A are the readings reduced to 0° C, while the rest are reduced readings to the sea level at lat. 45° and also to 0° C.

A. DIURNAL VARIATION OF ATMOSPHERIC PRESSURE.

The mean atmospheric pressures observed at our first class stations, where hourly observations are taken through day and night, are given in the following tables in millimetres, omitting the hundreds:

KUMAMOTO.

Month / Hour	Jan.	Feb.	Mar.	Apr.	May	June	July	Aug.	Sept.	Oct.	Nov.	Dec.	Year	Wint.	Spr.	Sum.	Aut.
1 am	67.47	66.26	65.08	62.93	59.61	58.15	58.11	57.41	58.00	62.69	67.15	67.08	62.49	66.91	62.54	57.89	62.61
2 am	67.50	66.25	64.93	62.73	59.46	58.02	57.98	57.35	57.85	62.65	67.10	67.08	62.41	66.91	62.37	57.78	62.53
3 am	67.34	66.17	64.74	62.63	59.39	57.88	57.92	57.25	57.78	62.54	67.03	67.06	62.31	66.86	62.25	57.68	62.45
4 am	67.21	66.05	64.61	62.57	59.43	57.90	57.87	57.22	57.73	62.50	66.99	66.98	62.26	66.76	62.21	57.66	62.41
5 am	67.09	66.11	64.69	62.69	59.54	57.96	57.96	57.27	57.80	62.62	67.03	66.90	62.30	66.70	62.31	57.73	62.48
6 am	67.21	66.21	64.83	62.88	59.71	58.08	58.05	57.46	57.88	62.72	67.18	67.01	62.43	66.81	62.47	57.86	62.59
7 am	67.31	66.53	65.06	63.16	59.92	58.20	58.20	57.61	58.10	62.96	67.45	67.29	62.65	67.04	62.71	58.00	62.84
8 am	67.61	66.80	65.37	63.23	59.95	58.34	58.28	57.66	58.26	63.16	67.66	67.55	62.82	67.32	62.85	58.09	63.03
9 am	67.97	66.93	65.50	63.29	59.92	58.30	58.28	57.61	58.40	63.22	67.81	67.75	62.91	67.52	62.90	58.06	63.14
10 am	68.17	66.82	65.46	63.34	59.79	58.23	58.26	57.56	58.17	63.09	67.80	67.90	62.91	67.63	62.86	58.02	63.12
11 am	67.95	66.64	65.24	63.16	59.60	58.05	58.11	57.41	58.26	62.83	67.47	67.60	62.69	67.40	62.67	57.86	62.85
Noon	67.33	66.16	64.86	62.85	59.29	57.82	57.88	57.14	58.02	62.43	66.90	66.98	62.30	66.82	62.33	57.61	62.45
1 pm	66.73	65.59	64.37	62.51	58.94	57.47	57.59	56.85	57.72	61.87	66.36	66.38	61.86	66.23	61.91	57.30	61.98
2 pm	66.40	65.11	64.00	62.20	58.63	57.27	57.33	56.07	57.42	61.50	66.01	65.98	61.53	65.83	61.61	57.06	61.64
3 pm	66.39	64.85	63.71	61.72	58.27	57.07	57.12	56.32	57.33	61.17	65.91	65.93	61.34	65.72	61.23	56.81	61.57
4 pm	66.51	64.80	63.68	61.57	58.12	56.96	57.01	56.22	57.35	61.57	66.04	66.05	61.32	65.79	61.12	56.73	61.65
5 pm	66.75	65.07	63.79	61.70	58.09	56.94	56.95	56.26	57.53	61.77	66.23	66.25	61.41	66.02	61.19	56.72	61.84
6 pm	67.01	65.42	64.10	61.91	58.21	57.01	57.03	56.39	57.66	62.04	66.52	66.55	61.66	66.33	61.41	56.82	62.07
7 pm	67.29	65.84	64.45	62.23	58.54	57.34	57.29	56.66	57.86	62.42	66.89	66·85	61.97	66.66	61.74	57.10	62.39
8 pm	67.55	46.14	64.73	62.67	58.90	57.68	57.60	57.06	58.23	62.65	67.10	67.12	62.29	66.94	62.10	57.45	62.66
9 pm	67.57	66.30	65.01	63.01	59.25	57.92	57.89	57.38	58.42	62.88	67.28	67.26	62.51	67.04	62.42	57.73	62.86
10 pm	67.57	66.39	65.14	63.18	59.56	58.23	58.18	57.54	58.36	63.02	67.45	67.32	62.66	67.09	62.63	57.98	62.94
11 pm	67.50	66.36	65.15	63.18	59.58	58.28	58.25	57.58	58.30	63.01	67.44	67.28	62.66	67.05	62.61	58.01	62.92
M. N.	67.41	66.17	65.05	63.20	59.55	58.19	58.16	57.56	58.19	62.92	67.41	67.17	62.58	66.92	62.60	57.97	62.84
Mean	67.29	66.05	64.74	62.69	59.21	57.80	57.81	57.14	57.95	62.52	67.01	66.97	62.26	66.77	62.21	57.58	62.49

MATSUYAMA.

1 am	62.99	62.69	61.70	59.78	56.15	54.98	55.19	54.31	55.85	59.62	63.40	63.16	59.15	62.95	59.21	54.83	59.62
2 am	63.10	62.70	61.55	59.57	56.02	54.83	55.09	54.21	55.73	59.55	63.31	63.14	59.07	62.98	59.05	54.71	59.52
3 am	63.00	62.59	61.32	59.48	56.05	54.77	55.04	54.17	55.60	59.42	63.29	63.14	58.99	62.91	58.95	54.66	59.41
4 am	62.77	62.46	61.24	59.39	56.05	54.85	55.02	54.17	55.53	59.39	63.24	62.99	58.92	62.74	58.89	54.68	59.39
5 am	62.74	62.60	61.31	59.56	56.23	55.05	55.11	54.25	55.58	59.55	63.30	62.97	59.02	62.77	59.03	54.80	59.48
6 am	62.81	62.63	61.47	59.85	56.47	55.13	55.32	54.50	55.76	59.69	63.44	63.13	59.18	62.86	59.26	54.98	59.63
7 am	63.09	62.90	61.73	60.15	56.65	55.32	55.48	54.62	55.99	59.89	63.79	63.15	59.42	63.15	59.51	55.14	59.89

MATSUYAMA.

Month / Hour	Jan.	Feb.	Mar.	Apr.	May	June	July	Aug.	Sept.	Oct.	Nov.	Dec.	Year	Wint.	Spr.	Sum.	Aut.
8 am	63,29	63,18	61,99	60,27	56,77	55,39	55,59	54,69	56,12	60,19	64,06	63,72	59,60	63,40	59,68	55,22	60,12
9 am	63,53	63,29	62,09	60,35	56,75	55,37	55,57	54,69	56,23	60,26	64,19	64,01	59,69	63,61	59,73	55,21	60,23
10 am	63,69	63,28	62,06	60,40	56,68	55,27	55,57	54,66	56,17	60,08	64,12	64,09	59,67	63,69	59,71	55,17	60,12
11 am	63,35	63,07	61,89	60,21	56,53	55,25	55,45	54,47	55,91	59,82	63,78	63,71	59,45	63,39	59,51	55,06	59,83
Noon	62.70	62,67	61,62	59,86	56,29	54,98	55,28	54,14	55,65	59,29	63,47	63,01	59,05	62,80	59,26	54,80	59,37
1 pm	62,18	62,08	61,16	59,55	55,92	54,73	54,94	53,91	55,28	58,79	62,72	62,51	58,65	62,27	58,88	54,53	58,93
2 pm	62,03	61,83	60,78	59,24	55,66	54,50	54,74	53,65	55,15	58,61	62,49	62,24	58,42	62,03	58,56	54,30	58,75
3 pm	62,10	61,77	60,61	58,95	55,37	54,24	54,54	53,38	55,05	58,57	62,46	62,25	58,27	62,01	58,31	54,05	58,69
4 pm	62,28	61,68	60,61	58,76	55,29	54,16	54,41	53,31	55,07	58,72	62,61	62,38	58,27	62,11	58,23	53,96	58,80
5 pm	62,49	61,87	60,75	58,80	55,18	54,09	54,38	53,31	55,33	58,87	62,83	62,58	58,37	62,31	58,24	53,93	59,01
6 pm	62,76	62,17	61,05	59,00	55,22	54,23	54,42	53,39	55,50	59,18	63,09	62,88	58,57	62,60	58,42	54,01	59,26
7 pm	63,05	62,31	61,27	59,27	55,47	54,50	54,61	53,68	55,74	59,45	63,37	63,07	58,82	62,82	58,67	54,27	59,52
8 pm	63,21	62,48	61,62	59,68	55,76	54,69	54,85	54,01	56,08	59,68	63,51	63,28	59,07	62,99	59,02	54,52	59,77
9 pm	63,33	62,64	61,79	60,10	56,09	54,97	55,13	54,36	56,25	59,90	63,70	63,41	59,30	63,13	59,33	54,82	59,95
10 pm	63,35	62,71	61,86	60,15	56,23	55,18	55,35	54,41	56,21	59,96	63,78	63,46	59,39	63,17	59,45	54,98	59,99
11 pm	63,27	62,68	61,81	60,15	56,30	55,20	55,40	54,43	56,19	59,88	63,75	63,41	59,37	63,12	59,42	55,01	59,94
M. N.	63,08	62,68	61,72	60,06	56,11	55,02	55,29	54,40	56,02	59,77	63,67	63,21	59,26	63,00	59,31	54,90	59,82
Mean	62,92	62,54	61,46	59,69	56,06	54,86	55,08	54,13	55,75	59,51	63,38	63,14	59,01	62,88	59,07	54,69	59,55

HIROSHIMA.

1 am	66,75	65,75	65,23	62,87	59,75	57,52	57,28	57,00	59,02	62,97	66,73	66,58	62,28	66,36	62,62	57,27	62,91
2 am	66,84	65,69	65,08	62,69	59,60	57,36	57,15	56,89	58,92	62,92	66,71	66,65	62,21	66,39	62,46	57,13	62,85
3 am	66,72	65,54	64,87	62,57	59,55	57,26	57,02	56,86	58,83	62,80	66,65	66,63	62,12	66,30	62,33	57,05	62,76
4 am	66,56	65,48	64,82	62,57	59,59	57,31	57,17	56,88	58,80	62,77	66,62	66,52	62,09	66,19	62,33	57,13	62,73
5 am	66,51	65,54	64,94	62,67	59,76	57,51	57,25	57,01	58,86	62,94	66,70	66,57	62,19	66,21	62,46	57,26	62,83
6 am	66,65	65,68	65,14	62,92	60,02	57,61	57,40	57,22	59,04	63,09	66,87	66,73	62,37	66,35	62,69	57,42	63,00
7 am	66,90	65,98	65,39	63,23	60,23	57,82	57,61	57,40	59,23	63,36	67,16	67,04	62,61	66,63	62,95	57,61	63,25
8 am	67,11	66,25	65,60	63,36	60,35	57,92	57,71	57,47	59,35	63,61	67,13	67,28	62,78	66,88	63,10	57,70	63,47
9 am	67,39	66,41	65,73	63,38	60,27	57,84	57,71	57,39	59,43	63,71	67,55	67,51	62,86	67,10	63,13	57,65	63,56
10 am	67,49	66,31	65,66	63,33	60,11	57,74	57,65	57,31	59,39	63,46	67,40	67,51	62,78	67,10	63,03	57,57	63,42
11 am	67,13	66,15	65,37	63,07	59,88	57,60	57,52	57,10	59,08	63,13	67,08	67,10	62,51	66,79	62,77	57,41	63,10
Noon	66,45	65,64	64,98	62,65	59,51	57,30	57,30	56,77	58,70	62,58	66,35	66,35	62,05	66,15	62,39	57,12	62,54
1 pm	65,90	65,04	64,42	62,26	59,17	56,99	57,01	56,47	58,30	62,01	65,79	65,79	61,59	65,58	61,95	56,82	62,03
2 pm	65,68	64,66	63,97	61,93	58,89	56,71	56,79	56,20	58,05	61,68	65,50	65,49	61,30	65,28	61,60	56,58	61,74
3 pm	65,69	64,54	63,72	61,61	58,52	56,51	56,61	55,88	57,91	61,69	65,46	65,51	61,14	65,26	61,28	56,31	61,70
4 pm	65,88	64,59	63,74	61,45	58,38	56,42	56,50	55,80	57,96	61,80	65,61	65,70	61,15	65,39	61,19	56,24	61,80
5 pm	66,15	64,79	63,90	61,56	58,36	56,37	56,14	55,81	58,13	61,95	65,84	65,96	61,27	65,63	61,27	56,21	61,97
6 pm	66,33	65,09	64,17	61,80	58,54	56,55	56,54	55,90	58,37	62,33	66,19	66,21	61,51	65,88	61,50	56,33	62,30
7 pm	66,71	65,11	64,18	62,11	58,87	56,85	56,78	56,25	58,69	62,67	66,52	66,52	61,82	66,20	61,82	56,63	62,63
8 pm	66,91	65,62	64,85	62,49	59,26	57,11	57,07	56,61	59,14	62,98	66,78	66,72	62,12	66,42	62,20	56,93	62,95
9 pm	67,03	65,77	65,11	62,86	59,58	57,37	57,21	56,90	59,31	63,11	66,84	66,76	62,32	66,52	62,52	57,17	63,09
10 pm	67,05	65,95	65,22	63,06	59,90	57,69	57,69	57,07	59,37	63,41	66,98	66,84	62,48	66,61	62,73	57,48	63,25
11 pm	66,98	65,81	65,19	63,06	59,91	57,71	57,53	57,11	59,37	63,15	66,95	66,81	62,46	66,53	62,72	57,45	63,16
M. N.	66,85	65,77	65,17	63,01	59,81	57,58	57,47	57,06	59,25	63,08	66,91	66,68	62,38	66,43	62,66	57,37	63,08
Mean	66,65	65,56	64,86	62,60	59,49	57,29	57,18	56,76	58,85	62,80	66,61	66,55	62,10	66,25	62,32	57,08	62,75

OZAKA.

1 am	65,64	65,08	64,80	62,52	59,61	57,14	57,67	57,56	59,60	63,19	65,83	65,91	62,05	65,54	62,31	57,46	62,87
2 am	65,74	65,01	64,82	62,36	59,51	56,99	57,53	57,53	59,46	63,05	65,76	65,96	61,98	65,57	62,23	57,35	62,76
3 am	65,67	64,83	64,18	62,29	59,43	56,95	57,49	57,55	59,43	62,96	65,71	65,91	61,89	65,47	62,07	57,33	62,70
4 am	65,51	64,77	64,50	62,28	59,16	57,05	57,52	57,57	59,42	62,96	65,68	65,82	61,88	65,37	62,08	57,38	62,69
5 am	65,45	64,82	64,64	62,39	59,57	57,18	57,64	57,71	59,59	63,08	65,72	65,83	61,97	65,37	62,20	57,51	62,80
6 am	65,58	64,94	64,87	62,65	59,78	57,35	57,77	57,91	59,69	63,24	65,90	66,05	62,11	65,52	62,43	57,68	62,94
7 am	65,78	65,18	65,09	62,91	59,92	57,53	57,95	58,07	59,83	63,49	66,13	66,31	62,35	65,76	62,64	57,85	63,15
8 am	65,99	65,14	65,29	63,01	60,01	57,57	58,07	58,14	59,80	63,77	66,40	66,57	62,50	66,00	62,77	57,93	63,32

OZAKA.

Hour \ Month	Jan.	Feb.	Mar.	Apr.	May	June	July	Aug.	Sept.	Oct.	Nov.	Dec.	Year	Wint.	Spr.	Sum.	Aut.
9 am	66,16	65,57	65,35	63,22	60,03	57,57	58,04	58,15	59,99	63,80	66,43	66,85	62,60	66,19	62,87	57,92	63,41
10 am	66,20	65,50	65,26	63,01	59,97	57,43	58,00	58,11	59,98	63,62	66,31	66,83	62,52	66,18	62,75	57,85	63,30
11 am	65,82	65,29	65,04	62,75	59,77	57,32	57,92	57,94	59,73	63,35	65,92	66,39	62,27	65,83	62,52	57,73	63,00
Noon	65,13	64,78	64,59	62,38	59,18	57,06	57,66	57,63	59,48	62,84	65,29	65,74	61,84	65,22	62,15	57,45	62,54
1 pm	64,58	64,23	64,07	62,03	59,12	56,77	57,39	57,34	59,00	62,25	64,80	65,24	61,40	64,68	61,74	57,17	62,02
2 pm	64,36	63,94	63,70	61,67	58,84	56,54	57,12	57,07	58,68	62,03	64,58	65,02	61,13	64,14	61,40	56,94	61,76
3 pm	64,46	63,88	63,54	61,37	58,84	56,35	56,90	56,83	58,51	61,99	64,58	65,08	61,02	64,47	61,24	56,69	61,69
4 pm	64,65	63,99	63,79	61,30	58,17	56,20	56,79	56,77	58,54	62,07	64,71	65,21	61,01	64,62	61,19	56,59	61,77
5 pm	64,94	64,22	63,55	61,38	58,18	56,15	56,83	56,77	58,69	62,23	64,95	65,35	61,13	64,83	61,14	56,58	61,96
6 pm	65,26	64,54	63,85	61,62	58,60	56,29	56,97	56,83	58,83	62,56	65,28	65,67	61,36	65,16	61,36	56,70	62,22
7 pm	65,49	64,87	64,19	61,96	58,88	56,53	57,17	57,10	59,14	62,94	65,59	65,90	61,65	65,12	61,68	56,93	62,56
8 pm	65,06	65,02	64,56	62,43	59,23	56,78	57,41	57,49	59,59	63,18	65,77	66,06	61,93	65,58	62,07	57,23	62,85
9 pm	65,78	65,16	64,78	62,76	59,55	57,08	57,69	57,72	59,78	63,36	65,96	66,12	62,14	65,69	62,36	57,50	63,03
10 pm	65,78	65,21	64,90	62,86	59,70	57,29	57,82	57,84	59,87	63,42	66,00	66,16	62,23	65,72	62,49	57,64	63,10
11 pm	65,72	65,13	64,94	62,84	59,69	57,30	57,85	57,84	59,84	63,35	65,94	66,13	62,24	65,66	62,17	57,65	63,04
M. N.	65,64	65,09	64,82	62,75	59,64	57,21	57,77	57,69	59,65	63,32	65,90	66,01	62,12	65,58	62,39	57,56	62,96
Mean	65,45	64,85	64,56	62,36	59,39	56,99	57,51	57,55	59,12	63,00	65,64	65,92	61,89	65,44	62,09	57,36	62,69

WAKAYAMA.

1 am	64,83	64,52	63,84	61,60	58,65	56,17	56,75	56,77	58,53	62,24	64,90	65,35	61,18	64,90	61,36	56,56	61,89
2 am	64,85	64,40	63,74	61,38	58,17	56,05	56,64	56,67	58,45	62,13	64,85	65,27	61,07	64,84	61,20	56,45	61,84
3 am	64,78	64,27	63,57	61,32	58,39	55,98	56,59	56,65	58,39	62,02	64,78	65,20	60,99	64,75	61,09	56,11	61,73
4 am	64,67	64,23	63,58	61,30	58,42	56,06	56,60	56,67	58,39	62,02	64,71	65,14	60,98	64,68	61,10	56,44	61,71
5 am	64,66	64,27	63,70	61,40	58,55	56,21	56,62	56,78	58,48	62,15	64,79	65,16	61,06	64,70	61,22	56,54	61,81
6 am	64,78	64,40	63,89	61,61	58,76	56,41	56,83	56,96	58,63	62,27	64,93	65,35	61,23	64,84	61,42	56,73	61,94
7 am	65,02	64,64	64,16	61,91	58,91	56,57	57,04	57,16	58,80	62,48	65,15	65,60	61,45	65,09	61,66	56,92	62,14
8 am	65,19	64,90	64,36	62,01	59,02	56,68	57,12	57,22	58,91	62,74	65,39	65,82	61,61	65,30	61,84	57,01	62,35
9 am	65,41	65,06	64,40	62,07	59,02	56,59	57,11	57,22	59,03	62,81	65,48	66,08	61,69	65,52	61,83	56,97	62,44
10 am	65,48	64,94	64,31	61,99	58,97	56,50	57,09	57,19	59,03	62,70	65,40	66,12	61,65	65,51	61,77	56,93	62,38
11 am	65,10	64,78	64,09	61,80	58,85	56,40	56,99	57,02	58,76	62,42	65,09	65,70	61,42	65,19	61,58	56,80	62,09
Noon	64,38	64,35	63,69	61,45	58,54	56,13	56,76	56,71	58,41	61,98	64,49	65,10	61,00	64,61	61,23	56,53	61,63
1 pm	63,86	63,80	63,18	61,12	58,21	55,94	56,48	56,45	58,07	61,50	64,03	64,59	60,60	64,08	60,85	56,29	61,20
2 pm	63,67	63,46	62,77	60,80	57,98	55,55	56,21	56,21	57,71	61,20	63,77	64,36	60,31	63,83	60,52	56,00	60,89
3 pm	63,77	63,40	62,53	60,48	57,69	55,54	56,01	55,95	57,54	61,12	63,76	64,40	60,18	63,86	60,23	55,84	60,81
4 pm	63,93	63,47	62,14	60,39	57,58	56,37	55,96	55,88	57,51	61,16	63,88	64,52	60,18	63,97	60,14	55,74	60,86
5 pm	64,17	63,67	62,53	60,17	57,56	55,31	55,96	55,87	57,67	61,30	64,03	64,65	60,27	64,16	60,19	55,71	61,00
6 pm	64,48	63,96	62,79	60,66	57,69	55,47	56,11	55,92	57,84	61,58	64,35	64,95	60,48	64,46	60,38	55,83	61,26
7 pm	64,74	64,29	63,10	60,98	58,00	55,69	56,33	56,21	58,17	61,95	64,63	65,19	60,77	64,74	60,69	56,08	61,58
8 pm	64,92	64,52	63,47	61,41	58,31	55,90	56,57	56,55	58,62	62,21	64,83	65,33	61,05	64,92	61,06	56,34	61,89
9 pm	65,01	64,68	63,70	61,75	58,63	56,16	56,84	56,82	58,81	62,40	65,00	65,44	61,27	65,05	61,36	56,61	62,07
10 pm	65,05	64,79	63,83	61,86	58,79	56,38	56,96	56,86	58,86	62,48	65,04	65,52	61,37	65,12	61,49	56,73	62,13
11 pm	65,02	64,75	63,84	61,83	58,75	56,38	56,95	56,84	58,84	62,42	65,05	65,49	61,35	65,09	61,47	56,72	62,10
M. N.	64,90	64,67	63,75	61,75	58,67	56,28	56,88	56,71	58,76	62,35	64,97	65,35	61,25	64,97	61,39	56,62	62,03
Mean	64,70	64,34	63,55	61,39	58,43	56,07	56,64	56,64	58,43	62,07	64,72	65,24	61,02	64,76	61,13	56,45	61,74

NAGANO.

1 am	25,95	25,31	26,51	24,88	23,66	21,28	22,04	21,20	23,14	25,61	28,51	25,79	24,50	25,68	25,02	21,51	25,76
2 am	26,05	25,28	26,40	24,69	23,58	21,18	21,94	21,12	23,13	25,59	28,53	25,87	24,45	25,73	24,89	21,11	25,75
3 am	25,97	25,15	26,34	24,61	23,58	21,25	21,96	21,12	23,13	25,50	28,43	25,86	24,40	25,66	24,83	21,44	25,69
4 am	25,89	25,11	26,28	24,61	23,61	21,34	22,02	21,17	23,21	25,61	28,46	25,77	24,43	25,59	24,84	21,51	25,76
5 am	25,96	25,17	26,37	24,84	23,76	21,56	22,17	21,32	23,33	25,76	28,52	25,83	24,51	25,65	24,99	21,68	25,87
6 am	26,14	25,27	26,57	25,10	23,87	21,71	22,35	21,48	23,55	25,93	28,70	26,07	24,73	25,83	25,18	21,85	26,06
7 am	26,38	25,63	26,76	25,31	24,01	21,85	22,46	21,58	23,75	26,17	28,99	26,34	24,93	26,08	25,36	21,96	26,30
8 am	26,61	25,76	26,87	25,27	23,93	21,78	22,51	21,59	23,78	26,43	29,25	26,60	25,03	26,32	25,36	21,96	26,49
9 am	26,76	25,72	27,09	25,17	23,79	21,63	22,45	21,55	23,83	26,40	29,26	26,82	25,01	26,43	25,35	21,88	26,50

NAGANO.

Month / Hour	Jan.	Feb.	Mar.	Apr.	May	June	July	Aug.	Sept.	Oct.	Nov.	Dec.	Year	Wint.	Spr.	Sum.	Aut.
10 am	26,69	25,56	26,74	25,01	23,56	21,47	22,29	21,40	23,57	26,11	29,02	26,75	24,85	26,33	25,10	21,72	26,23
11 am	26,23	25,27	26,32	24,52	23,27	21,17	22,07	21,08	23,39	25,69	28,54	26,18	24,48	25,89	24,70	21,44	25,87
Noon	25,57	24,71	25,97	23,98	22,79	20,79	21,76	20,71	22,84	25,13	27,96	25,50	23,98	25,26	24,25	21,09	25,31
1 pm	25,07	24,15	25,44	23,62	22,41	20,45	21,47	20,39	22,45	24,57	27,55	24,96	23,54	24,73	23,82	20,77	24,86
2 pm	24,99	23,96	25,22	23,34	22,02	20,21	21,23	20,16	22,17	24,87	27,41	24,85	23,33	24,60	23,53	20,53	24,65
3 pm	25,17	24,04	25,28	23,15	21,90	20,12	21,09	20,03	22,10	24,40	27,60	25,02	23,32	24,74	23,43	20,41	24,70
4 pm	25,42	24,15	25,20	23,27	21,95	20,03	21,03	20,12	22,18	24,52	27,73	25,19	23,40	24,92	23,47	20,39	24,81
5 pm	25,63	24,43	25,46	23,51	22,08	20,13	21,10	20,21	22,44	24,70	27,97	25,41	23,59	25,16	23,68	20,48	25,01
6 pm	25,87	24,75	25,80	23,92	22,36	20,43	21,38	20,40	22,68	25,09	28,29	25,66	23,89	25,43	24,03	20,74	25,35
7 pm	26,02	25,10	26,10	24,37	22,81	20,80	21,74	20,75	23,05	25,36	28,50	25,84	24,20	25,65	24,43	21,10	25,61
8 pm	26,10	25,28	26,48	24,78	23,26	21,13	22,01	21,15	23,35	25,59	28,57	25,98	24,47	25,79	24,84	21,43	25,81
9 pm	26,07	25,39	26,53	24,90	23,62	21,45	22,32	21,38	23,21	25,75	28,69	26,04	24,61	25,83	25,02	21,72	25,88
10 pm	26,05	25,41	26,70	25,00	23,72	21,56	22,37	21,43	23,35	25,73	28,71	26,09	24,67	25,85	25,14	21,79	25,93
11 pm	25,94	25,36	26,64	25,05	23,75	21,57	22,33	21,42	23,30	25,64	28,67	25,96	24,63	25,75	25,15	21,77	25,87
M. N.	25,85	25,33	26,58	24,94	23,68	21,44	22,19	21,31	23,25	25,56	28,61	25,74	24,57	25,61	25,07	21,65	25,81
Mean	25,93	25,05	26,23	24,49	23,21	21,10	21,93	21,00	23,09	25,47	28,44	25,84	24,32	25,61	24,64	21,31	25,67

TOKIO.

1 am	60,77	60,74	61,09	60,10	57,93	55,35	56,62	56,79	58,13	61,30	62,39	60,77	59,33	60,76	59,71	56,25	60,61
2 am	60,84	60,69	60,91	59,94	57,81	55,23	56,47	56,71	58,02	61,16	62,34	60,82	59,24	60,78	59,55	56,14	60,54
3 am	60,73	60,57	60,74	59,87	57,81	55,25	56,44	56,60	57,99	61,07	62,28	60,73	59,18	60,68	59,47	56,13	60,45
4 am	60,63	60,62	60,82	59,93	57,87	55,35	56,48	56,60	58,01	61,14	62,27	60,67	59,21	60,61	59,54	56,17	60,47
5 am	60,69	60,72	61,01	60,10	58,07	55,46	56,64	56,80	58,12	61,28	62,41	60,81	59,35	60,74	59,73	56,33	60,60
6 am	60,96	61,00	61,23	60,39	58,31	55,64	56,77	57,09	58,36	61,43	62,63	61,10	59,58	61,02	59,98	56,50	60,81
7 am	61,14	61,24	61,51	60,59	58,46	55,86	56,94	57,24	58,54	61,69	62,93	61,32	59,79	61,23	60,19	56,68	61,05
8 am	61,34	61,52	61,72	60,69	58,53	55,97	56,99	57,30	58,65	61,95	63,18	61,60	59,95	61,49	60,31	56,75	61,26
9 am	61,56	61,57	61,76	60,72	58,47	55,87	56,97	57,32	58,76	61,95	63,19	61,82	60,00	61,65	60,32	56,72	61,30
10 am	61,39	61,44	61,59	60,61	58,41	55,80	56,92	57,23	58,66	61,74	63,04	61,68	59,88	61,50	60,20	56,65	61,15
11 am	60,70	61,06	61,19	60,25	58,17	55,65	56,72	56,96	58,33	61,35	62,49	61,01	59,49	60,92	59,87	56,44	60,72
Noon	59,84	60,34	60,73	59,79	57,78	55,36	56,46	56,65	57,99	60,82	61,75	60,25	58,98	60,14	59,43	56,16	60,19
1 pm	59,31	59,74	60,11	59,37	57,45	55,09	56,21	56,36	57,64	60,36	61,30	59,76	58,56	59,60	58,98	55,89	59,77
2 pm	59,14	59,15	59,78	58,98	57,17	54,86	55,99	56,07	57,38	60,20	61,12	59,62	58,31	59,40	58,64	55,64	59,57
3 pm	59,29	59,47	59,75	58,73	56,89	54,67	55,83	55,95	57,31	60,29	61,22	59,81	58,27	59,52	58,46	55,48	59,61
4 pm	59,52	59,57	59,81	58,72	56,83	54,57	55,76	56,00	57,40	60,39	61,41	60,00	58,33	59,70	58,45	55,44	59,73
5 pm	59,85	59,84	60,06	58,94	56,89	54,56	55,78	56,02	57,59	60,60	61,73	60,34	58,52	60,01	58,63	55,45	59,97
6 pm	60,27	60,30	60,44	59,23	57,13	54,75	55,98	56,21	57,78	61,01	62,09	60,63	58,82	60,40	58,93	55,65	60,29
7 pm	60,58	60,62	60,77	59,60	57,50	55,07	56,28	56,58	58,13	61,29	62,31	60,90	59,14	60,70	59,29	55,98	60,58
8 pm	60,78	60,82	61,06	60,04	57,83	55,31	56,60	56,96	58,49	61,48	62,41	61,03	59,40	60,88	59,64	56,29	60,80
9 pm	60,87	61,01	61,28	60,25	58,20	55,62	56,89	57,17	58,52	61,65	62,55	61,08	59,60	60,99	59,91	56,56	60,93
10 pm	60,90	61,04	61,35	60,27	58,25	55,68	56,96	57,17	58,51	61,61	62,53	61,08	59,61	61,01	59,96	56,60	60,88
11 pm	60,82	60,99	61,28	60,23	58,20	55,60	56,88	57,14	58,42	61,51	62,41	61,03	59,54	60,95	59,90	56,54	60,79
M. N.	60,66	60,92	61,19	60,18	58,06	55,51	56,75	57,04	58,29	61,43	62,34	60,83	59,43	60,80	59,81	56,43	60,69
Mean	60,52	60,64	60,88	59,90	57,83	55,34	56,51	56,76	58,13	61,20	62,27	60,78	59,23	60,65	59,54	56,20	60,53

HAKODATE.

1 am	60,44	61,54	62,26	60,68	57,97	57,27	56,88	57,58	59,51	61,10	63,19	61,64	60,01	61,21	60,30	57,24	61,28
2 am	60,58	61,52	62,11	60,60	57,84	57,20	56,78	57,49	59,46	60,98	63,26	61,76	59,96	61,29	60,18	57,16	61,23
3 am	60,45	61,38	61,96	60,54	57,77	57,11	56,71	57,41	59,30	60,91	63,18	61,61	59,86	61,15	60,09	57,09	61,13
4 am	60,33	61,30	62,00	60,50	57,74	57,22	56,78	57,52	59,31	60,97	63,13	61,54	59,86	61,06	60,08	57,17	61,11
5 am	60,32	61,29	62,08	60,63	57,88	57,39	56,92	57,69	59,41	61,08	63,18	61,50	59,95	61,04	60,20	57,33	61,22
6 am	60,44	61,30	62,17	60,75	57,93	57,46	56,99	57,84	59,55	61,17	63,24	61,69	60,01	61,14	60,28	57,43	61,32
7 am	60,57	61,37	62,22	60,87	57,97	57,57	57,06	57,89	59,56	61,36	63,53	61,86	60,15	61,27	60,35	57,51	61,48
8 am	60,73	61,51	62,29	60,81	57,94	57,63	57,09	57,88	59,58	61,43	63,66	62,07	60,22	61,44	60,35	57,53	61,56
9 am	60,87	61,39	62,31	60,83	57,80	57,59	57,04	57,86	59,68	61,39	63,63	62,25	60,22	61,50	60,32	57,50	61,57
10 am	60,87	61,32	62,25	60,79	57,68	57,53	56,99	57,80	59,57	61,20	63,48	62,21	60,11	61,47	60,24	57,44	61,42

HAKODATE.

Hour / Month	Jan.	Feb.	Mar.	Apr.	May	June	July	Aug.	Sept.	Oct.	Nov.	Dec.	Year	Wint.	Spr.	Sum.	Aut.
11 am	60,52	61,14	62,08	60,57	57,50	57,49	56,90	57,61	59,31	60,90	63,13	61,77	59,92	61,14	60,05	57,33	61,15
Noon	60,04	60,66	61,89	60,30	57,24	57,29	56,72	57,43	59,16	60,51	62,62	61,27	59,59	60,65	59,81	57,15	60,76
1 pm	59,72	60,28	61,58	60,14	57,02	57,18	56,51	57,23	59,00	60,22	62,42	60,97	59,36	60,32	59,58	56,98	60,55
2 pm	59,76	60,21	61,45	59,92	56,80	57,09	56,38	57,05	58,90	60,16	62,35	60,95	59,25	60,30	59,39	56,84	60,47
3 pm	59,99	60,35	61,51	59,80	56,72	56,99	56,32	56,94	58,94	60,37	62,61	61,21	59,31	60,52	59,34	56,75	60,64
4 pm	60,22	60,50	61,62	59,88	56,84	56,96	56,30	56,96	59,06	60,50	62,74	61,35	59,41	60,69	59,44	56,74	60,77
5 pm	60,53	60,82	61,84	60,06	56,87	56,93	56,29	57,01	59,23	60,74	62,92	61,50	59,56	60,95	59,59	56,74	60,95
6 pm	60,75	61,17	62,14	60,25	57,09	57,08	56,46	57,16	59,36	60,97	63,06	61,59	59,76	61,17	59,83	56,90	61,13
7 pm	60,87	61,33	62,42	60,60	57,40	57,29	56,69	57,43	59,70	61,15	63,19	61,71	59,98	61,30	60,14	57,11	61,35
8 pm	60,89	61,41	62,60	60,93	57,74	57,50	56,94	57,77	59,90	61,29	63,23	61,75	60,16	61,35	60,12	57,40	61,47
9 pm	60,87	61,49	62,69	61,00	57,94	57,68	57,08	57,82	59,82	61,32	63,23	61,78	60,22	61,38	60,53	57,53	61,46
10 pm	60,82	61,50	62,83	60,92	57,97	57,63	57,08	57,81	59,75	61,32	63,21	61,79	60,22	61,37	60,57	57,52	61,44
11 pm	60,63	61,45	62,67	60,87	57,98	57,60	57,08	57,82	59,72	61,27	63,23	61,73	60,17	61,27	60,51	57,50	61,41
M. N.	60,45	61,38	62,57	60,83	57,90	57,51	56,97	57,77	59,68	61,20	63,25	61,54	60,08	61,12	60,43	57,42	61,38
Mean	60,48	61,15	62,13	60,51	57,56	57,31	56,79	57,53	59,41	60,99	63,11	61,63	59,89	61,08	60,08	57,22	61,18

SAPPORO.

Hour / Month	Jan.	Feb.	Mar.	Apr.	May	June	July	Aug.	Sept.	Oct.	Nov.	Dec.	Year	Wint.	Spr.	Sum.	Aut.
1 am	58,17	59,77	60,63	58,55	55,88	55,62	55,08	55,99	58,19	59,31	61,14	59,63	58,19	59,29	58,35	55,56	59,55
2 am	58,53	59,77	60,48	58,47	55,81	55,52	55,03	55,94	58,10	59,22	61,21	59,78	58,15	59,36	58,25	55,50	59,51
3 am	58,44	59,65	60,35	58,16	55,77	55,51	55,00	55,94	58,02	59,18	61,13	59,65	58,09	59,25	58,19	55,18	59,44
4 am	58,33	59,61	60,35	58,13	55,79	55,58	55,09	56,01	58,03	59,23	61,14	59,55	58,10	59,17	58,19	55,56	59,47
5 am	58,33	59,67	60,42	58,50	55,90	55,69	55,22	56,20	58,11	59,31	61,16	59,56	58,17	59,19	58,27	55,70	59,53
6 am	58,45	59,73	60,43	58,68	55,94	55,73	55,25	56,33	58,19	59,39	61,23	59,69	58,25	59,29	58,35	55,77	59,60
7 am	58,65	59,93	60,52	58,71	55,92	55,78	55,27	56,39	58,23	59,51	61,48	59,84	58,35	59,47	58,38	55,81	59,71
8 am	58,83	60,03	60,56	58,70	55,82	55,75	55,20	56,31	58,23	59,60	61,56	60,02	58,38	59,63	58,36	55,75	59,80
9 am	58,93	60,00	60,51	58,68	55,61	55,64	55,10	56,28	58,27	59,49	61,52	60,08	58,34	59,67	58,27	55,67	59,76
10 am	58,92	59,88	60,37	58,61	55,48	55,55	54,97	56,17	58,12	59,31	61,17	60,09	58,24	59,63	58,15	55,56	59,63
11 am	58,56	59,64	60,16	58,15	55,27	55,41	54,80	55,90	57,87	59,06	61,05	59,67	57,99	59,29	57,96	55,37	59,33
Noon	58,16	59,16	59,89	58,26	55,05	55,20	54,59	55,69	57,65	58,74	60,63	59,23	57,69	58,85	57,73	55,16	59,01
1 pm	57,93	58,80	59,61	58,14	54,82	55,05	54,41	55,53	57,43	58,50	60,10	58,98	57,17	58,57	57,53	55,00	58,78
2 pm	57,98	58,74	59,52	57,97	54,70	54,99	54,29	55,37	57,34	58,49	60,10	58,99	57,10	58,56	57,10	54,88	58,74
3 pm	58,17	58,81	59,63	57,90	54,62	54,98	54,25	55,33	57,40	58,66	60,56	59,18	57,16	58,73	57,38	54,85	58,87
4 pm	58,34	59,03	59,78	57,97	54,77	55,01	54,25	55,42	57,54	58,77	60,76	59,32	57,58	58,90	57,51	54,89	59,02
5 pm	58,52	59,22	60,01	58,10	54,86	55,10	54,30	55,47	57,71	58,93	60,90	59,48	57,72	59,07	57,66	54,96	59,18
6 pm	58,70	59,50	60,27	58,26	55,09	55,34	54,50	55,66	57,88	59,26	61,07	59,63	57,93	59,28	57,87	55,17	59,40
7 pm	58,81	59,59	60,53	58,51	55,36	55,55	54,70	55,91	58,20	59,42	61,10	59,72	58,12	59,37	58,13	55,40	59,57
8 pm	58,87	59,68	60,79	58,80	55,65	55,80	55,01	56,24	58,40	59,50	61,18	59,74	58,30	59,43	58,41	55,68	59,69
9 pm	58,87	59,73	60,94	58,84	55,80	55,97	55,28	56,28	58,40	59,56	61,25	59,77	58,39	59,46	58,53	55,84	59,74
10 pm	58,84	59,71	61,02	58,78	55,84	55,92	55,18	56,30	58,41	59,54	61,20	59,75	58,37	59,42	58,55	55,80	59,71
11 pm	58,66	59,69	61,01	58,72	55,87	55,88	55,15	56,29	58,31	59,42	61,20	59,67	58,32	59,34	58,53	55,77	59,64
M. N.	58,47	59,60	60,93	58,68	55,80	55,81	55,09	56,17	58,26	59,35	61,16	59,49	58,23	59,19	58,15	55,69	59,59
Mean	58,50	59,54	60,36	58,46	55,48	55,51	54,88	55,97	58,01	59,20	61,08	59,61	58,05	59,22	58,10	55,45	59,43

NEMURO.

Hour / Month	Jan.	Feb.	Mar.	Apr.	May	June	July	Aug.	Sept.	Oct.	Nov.	Dec.	Year	Wint.	Spr.	Sum.	Aut.
1 am	56,19	57,63	59,25	57,74	55,29	55,92	55,09	56,08	58,04	57,81	59,02	57,66	57,14	57,16	57,43	55,70	58,29
2 am	56,17	57,55	59,08	57,68	55,23	55,83	54,97	55,98	57,94	57,74	59,01	57,80	57,08	57,17	57,33	55,59	58,23
3 am	56,09	57,52	59,05	57,60	55,19	55,76	54,94	55,98	57,90	57,76	58,93	57,62	57,03	57,08	57,28	55,56	58,20
4 am	55,99	57,55	59,11	57,67	55,26	56,00	55,08	56,07	57,93	57,90	59,00	57,62	57,09	57,05	57,36	55,68	58,28
5 am	56,04	57,59	59,21	57,82	55,40	55,96	55,20	56,27	58,02	57,96	59,14	57,69	57,19	57,11	57,48	55,81	58,37
6 am	56,16	57,69	59,35	57,94	55,51	56,03	55,35	56,35	58,13	58,12	59,27	57,83	57,31	57,23	57,60	55,91	58,51
7 am	56,30	57,94	59,45	57,93	55,62	56,11	55,41	56,41	58,06	58,32	59,55	58,08	57,43	57,44	57,67	55,98	58,64
8 am	56,45	58,02	59,47	57,91	55,58	56,08	55,37	56,42	58,07	58,41	59,65	58,28	57,48	57,58	57,65	55,96	58,71
9 am	56,50	57,99	59,56	57,98	55,53	56,01	55,44	56,51	58,07	58,34	59,62	58,40	57,50	57,63	57,69	55,99	58,68
10 am	56,43	57,94	59,43	57,79	55,41	55,99	55,36	56,11	57,90	58,16	59,47	58,13	57,37	57,49	57,55	55,93	58,51
11 am	55,92	57,62	59,24	57,53	55,25	55,90	55,25	56,27	57,68	57,87	59,03	57,65	57,10	57,06	57,31	55,81	58,20

NEMURO.

Month / Hour	Jan.	Feb.	Mar.	Apr.	May	June	July	Aug.	Sept.	Oct.	Nov.	Dec.	Year	Wint.	Spr.	Sum.	Aut.
Noon	55,47	57,21	58,90	57,29	55,07	55,74	55,13	56,12	57,50	57,52	58,60	57,22	56,81	56,63	57,09	55,66	57,87
1 pm	55,34	56,82	58,57	57,12	54,87	55,66	54,97	55,97	57,31	57,23	58,39	57,01	56,61	56,40	56,85	55,53	57,64
2 pm	55,48	56,77	58,48	56,97	54,70	55,59	54,83	55,82	57,24	57,19	58,10	57,19	56,56	56,48	56,72	55,41	57,61
3 pm	55,74	56,90	58,53	56,94	54,68	55,59	54,75	55,81	57,32	57,32	58,57	57,39	56,63	56,68	56,72	55,38	57,74
4 pm	55,94	57,04	58,60	57,02	54,77	55,62	54,71	55,87	57,45	57,42	58,69	57,51	56,72	56,94	56,80	55,40	57,85
5 pm	56,16	57,25	58,82	57,16	54,83	55,74	54,77	55,92	57,61	57,60	58,91	57,63	56,87	57,01	56,94	55,48	58,01
6 pm	56,31	57,42	59,11	57,41	55,06	55,90	54,93	56,09	57,83	57,88	59,06	57,79	57,06	57,17	57,19	55,61	58,26
7 pm	56,35	57,44	59,32	57,78	55,28	56,01	55,01	56,35	58,14	58,02	59,09	57,82	57,22	57,20	57,46	55,81	58,42
8 pm	56,40	57,57	59,51	58,01	55,51	56,36	55,32	56,59	58,24	58,11	59,13	57,83	57,38	57,27	57,70	56,09	58,49
9 pm	56,36	57,61	59,62	58,05	55,62	56,42	55,37	56,57	58,24	58,15	59,17	57,82	57,42	57,26	57,76	56,12	58,54
10 pm	56,31	57,53	59,69	58,04	55,55	56,37	55,31	56,56	58,21	58,14	59,15	57,77	57,38	57,20	57,76	56,08	58,50
11 pm	56,21	57,45	59,66	58,02	55,47	56,25	55,20	56,52	58,13	58,06	59,06	57,56	57,30	57,07	57,72	55,99	58,42
M. N.	56,13	57,36	59,65	57,90	55,32	56,11	55,06	56,35	58,02	58,03	58,98	57,49	57,19	56,96	57,62	55,84	58,31
Mean	56,11	57,47	59,20	57,64	55,25	55,95	55,12	56,22	57,88	57,88	59,01	57,70	57,12	57,09	57,36	55,76	58,27

Thus there are generally two maxima and minima of the atmospheric pressure in a day (see the Plate II). Of these, the chief maximum appears at about 9 o'clock am and the chief minimum at about 3 o'clock pm. In winter, however, we have another pair of maximum and minimum pressures after midnight, besides those just mentioned.

Now if the stations in Hokkaido be grouped into one and called "the northern" while the rest be called "the southern," the mean times (local time) of the maximum and the minimum pressures are as follows (The underlined in the table are the chief maximum and the chief minimum.):--

TIME OF OCCURRENCE OF MAXIMUM AND MINIMUM OF AIR PRESSURE (LOCAL TIME).

		1st Min.	1st Max.	2nd Min.	2nd Max.	3rd Min.	3rd Max.
		h	h	h	h	h	h
Winter	Northern part	0,5a	2,3a	4,6a	9,5a	1,5p	9,3p
	Southern part	0,5a	1,7a	4,2a	9,6a	2,4p	10,0p
Spring	Northern part			3,6a	8,5a	3,0p	10,5p
	Southern part			3,6a	8,9a	4,1p	10,5p
Summer	Northern part			3,5a	8,1a	4,0p	9,8p
	Southern part			3,2a	8,1a	4,6p	10,6p
Autumn	Northern part			3,8a	9,0a	2,3p	10,0p
	Southern part			3,6a	9,0a	2,8p	9,9p
Year	Northern part			3,8a	9,0a	2,5p	9,9p
	Southern part			3,7a	9,2a	3,2p	10,1p

It seems that there is no appreciable difference between the northern and the southern stations in their daily variations of the atmospheric pressure. With regard to different seasons, we observe that the chief maximum comes somewhat later in autumn and winter than in spring and summer, while on the contrary, the chief minimum comes far earlier. In the mean of the year, the chief maximum occurs at 9 o'clock am and the chief minimum at 3 o'clock pm and the secondary maximum at 10 o'clock pm and the secondary minimum at 4 o'clock am.

B. ANNUAL VARIATION OF ATMOSPHERIC PRESSURE.

The monthly mean pressures, reduced to sea level and to standard gravity, are given in the following table in millimetres, omitting the hundreds:

Mean air pressure.

Locality.	Jan.	Feb.	Mar.	Apr.	May	June	July	Aug.	Sept.	Oct.	Nov.	Dec.	Year	Wint.	Spr.	Sum.	Aut.
Naha	61.8	64.3	62.7	60.7	57.7	55.1	55.8	55.6	51.7	59.2	63.5	64.5	59.9	64.5	60.4	55.5	59.1
Kagoshima	65.8	65.2	63.5	61.3	58.9	56.8	57.2	56.7	58.2	61.9	65.0	65.8	61.4	65.6	61.2	56.9	61.7
Miyazaki	65.6	65.2	63.5	61.6	59.2	57.0	57.6	57.2	58.7	62.3	65.1	65.7	61.6	65.5	61.4	57.3	62.0
Kochi	64.7	64.4	63.1	61.5	58.9	56.8	57.7	57.4	59.0	62.1	64.5	64.9	61.3	64.7	61.2	57.3	61.9
Wakayama	64.7	64.7	63.3	61.5	58.9	56.6	57.1	57.0	59.0	62.3	64.2	64.6	61.2	64.7	61.2	56.9	61.8
Oita	65.7	65.3	63.6	61.4	58.7	56.3	56.7	56.6	58.4	62.3	65.0	65.5	61.3	65.5	61.2	56.5	61.9
Yamaguchi	65.9	65.7	63.8	61.6	59.0	56.6	56.9	56.8	58.6	62.6	65.2	65.1	61.5	65.6	61.5	56.8	62.1
Hiroshima	65.9	65.5	64.0	61.7	58.8	56.3	56.9	56.6	58.6	62.6	64.9	65.2	61.4	65.5	61.5	56.6	62.0
Matsuyama	65.6	65.1	63.7	61.5	58.9	56.6	57.1	57.0	58.5	62.5	64.9	65.3	61.5	65.3	61.4	56.9	62.0
Okayama	65.3	65.3	63.7	61.9	58.8	56.5	57.1	57.1	59.1	62.9	65.0	65.3	61.5	65.3	61.5	56.9	62.3
Ozaka	64.6	64.6	63.3	61.7	59.1	56.8	57.4	57.3	59.2	62.5	64.7	64.7	61.3	64.6	61.4	57.2	62.1
Kioto	64.5	64.8	63.6	61.8	59.2	56.8	57.4	57.5	59.4	62.7	65.2	65.5	61.4	64.9	61.5	57.2	62.4
Kumamoto	66.7	66.0	64.0	61.6	59.0	56.7	56.9	56.6	58.4	62.5	65.6	66.3	61.7	66.3	61.5	56.7	62.2
Saga	66.1	65.7	63.8	61.2	58.7	56.1	56.3	56.1	58.2	62.3	65.0	66.0	61.3	65.9	61.2	56.2	61.8
Nagasaki	66.6	65.8	64.2	61.6	58.8	56.6	56.6	56.5	58.4	62.5	65.5	66.3	61.6	66.2	61.5	56.6	62.1
Fukuoka	66.3	65.9	64.0	61.6	58.8	56.3	56.4	56.3	58.6	62.7	65.5	66.0	61.5	66.1	61.5	56.3	62.3
Itsugahara	66.4	66.3	64.3	61.9	59.0	56.5	56.7	56.7	59.4	63.3	65.6	66.6	61.9	66.4	61.7	56.6	62.8
Akamagaseki	66.1	66.0	64.0	61.8	59.3	56.7	57.1	57.0	59.0	63.0	65.6	65.9	61.8	66.0	61.7	56.9	62.5
Sakai	65.1	65.4	63.6	61.7	59.0	56.5	57.0	57.0	59.2	63.0	64.9	64.6	61.4	65.0	61.4	56.8	62.4
Tsu	64.3	64.3	63.2	62.0	59.4	57.2	57.8	57.7	59.0	62.4	64.3	64.3	61.3	64.3	61.5	57.6	61.9
Nagoya	63.7	63.9	62.8	61.7	59.0	57.0	57.4	57.4	59.3	62.3	64.0	63.9	61.0	63.8	61.2	57.3	61.9
Gifu	64.1	64.1	63.1	62.0	59.4	57.2	57.9	58.0	59.7	62.7	64.4	64.2	61.4	64.1	61.5	57.7	62.3
Hamamatsu	62.7	62.6	61.9	61.3	59.0	57.1	57.9	57.9	59.4	61.9	63.3	63.1	60.7	62.8	60.7	57.6	61.5
Numazu	61.5	61.7	61.2	61.0	58.8	57.0	57.7	57.9	59.3	61.9	62.7	61.7	60.2	61.6	60.3	57.5	61.3
Tokio	62.3	62.7	61.7	61.7	59.0	57.1	57.8	58.1	59.7	62.5	62.8	61.9	60.6	62.3	60.8	57.7	61.7
Utsunomiya	62.0	62.6	62.0	61.5	58.9	57.4	57.6	57.9	60.0	62.5	62.8	61.9	60.6	62.2	60.8	57.6	61.8
Choshi	62.1	62.7	61.6	61.8	59.1	57.3	58.0	58.2	59.8	62.1	62.4	61.7	60.6	62.2	70.8	57.8	61.4
Kanazawa	64.9	65.2	63.5	61.9	59.2	57.2	57.6	57.7	59.6	63.0	64.7	63.9	61.6	64.7	61.5	57.5	62.4
Fushiki	64.6	65.1	63.6	62.0	59.1	56.8	57.3	57.3	59.4	62.8	64.2	63.8	61.3	64.5	61.6	57.1	62.1
Nagano	64.2	64.2	62.6	61.7	58.7	56.4	57.1	57.4	59.4	63.7	64.9	63.4	61.1	63.9	61.0	57.0	62.7
Niigata	63.1	63.7	62.8	61.8	59.0	56.8	57.4	57.6	59.6	62.9	63.7	62.9	60.9	63.2	61.2	57.3	62.1
Yamagata	62.3	63.2	62.5	61.5	58.8	56.8	57.4	57.9	59.9	62.9	63.7	62.1	60.7	62.5	60.9	57.4	62.2
Akita	62.5	63.2	62.5	62.2	59.5	57.6	58.1	58.2	59.9	63.3	63.9	61.9	61.1	62.5	61.4	58.0	62.4
Fukushima	61.7	62.7	62.2	61.3	58.7	56.9	57.7	58.0	60.0	62.7	63.5	61.9	60.6	62.1	60.7	57.5	62.1
Nobiru	61.5	62.2	61.6	61.4	58.8	57.0	57.7	58.1	59.8	62.6	63.2	61.5	60.4	61.7	60.6	57.6	61.9
Miyako	60.9	61.0	61.3	61.3	59.1	57.7	58.1	58.4	60.1	62.8	62.9	60.5	60.3	60.8	60.6	58.1	61.9
Aomori	61.4	62.2	61.3	60.9	58.2	57.0	57.5	58.0	59.7	62.7	63.1	60.7	60.2	61.4	60.1	57.5	61.8
Hakodate	60.8	61.7	60.8	60.7	58.6	57.5	57.3	57.8	59.7	62.1	61.8	60.5	59.9	61.0	60.0	57.5	61.2
Suttsu	60.1	61.2	60.2	60.2	57.9	57.0	56.9	57.4	59.5	61.8	61.6	58.8	59.4	60.0	59.4	57.1	61.0
Sapporo	59.7	61.1	59.8	60.1	57.7	56.8	56.6	57.2	59.3	61.5	61.4	59.0	59.2	59.9	59.2	56.9	60.7
Kamikawa	59.8	61.5	60.3	60.3	58.0	57.4	57.3	58.0	60.0	62.0	61.8	59.0	59.6	60.1	59.5	57.6	61.3
Soya	59.3	63.0	59.3	59.0	57.4	57.1	56.5	57.3	59.2	60.7	60.2	57.4	58.9	59.9	58.6	57.1	60.0
Abashiri	58.6	60.6	60.7	59.3	57.4	57.4	56.7	57.5	59.7	61.1	60.3	57.6	58.9	58.9	59.1	57.2	60.4
Nemuro	57.7	59.8	59.5	59.9	58.4	58.0	57.7	58.3	60.0	61.5	60.6	57.6	59.1	58.4	59.3	58.0	60.7
Erimo	58.6	60.2	59.9	60.1	58.4	57.5	57.8	58.4	59.9	61.7	61.0	57.6	59.2	58.8	59.5	57.9	60.9

A glance at this table clearly shows the manner of the annual variation. Thus the pressure is higher in winter and lower in summer throughout the whole country. Or precisely speaking, the atmospheric pressure is generally lower than the mean value of the year during the five months from May to September, and higher during the remaining seven months from October to April. The highest pressure is generally in January or February, while the lowest is in June. The amplitude, or the difference between the maximum and the minimum, gets less and less as we proceed from the western part of this country toward the eastern. Thus at Okinawa and the western part of Kiushu, the amplitude is generally about 10 mm. It diminishes to about 5 mm at the eastern coast of Nippon, and to only 4 mm or even less, at the eastern extremity of Hokkaido.

Again the last table shows us that in order to study the annual variation of atmospheric pressure, it is necessary to divide the whole country into the following three territrial divisions: (1) The west of Kioto (2) The east of Kioto (3) Hokkaido.

The mean monthly pressures for each of these divisions are as follows:--

	Jan.	Feb.	March	April	May	June	July	August	Sept.	Oct.	Nov.	Dec.
West of Kioto	765.6	765.3	763.6	761.5	758.9	756.5	757.0	756.8	758.5	762.4	764.9	765.4
East of Kioto	762.8	763.3	762.1	761.6	759.0	757.1	757.6	757.9	759.7	762.7	763.7	762.7
Hokkaido	759.5	761.3	760.2	760.0	758.0	757.3	757.1	757.8	759.7	761.7	761.3	759.9

Thus in the first division lying west of Kioto, the mean maximum pressure 765.6 mm is in January. It quickly decreases toward summer and reaches its minimum value of 756.5 mm in June. In July, it increases slightly and a secondary maximum of 757.0 mm is formed. Again in August, we have a secondary minimum of 756.8 mm. After that month, it increases steadily and ultimately attains the maximum of January. In this division, therefore, we have a pair of maximum and minimum pressures in its annual variation. Also its amplitude 9.1 mm far exceeds those in any other divisions.

In the second division lying east of Kioto, the chief maximum pressure 763.7 mm is in November. The pressure decreases continually and has a minimum in December, or January and a maximum in February. Then it decreases very quickly till it reaches the chief minimum 757.1 mm in June. After that, it increases slowly until September when it begins to increase suddenly and ultimately reaches to the maximum of November. The annual variation in this division has thus two maxima and minima, and the greatest amplitude is 6.6 mm.

Lastly in Hokkaido, the variation is similar to the last division for having two maxima and minima. The chief maximum is in October and the chief minimum in July, the amplitude being 4.6 mm. The secondary maximum is in February and the secondary minimum in January. Both of these secondary maximum and minimum are very much conspicuous when compared with those of the last division.

Thus we summarize that:--

In winter, the atmospheric pressure remains high in the first division lying west of Kioto, but there is a minimum in Hokkaido. In summer, we have a secondary maximum in the first division (in July) but is absent in Hokkaido. The variation in the second division lying east of Kioto is intermediate between them. (For the detail, see the Plate III.)

C. ABSOLUTE MAXIMUM AND MINIMUM OF THE ATMOSPHERIC PRESSURE.

The extreme of the highest atmospheric pressure ever observed in this country is 779.8 mm at Nemuro on the 4th of February, 1882; and the extreme of the lowest pressure is 713.1 mm at Nagasaki on the 14th of September, 1891. The difference between these two extremes is 66.7 mm. Though this last is the difference between the extremes on the whole country, yet sometimes such great difference occurs at some of stations. Thus the difference of the extremes observed at Nemuro was 60 mm. The extreme values of the highest and the lowest pressures are:--

Extremes of air pressure.

Locality.	Max.	Day	Month	Year	Min.	Day	Month	Year	Range
Naha	772,6	26	XII	1891	736,2	4	VIII	1891	36,4
Kagoshima	776,1	21	XII	1883	721,4	25	VIII	1884	54,7
Miyazaki	775,9	22;28	XI;XII	1883;91	725,8	10	IX	1886	50,1
Kochi	775,2	22;28	XI;XII	1883;91	730,1	16	VIII	1891	45,1
Wakayama	776,7	29	I	1882	717,0	30	VIII	1888	59,7
Oita	776,0	28	XI	1891	731,6	14	IX	1891	44,4
Yamaguchi	776,8	28	XI	1891	732,6	14	IX	1891	44,2
Hiroshima	776,7	28	XI	1891	727,4	25	VIII	1884	49,3
Matsuyama	776,2	28	XI	1891	737,1	14	IX	1891	39,1
Okayama	776,1	7	XII	1891	737,0	16	VIII	1891	39,1
Ozaka	776,0	13	II	1883	725,8	31	VIII	1888	50,2
Kioto	776,6	29	I	1882	729,6	1	VII	1885	47,0
Kumamoto	776,6	27	XI	1891	723,7	14	IX	1891	52,9
Saga	775,7	28	XI	1891	720,4	11	IX	1891	55,3
Nagasaki	776,7	21	XII	1883	713,1	11	IX	1891	63,6
Fukuoka	776,1	28	XI	1891	725,8	11	IX	1891	50,3
Itsugahara	776,3	27	XI	1891	733,8	11	IX	1891	42,5
Akamagaseki	776,5	28	XI	1891	725,8	11	IX	1891	50,7
Sakai	777,3	28	XI	1891	729,0	25	VIII	1884	48,3
Tsu	775,2	7	XII	1891	716,7	11	IX	1889	58,5
Nagoya	774,9	8;7	XI;XII	1890;91	733,7	30	IX	1891	41,2
Gifu	775,0	8	XI	1890	729,1	11	IX	1889	45,9
Hamamatsu	774,2	8	XI	1890	716,0	15	IX	1884	58,2
Numazu	774,7	16	XI	1885	726,4	24	V	1884	48,3
Tokio	777,5	12	IV	1878	730,1	4	X	1880	47,4
Utsunomiya	775,6	9	XI	1890	732,6	30	IX	1890	43,0
Choshi	775,3	17	I	1887	733,9	9	III	1887	41,4
Kanazawa	777,1	17	I	1887	732,9	1	VII	1885	44,2
Fushiki	776,6	15	XI	1889	733,2	11	IX	1889	43,4
Nagano	777,5	7	XII	1891	731,3	11	IX	1889	46,2
Niigata	777,1	29	I	1882	733,0	11	IX	1889	44,1
Yamagata	777,9	9	XI	1890	734,6	12	IX	1889	43,6
Akita	777,5	9	XI	1890	736,0	12	IX	1889	41,5
Fukushima	776,1	9	XI	1890	736,0	12	IX	1889	40,3
Ishinomaki	776,1	9	XI	1890	737,1	12	IX	1889	39,0
Miyako	776,2	9	XI	1890	733,5	16	III	1883	42,7
Aomori	779,3	4	II	1882	735,3	21	X	1889	44,0
Hakodate	779,0	4	II	1882	733,3	21	X	1889	45,7
Suttsu	776,1	9	XI	1890	729,3	21	X	1889	46,8
Sapporo	779,1	24	I	1876	727,3	22	X	1889	51,8
Kamikawa	777,8	29	I	1891	728,5	22	X	1889	49,3
Soya	777,3	9	XI	1890	733,4	21	X	1889	43,9
Abashiri	776,8	27	III	1890	729,3	22	X	1889	47,5
Nemuro	779,8	4	II	1882	719,4	3	IX	1879	60,4
Erimo	775,2	27	III	1890	731,6	22	X	1889	43,6

Thus it will be clear that the absolute maximum happens always in winter, and the absolute minimum generally in early autumn. Again it will be easily noticed that though the mean pressure is higher in the southwestern part of this country as mentioned above, yet the absolute maximum is apt to occur frequently in the northeastern part.

D. THE DISTRIBUTION OF ATMOSPHERIC PRESSURE.

The distribution of the atmospheric pressure in this country is shown in the Plate XIV. The essential points to be remarked are as follows:—

(1) On the mean of the year, the pressure is high on Corea, and low on the Sea of Okhotsk. The isobars run nearly along the meridians and they are concave toward west.

(2) In winter, the pressure is high on Corea and Siberia, and low on the Pacific near the eastern Hokkaido. Though the isobars run nearly along the meridians, yet they bend toward west in the southern regions, and toward east in the northern. The distribution peculiar to winter, and very much different to any other seasons, is that the isobars are very closely packed together. In this season, for every one degree of longitude eastward, the atmospheric pressure decreases about 1 mm. or more.

(3) In spring, the high pressure is on Corea and the western part of the Sea of Japan, and the low pressure on the Sea of Okhotsk and that part of the Pacific lying east of Japan. All the isobars are so very much concave toward west, that they are elliptical in their forms and enclose the Sea of Japan.

(4) In summer, the distribution is just opposite to the two last cases, winter and spring. Now the pressure is low on Corea and Siberia, and high on the Pacific, east of Japan. The isobars are generally concave toward southwest, and the area of low pressure includes far into the Sea of Japan in tongue-like form.

(5) In autumn, the pressure distribution is restored to the conditions of winter and spring. The pressure is high on Corea and Siberia, and low on the Pacific to our east. The isobars are bend toward west, so that they run along the parallels of latitudes, on the southern part, but along the meridians on the northern.

Owing to the above mentioned facts, there is a considerable difference of the prevailing winds between winter and summer. In autuman, winter, and spring, the north wind predominates, and the wind in winter is generally violent. For minutes informations, refer to our chapter III on winds.

CHAPTER III.

WIND.

Of the observations in wind at our meteorological stations, its direction is distinguished into the sixteen points, viz., N, NNE, NE, ENE, E, ESE, SE, SSE, S, SSW, SW, WSW, W, WNW, NW, and NNW; and its velocity is measured by Robinson's anemometer. At those stations, where hourly observations are taken, the mean velocity of wind during one hour previous to the time of observation is taken as the velocity at that time: and at those where six observations are taken in a day, the mean velocity during each ten minutes both preceding and succeeding to the time of observation is taken as the velocity at the observed time.

A. DIURNAL VARIATION OF THE VELOCITY OF WIND.

The mean velocity of wind hourly observed at our first class stations are given in metres per second:

KUMAMOTO.

Month / Hour	Jan.	Feb.	Mar.	Apr.	May	June	July	Aug.	Sept.	Oct.	Nov.	Dec.	Year	Wint.	Spr.	Sum.	Aut.
1 am	0,69	1,01	1,31	1,29	1,01	0,78	0,95	0,61	1,32	1,12	1,09	1,09	1,03	0,93	1,21	0,79	1,18
2 am	0,85	1,05	1,33	1,39	1,02	0,86	0,98	0,71	1,28	1,06	1,09	1,21	1,07	1,04	1,25	0,85	1,14
3 am	0,76	1,01	1,38	1,32	1,01	0,87	0,92	0,61	1,36	1,15	1,07	1,10	1,05	0,96	1,27	0,80	1,19
4 am	0,84	1,14	1,29	1,55	1,13	0,95	0,86	0,62	1,27	1,28	1,13	0,99	1,09	0,99	1,32	0,81	1,23
5 am	0,96	1,02	1,26	1,66	1,08	1,03	0,95	0,58	1,23	1,33	1,13	1,10	1,11	1,03	1,33	0,85	1,23
6 am	1,04	1,08	1,24	1,51	1,26	1,05	1,04	0,73	1,30	1,32	1,13	1,08	1,15	1,07	1,34	0,94	1,25
7 am	1,05	1,02	1,44	1,45	1,28	1,13	1,11	0,79	1,37	1,43	1,06	1,09	1,18	1,05	1,39	1,01	1,29
8 am	1,12	1,13	1,53	1,68	1,22	1,24	1,28	0,96	1,57	1,46	1,03	1,08	1,27	1,11	1,48	1,16	1,35
9 am	1,16	1,08	1,75	1,86	1,29	1,11	1,68	1,33	1,89	1,53	1,18	1,19	1,45	1,14	1,63	1,18	1,53
10 am	1,50	1,52	2,01	2,10	1,67	1,79	2,12	1,58	2,18	1,78	1,20	1,37	1,73	1,46	1,93	1,83	1,72
11 am	1,70	1,82	2,11	2,55	1,92	2,09	2,29	1,89	2,62	2,01	1,51	1,55	2,03	1,69	2,29	2,09	2,06
Noon	1,94	1,95	2,62	2,89	2,16	2,13	2,73	2,37	2,84	2,27	2,01	1,80	2,33	1,90	2,56	2,51	2,37
1 pm	2,75	2,35	2,96	3,35	2,67	2,91	3,08	2,85	2,95	2,63	2,31	2,32	2,76	2,47	2,99	2,95	2,61
2 pm	3,02	2,49	3,19	3,33	2,83	2,99	3,19	3,12	3,09	2,81	2,63	2,58	2,94	2,70	3,12	3,10	2,85
3 pm	3,05	2,78	3,32	3,39	2,99	3,07	3,09	3,17	3,02	2,97	2,68	2,53	3,00	2,79	3,23	3,11	2,89
4 pm	2,43	2,71	3,30	3,34	3,09	3,00	3,12	2,98	2,94	2,85	2,51	2,58	2,91	2,57	3,21	3,03	2,78
5 pm	2,18	2,59	3,03	3,29	2,89	2,99	2,81	2,77	2,67	2,49	2,12	1,93	2,67	2,33	3,07	2,87	2,43
6 pm	1,88	2,09	2,68	2,96	2,54	2,71	2,49	2,54	2,43	1,76	1,38	1,35	2,23	1,77	2,73	2,58	1,86
7 pm	1,29	1,60	2,07	2,38	2,01	2,34	2,28	1,90	2,02	1,51	1,17	1,25	1,82	1,38	2,15	2,17	1,58
8 pm	1,19	1,37	1,57	1,85	1,54	1,94	1,67	1,49	1,81	1,51	1,53	1,01	1,54	1,20	1,65	1,69	1,62
9 pm	1,07	1,16	1,41	1,69	1,23	1,33	1,24	1,09	1,59	1,36	1,01	0,91	1,26	1,05	1,45	1,22	1,32
10 pm	0,76	1,12	1,18	1,57	1,19	1,13	1,11	0,91	1,56	1,37	1,07	1,14	1,20	1,01	1,11	1,05	1,33
11 pm	0,69	1,09	1,33	1,17	0,94	1,00	1,05	0,70	1,47	1,22	0,97	1,01	1,08	0,94	1,21	0,92	1,22
M. N.	0,59	1,13	1,43	1,31	0,94	0,93	1,00	0,62	1,27	1,12	1,02	1,15	1,04	0,96	1,23	0,85	1,14
Mean	1,45	1,56	1,98	2,14	1,70	1,75	1,79	1,54	1,96	1,73	1,44	1,44	1,71	1,48	1,94	1,69	1,71

MATSUYAMA.

1 am	1,86	1,71	1,46	1,14	0,93	1,02	0,89	0,86	0,72	0,91	1,01	1,79	1,20	1,80	1,17	0,92	0,89
2 am	1,89	1,80	1,21	0,96	0,93	0,90	0,94	0,79	0,91	1,05	0,95	1,93	1,19	1,87	1,03	0,88	0,98
3 am	2,11	1,51	1,36	1,01	1,02	0,92	0,92	0,70	0,82	1,10	1,06	1,91	1,20	1,84	1,14	0,85	0,99
4 am	2,07	1,72	1,34	1,03	0,89	0,87	1,00	0,67	0,99	1,14	1,10	2,17	1,25	1,99	1,09	0,85	1,08
5 am	2,06	1,50	1,57	1,09	0,98	0,72	0,87	0,69	0,92	1,18	1,11	1,96	1,22	1,81	1,21	0,76	1,07
6 am	2,12	1,38	1,55	1,21	0,88	0,87	0,86	0,73	0,92	0,91	1,13	2,02	1,21	1,84	1,21	0,82	0,99
7 am	2,23	1,41	1,59	1,35	1,01	1,06	0,97	0,79	0,94	1,23	1,17	2,23	1,33	1,96	1,32	0,94	1,13
8 am	2,21	1,57	1,67	1,28	1,27	1,41	1,12	1,08	0,96	1,25	1,25	2,03	1,42	1,94	1,41	1,20	1,15
9 am	2,59	1,84	2,16	2,14	1,69	2,33	1,80	1,57	1,33	1,52	1,18	2,37	1,90	2,27	2,00	1,90	1,44
10 am	3,02	2,33	3,01	2,96	2,81	3,31	2,71	2,39	1,98	2,02	2,00	2,71	2,60	2,70	2,93	2,80	2,00

MATSUYAMA.

Month / Hour	Jan.	Feb.	Mar.	Apr.	May	June	July	Aug.	Sep.	Oct.	Nov.	Dec.	Year	Wint.	Spr.	Sum.	Aut.
11 am	3,62	3,01	3,74	3,64	3,51	4,07	3,23	3,18	2,75	2,60	2,94	3,20	3,30	3,31	3,63	3,49	2,76
Noon	4,45	3,89	4,58	3,86	3,86	4,30	3,57	3,67	3,24	3,08	3,51	3,89	3,83	4,08	4,10	3,85	3,29
1 pm	4,86	4,40	5,05	4,14	3,70	4,20	3,97	3,85	3,47	3,53	3,95	4,05	4,09	4,11	4,30	4,01	3,65
2 pm	4,97	4,40	4,56	4,16	3,57	4,21	3,75	3,59	3,18	3,58	4,05	4,22	4,01	4,53	4,10	3,85	3,70
3 pm	4,93	4,58	4,67	4,04	3,30	4,10	3,49	3,72	3,27	3,33	3,75	4,16	3,94	4,56	4,00	3,77	3,45
4 pm	4,60	3,81	3,91	3,81	3,14	4,02	3,01	3,35	2,98	2,96	3,25	3,63	3,56	4,01	3,72	3,46	3,05
5 pm	3,61	3,37	3,60	3,40	2,87	3,34	2,91	3,01	2,31	2,08	2,25	2,95	2,97	3,31	3,29	3,09	2,21
6 pm	3,06	2,39	2,73	2,75	2,37	2,88	2,55	2,55	1,56	1,29	1,95	2,53	2,39	2,66	2,62	2,66	1,63
7 pm	2,49	2,00	1,65	1,81	1,54	2,07	1,74	1,53	0,96	1,18	1,77	2,41	1,76	2,30	1,67	1,78	1,30
8 pm	2,13	1,88	1,58	1,45	1,19	1,45	1,33	1,16	0,98	1,23	1,94	2,39	1,55	2,13	1,34	1,31	1,38
9 pm	2,05	1,78	1,29	1,36	1,07	1,21	1,15	1,13	1,01	1,15	1,79	2,22	1,43	2,02	1,24	1,17	1,32
10 pm	2,01	2,16	1,28	1,36	1,12	1,25	1,08	1,11	0,97	1,06	1,86	2,07	1,41	2,08	1,25	1,15	1,30
11 pm	1,71	1,74	1,23	1,20	0,99	1,27	0,99	1,11	0,80	0,98	1,51	1,79	1,28	1,75	1,14	1,12	1,10
M. N.	1,84	1,61	1,26	1,28	0,90	1,01	0,90	0,80	0,75	1,02	1,32	1,83	1,22	1,77	1,15	0,91	1,03
Mean	2,85	2,41	2,11	2,18	1,91	2,20	1,91	1,84	1,63	1,73	2,01	2,61	2,14	2,63	2,17	1,87	1,79

HIROSHIMA.

1 am	2,11	1,77	2,61	2,11	2,28	1,60	1,41	2,13	2,20	2,94	3,15	2,17	2,21	2,02	2,14	1,72	2,76
2 am	2,10	1,90	2,70	2,37	2,40	1,59	1,55	2,29	2,28	2,91	3,12	2,15	2,27	2,05	2,49	1,78	2,75
3 am	2,11	1,97	2,87	2,35	2,68	1,66	1,71	2,29	2,26	2,91	3,28	2,38	2,38	2,15	2,63	1,89	2,83
4 am	2,15	2,00	2,76	2,52	2,83	1,88	1,77	2,37	2,45	2,99	3,22	2,56	2,45	2,21	2,70	2,01	2,89
5 am	2,27	1,96	2,79	2,57	2,87	1,98	1,69	2,16	2,44	3,29	3,28	2,56	2,51	2,26	2,71	2,01	3,00
6 am	2,18	1,78	2,81	2,71	2,88	1,99	1,77	2,29	2,48	3,31	3,25	2,49	2,50	2,15	2,81	2,02	3,01
7 am	2,30	1,83	2,85	2,86	2,99	2,23	1,77	2,35	2,65	3,53	3,31	2,59	2,60	2,21	2,90	2,12	3,16
8 am	2,30	1,69	2,81	2,70	2,47	1,90	1,56	2,30	2,49	3,50	3,17	2,15	2,15	2,15	2,66	1,92	3,05
9 am	2,27	1,74	2,31	2,31	1,76	1,60	1,23	1,88	2,52	3,31	3,19	2,48	2,22	2,16	2,14	1,57	3,01
10 am	2,26	1,78	2,00	2,16	1,84	1,61	1,26	1,87	2,11	2,92	2,77	2,55	2,09	2,20	2,00	1,58	2,60
11 am	2,32	1,92	2,10	2,55	2,16	1,71	1,69	2,12	2,01	2,38	2,31	2,22	2,16	2,15	2,27	1,95	2,21
Noon	2,47	2,17	2,68	2,86	2,60	2,08	2,22	2,90	1,98	2,17	2,08	1,87	2,31	2,17	2,71	2,10	2,08
1 pm	2,82	2,33	2,61	2,90	2,75	2,29	2,31	3,17	2,18	2,29	2,15	2,11	2,50	2,12	2,75	2,59	2,21
2 pm	2,85	2,56	2,81	3,06	2,88	2,41	2,57	3,32	2,13	2,52	2,32	2,20	2,67	2,54	2,92	2,79	2,12
3 pm	3,09	2,66	2,98	3,48	2,95	2,18	2,71	3,72	2,79	2,65	2,66	2,55	2,89	2,77	3,14	2,97	2,70
4 pm	2,76	2,65	3,25	3,09	2,84	2,47	2,54	3,49	2,39	2,48	2,31	2,35	2,71	2,59	3,05	2,81	2,39
5 pm	2,51	2,58	3,00	2,56	2,51	2,33	2,39	3,15	2,15	2,14	1,87	2,05	2,44	2,38	2,70	2,62	2,04
6 pm	1,84	1,99	2,31	2,27	2,15	2,47	2,21	2,72	1,68	1,66	1,41	1,63	2,00	1,82	2,24	2,37	1,58
7 pm	1,51	1,48	1,77	1,71	1,88	1,76	1,76	2,21	1,37	1,69	1,38	1,46	1,67	1,48	1,79	1,92	1,18
8 pm	1,57	0,88	1,48	1,54	1,55	1,44	1,50	2,25	1,41	2,03	1,76	1,17	1,58	1,31	1,52	1,73	1,73
9 pm	1,50	1,21	1,60	1,68	1,56	1,17	1,14	1,76	1,59	2,31	2,27	1,60	1,62	1,14	1,61	1,36	2,06
10 pm	1,73	1,38	1,99	1,71	1,72	1,24	1,19	1,74	1,86	2,76	2,59	1,76	1,84	1,62	1,82	1,39	2,40
11 pm	1,90	1,49	2,24	1,91	1,90	1,25	1,42	1,91	1,98	2,79	2,81	1,94	1,96	1,78	2,02	1,53	2,53
M. N.	1,92	1,52	2,37	2,24	2,32	1,38	1,47	1,91	1,95	2,90	2,94	2,01	2,08	1,82	2,31	1,59	2,60
Mean	2,20	1,88	2,49	2,16	2,37	1,84	1,80	2,41	2,15	2,68	2,64	2,15	2,25	2,08	2,41	2,03	2,48

OZAKA.

1 am	2,88	2,86	3,08	3,08	2,28	2,38	2,43	3,21	2,63	2,05	2,43	2,79	2,67	2,84	2,81	2,67	2,37
2 am	2,82	2,88	3,22	2,90	2,30	2,43	2,30	2,70	2,54	2,31	2,31	2,57	2,61	2,76	2,81	2,48	2,39
3 am	2,77	2,91	3,52	3,27	2,41	2,57	2,07	2,39	2,66	2,30	2,29	2,71	2,66	2,83	3,07	2,34	2,42
4 am	2,65	2,94	3,47	3,10	2,34	2,61	2,03	2,71	2,32	2,48	2,38	2,74	2,65	2,78	2,97	2,45	2,39
5 am	2,59	2,81	3,45	3,10	2,36	2,45	2,05	2,71	2,31	2,37	2,39	2,63	2,60	2,68	2,97	2,40	2,36
6 am	2,30	2,78	3,30	2,87	2,35	2,55	1,96	2,60	2,44	2,54	2,30	2,58	2,55	2,55	2,84	2,37	2,43
7 am	2,52	2,80	3,23	3,19	3,05	2,84	2,06	2,94	2,60	2,54	2,36	2,70	2,71	2,67	3,16	2,61	2,50
8 am	2,70	2,73	3,64	3,49	3,00	3,25	2,39	3,35	3,05	2,61	2,14	2,91	2,96	2,78	3,38	3,06	2,70
9 am	2,64	3,27	4,02	3,69	3,21	3,20	2,70	3,89	3,35	2,79	2,67	3,14	3,21	3,01	3,64	3,26	2,94
10 am	2,95	3,82	4,28	4,10	3,70	3,74	3,36	4,12	3,59	3,10	2,99	3,28	3,59	3,35	4,03	3,74	3,23
11 am	3,96	4,44	4,77	4,35	4,18	4,27	3,74	4,54	3,68	3,16	3,39	3,65	4,01	4,02	4,43	4,18	3,41

OZAKA.

Hour \ Month	Jan.	Feb.	Mar.	Apr.	May	June	July	Aug.	Sept.	Oct.	Nov.	Dec.	Year	Wint.	Spr.	Sum.	Au'.
Noon	4,48	5,48	4,85	4,74	4,54	4,50	4,29	4,85	4,08	3,51	3,86	4,24	4,36	4,40	4,71	4,55	3,80
1 pm	4,98	4,93	4,82	5,01	5,04	4,84	4,83	5,11	4,32	3,75	4,09	4,35	4,67	4,75	4,96	4,93	4,05
2 pm	5,60	5,17	5,27	5,10	5,66	5,13	5,16	5,23	4,75	3,99	4,24	4,49	4,99	5,09	5,31	5,21	4,33
3 pm	5,38	5,30	5,67	5,12	5,97	5,39	5,48	5,65	4,81	4,14	4,41	4,52	5,15	5,07	5,59	5,51	4,45
4 pm	5,08	5,13	5,46	5,25	5,62	5,48	5,41	5,70	4,51	4,08	4,18	4,08	5,00	4,76	5,11	5,53	4,26
5 pm	4,42	4,83	5,06	5,01	5,15	5,35	4,84	5,26	4,42	3,58	3,82	3,43	4,60	4,23	5,08	5,15	3,94
6 pm	4,03	4,00	4,38	4,18	4,58	4,93	4,26	4,65	3,83	3,21	3,37	3,22	4,05	3,75	4,38	4,61	3,47
7 pm	3,66	3,60	3,72	3,78	3,49	4,13	3,63	3,82	3,52	2,76	3,22	3,05	3,53	3,44	3,66	3,86	3,17
8 pm	3,69	3,34	3,59	3,28	3,19	3,60	3,15	3,46	3,35	2,62	2,98	2,84	3,26	3,29	3,35	3,40	2,98
9 pm	3,51	3,09	3,04	3,16	2,98	3,08	2,71	3,42	2,96	2,44	2,81	3,10	3,02	3,23	3,06	3,07	2,74
10 pm	3,53	3,61	3,10	3,38	2,68	2,91	2,79	3,36	2,90	2,26	2,85	3,13	3,05	3,43	3,05	3,03	2,67
11 pm	3,38	3,15	3,01	2,98	2,19	2,89	2,58	2,90	2,77	2,28	2,58	3,08	2,86	3,20	2,83	2,82	2,58
M. N.	3,26	3,12	2,88	3,17	2,22	2,93	2,57	2,87	2,51	2,19	2,68	2,72	2,76	3,03	2,76	2,79	2,46
Mean	3,57	3,68	3,95	3,81	3,53	3,65	3,28	3,82	3,33	2,87	3,05	3,25	3,48	3,50	3,76	3,58	3,08

WAKAYAMA.

1 am	2,71	2,36	2,19	2,07	1,83	1,61	1,66	1,87	1,87	2,55	2,77	2,99	2,24	2,69	2,03	1,72	2,40
2 am	2,80	2,56	2,18	2,32	1,86	1,61	1,70	2,01	1,96	2,47	2,88	3,06	2,28	2,81	2,12	1,77	2,44
3 am	2,74	2,51	2,20	2,22	2,00	1,80	1,86	2,06	2,22	2,76	2,98	3,08	2,37	2,79	2,14	1,91	2,65
4 am	2,74	2,50	2,36	2,21	2,10	1,90	1,91	2,02	2,24	2,75	3,03	2,92	2,39	2,72	2,22	1,94	2,67
5 am	2,62	2,58	2,29	2,24	2,16	1,76	1,90	2,12	2,37	2,81	2,87	3,06	2,40	2,75	2,23	1,93	2,68
6 am	2,34	2,32	2,28	1,97	2,03	1,89	1,84	2,05	2,33	2,65	2,77	2,89	2,28	2,52	2,09	1,93	2,58
7 am	2,59	2,57	2,58	2,37	2,29	2,18	1,98	2,41	2,68	2,93	3,07	3,27	2,58	2,81	2,41	2,19	2,89
8 am	2,49	2,51	2,64	2,48	2,13	2,23	2,10	2,30	2,76	2,98	3,11	3,04	2,56	2,68	2,42	2,21	2,95
9 am	2,71	2,96	2,73	2,68	2,01	2,11	2,16	2,30	2,57	2,84	2,98	3,11	2,60	2,93	2,48	2,19	2,80
10 am	3,04	3,09	2,82	2,84	2,43	2,48	2,46	2,71	2,58	2,72	2,97	3,27	2,78	3,13	2,70	2,55	2,76
11 am	3,30	3,13	2,88	3,02	2,87	2,83	2,85	3,25	2,96	2,72	2,83	3,12	2,98	3,18	2,92	2,98	2,84
Noon	3,66	3,34	3,10	3,39	2,99	3,26	3,26	3,63	3,34	2,95	3,01	3,35	3,28	3,45	3,16	3,38	3,11
1 pm	3,96	3,58	3,38	3,41	3,10	3,38	3,48	4,09	3,69	3,17	3,41	3,50	3,51	3,68	3,31	3,65	3,42
2 pm	3,91	3,56	3,25	3,61	3,13	3,42	3,54	3,98	3,51	3,03	3,36	3,62	3,50	3,70	3,34	3,65	3,30
3 pm	4,12	3,77	3,52	3,63	3,38	3,68	3,76	4,23	3,58	3,35	3,52	3,70	3,69	3,86	3,51	3,89	3,48
4 pm	3,77	3,43	3,29	3,60	3,06	3,54	3,48	4,07	3,47	2,89	3,07	3,28	3,41	3,49	3,32	3,70	3,14
5 pm	3,42	3,09	2,94	3,45	2,66	3,26	3,19	3,59	3,04	2,31	2,50	2,67	2,98	3,06	2,92	3,35	2,62
6 pm	2,97	2,63	2,59	2,76	2,33	2,83	2,87	3,35	2,47	1,74	2,10	2,36	2,58	2,65	2,56	3,02	2,10
7 pm	2,79	2,29	2,19	2,34	2,00	2,39	2,47	2,75	2,07	1,66	2,06	2,42	2,28	2,50	2,17	2,54	1,93
8 pm	2,65	2,15	1,91	2,19	1,52	2,07	1,96	2,45	1,99	2,01	2,31	2,77	2,17	2,52	1,87	2,16	2,11
9 pm	2,62	2,01	1,85	2,28	1,45	1,69	1,66	2,01	1,69	2,12	2,43	2,92	2,06	2,52	1,86	1,79	2,08
10 pm	2,76	2,19	2,14	2,48	1,55	1,62	1,71	1,99	1,79	2,58	2,95	3,22	2,25	2,72	2,04	1,78	2,41
11 pm	2,60	2,12	2,07	2,32	1,55	1,59	1,58	1,92	1,70	2,52	2,85	3,14	2,16	2,62	1,98	1,70	2,36
M. N.	2,61	2,11	2,03	2,17	1,68	1,52	1,48	1,87	1,71	2,49	2,90	3,12	2,14	2,61	1,96	1,62	2,37
Mean	3,00	2,72	2,56	2,66	2,26	2,36	2,37	2,71	2,52	2,62	2,87	3,08	2,64	2,93	2,49	2,18	2,67

NAGANO.

1 am	1,71	2,42	2,22	4,25	3,36	1,67	2,21	3,28	3,84	2,05	2,11	2,40	2,62	2,18	3,28	2,39	2,67
2 am	1,08	2,43	2,34	4,44	3,63	1,61	2,00	2,96	3,71	2,08	1,90	2,56	2,57	2,02	3,47	2,29	2,56
3 am	1,19	2,32	2,41	4,22	3,20	1,62	1,83	2,03	3,71	1,76	1,88	2,07	2,13	1,86	3,28	2,13	2,45
4 am	1,48	2,04	2,46	3,47	2,92	1,46	1,74	2,76	3,08	1,74	1,80	1,87	2,23	1,80	2,95	1,99	2,21
5 am	1,06	2,17	2,80	3,14	2,65	1,44	1,63	2,86	2,89	1,82	2,12	1,81	2,20	1,69	2,86	1,98	2,28
6 am	0,75	2,08	2,43	2,83	2,51	1,42	1,80	2,72	3,05	1,69	1,98	1,80	2,09	1,54	2,60	1,98	2,21
7 am	0,80	2,14	2,70	3,43	2,62	1,85	1,96	2,56	2,95	1,71	1,88	1,57	2,18	1,50	2,92	2,12	2,18
8 am	0,64	2,03	2,67	3,52	3,21	2,00	2,20	3,08	3,56	1,87	2,06	1,55	2,37	1,41	3,13	2,43	2,50
9 am	0,98	2,39	3,33	4,25	3,67	2,53	2,54	3,94	3,71	2,21	2,23	1,96	2,81	1,78	3,75	3,00	2,72
10 am	1,67	2,97	4,03	4,48	3,92	2,70	2,84	4,34	4,05	3,00	2,79	2,43	3,26	2,36	4,14	3,28	3,28
11 am	2,20	3,41	4,57	4,64	3,91	2,98	3,03	4,23	4,20	3,32	3,23	2,69	3,53	2,77	4,37	3,41	3,58
Noon	2,70	3,87	4,76	4,89	4,38	3,57	2,80	4,17	4,52	3,55	3,73	2,91	3,82	3,16	4,68	3,51	3,93

NAGANO.

Hour \ Month	Jan.	Feb.	Mar.	Apr.	May	June	July	Aug.	Sept.	Oct.	Nov.	Dec.	Year	Wint.	Spr.	Sum.	Aut.
1 pm	2,76	3,79	4,93	5,06	4,63	3,68	2,95	4,11	4,34	3,62	3,72	3,18	3,90	3,24	4,87	3,58	3,89
2 pm	3,05	4,00	5,32	5,13	4,91	4,45	2,96	4,47	4,71	4,10	4,19	3,49	4,23	3,51	5,12	3,96	4,33
3 pm	3,25	3,82	5,94	6,06	5,09	5,58	3,94	5,22	5,53	4,69	4,36	3,77	4,77	3,61	5,70	4,90	4,86
4 pm	2,70	3,88	5,83	6,01	5,70	6,31	4,69	5,88	6,03	5,16	4,37	3,31	4,99	3,30	5,85	5,63	5,19
5 pm	2,56	3,28	5,17	6,49	6,33	5,68	5,39	6,11	5,60	5,03	4,36	2,79	4,92	2,88	6,10	5,73	5,00
6 pm	2,02	2,98	4,91	6,12	6,01	4,81	4,06	5,45	4,82	4,13	3,36	2,83	4,29	2,61	5,69	4,77	4,10
7 pm	1,58	2,60	4,51	5,62	4,88	3,50	3,18	4,62	3,61	2,86	3,14	2,71	3,51	2,31	4,88	3,77	3,20
8 pm	1,31	2,12	3,25	4,75	4,32	2,75	2,56	3,76	3,10	2,32	2,90	2,89	3,01	2,11	4,11	3,02	2,80
9 pm	1,31	1,99	2,89	4,33	3,40	2,25	2,15	3,01	3,01	2,21	2,61	3,05	2,69	2,12	3,54	2,48	2,63
10 pm	1,92	2,33	3,35	5,15	4,33	2,51	2,79	3,55	3,61	2,75	2,77	2,92	3,47	2,39	4,28	2,95	3,05
11 pm	1,75	2,32	2,71	4,24	3,63	2,18	2,87	3,25	3,46	2,40	2,63	2,56	2,79	2,21	3,52	2,60	2,83
M. N.	1,17	2,22	2,50	4,26	3,31	2,31	2,38	3,27	3,71	2,15	2,31	2,52	2,68	1,97	3,36	2,65	2,73
Mean	1,74	2,73	3,67	4,61	4,02	2,95	2,75	3,86	3,95	2,81	2,86	2,57	3,21	2,35	4,10	3,19	3,22

TOKIO.

1 am	2,99	3,36	3,26	3,39	2,62	2,63	2,63	2,61	2,87	2,72	3,08	2,86	2,92	3,07	3,09	2,62	2,89
2 am	2,83	3,38	3,21	3,30	2,55	2,46	2,40	2,42	2,89	2,80	3,05	2,82	2,84	3,01	3,03	2,43	2,91
3 am	2,76	3,53	3,22	3,17	2,60	2,32	2,26	2,26	2,85	2,81	2,99	2,80	2,80	3,03	3,00	2,28	2,89
4 am	2,69	3,47	3,20	3,09	2,57	2,28	2,15	2,29	2,85	2,85	2,88	2,87	2,77	3,01	2,95	2,24	2,86
5 am	2,68	3,24	3,10	3,01	2,51	2,25	2,13	2,22	2,78	2,90	2,85	2,79	2,71	2,90	2,88	2,20	2,84
6 am	2,61	3,14	3,03	2,93	2,58	2,21	2,07	2,28	2,67	2,81	2,76	2,68	2,65	2,81	2,85	2,20	2,75
7 am	2,56	2,99	3,02	3,00	2,73	2,31	2,37	2,37	2,71	2,85	2,71	2,58	2,68	2,71	2,92	2,35	2,76
8 am	2,57	3,11	3,37	3,37	2,87	2,55	2,60	2,72	3,01	2,94	2,80	2,55	2,87	2,74	3,20	2,62	2,92
9 am	2,85	3,76	3,80	3,69	3,07	2,95	3,09	2,99	3,38	3,20	3,11	2,68	3,21	3,10	3,52	3,01	3,23
10 am	3,28	4,13	4,22	3,92	3,54	3,36	3,50	3,27	3,60	3,32	3,28	3,06	3,54	3,49	3,88	3,38	3,40
11 am	3,64	4,09	4,26	4,14	3,99	3,70	3,93	3,78	3,67	3,20	3,07	3,28	3,73	3,67	4,13	3,80	3,31
Noon	3,80	4,08	4,18	4,70	4,67	4,38	4,60	4,52	4,17	3,31	3,22	3,14	4,14	3,77	4,62	4,50	3,57
1 pm	3,95	4,25	4,79	5,13	5,19	4,95	5,10	5,19	4,50	3,47	3,31	3,55	4,45	3,92	5,04	5,08	3,76
2 pm	3,89	4,51	5,03	5,46	5,47	5,32	5,48	5,50	4,71	3,57	3,39	3,54	4,66	3,98	5,32	5,43	3,89
3 pm	3,86	4,57	5,08	5,64	5,56	5,51	5,60	5,58	4,82	3,55	3,33	3,48	4,71	3,97	5,43	5,56	3,90
4 pm	3,77	4,56	5,25	5,71	5,62	5,58	5,61	5,77	4,83	3,47	3,11	3,23	4,71	3,85	5,53	5,66	3,80
5 pm	3,37	4,13	5,14	5,29	5,28	5,32	5,35	5,47	4,51	2,99	2,65	2,65	4,35	3,38	5,24	5,38	3,38
6 pm	2,96	3,53	4,62	4,98	4,91	5,10	4,95	4,86	4,07	2,58	2,62	2,40	3,96	2,96	4,84	4,97	3,09
7 pm	2,91	3,39	4,08	4,39	4,32	4,52	4,34	4,15	3,66	2,35	2,70	2,50	3,63	2,93	4,28	4,31	2,90
8 pm	2,88	3,38	4,03	3,93	3,69	3,96	3,80	3,72	3,31	2,31	2,71	2,56	3,36	2,91	3,88	3,83	2,79
9 pm	2,76	3,23	3,71	3,83	3,31	3,63	3,59	3,49	3,15	2,27	2,69	2,66	3,20	2,88	3,63	3,57	2,70
10 pm	2,80	3,19	3,50	3,53	3,00	3,42	3,32	3,19	2,94	2,33	2,79	2,74	3,05	2,91	3,31	3,31	2,69
11 pm	2,90	3,23	3,47	3,51	2,79	3,35	3,08	2,96	2,88	2,50	2,93	2,89	3,04	3,01	3,26	3,13	2,77
M. N.	2,91	3,31	3,32	3,42	2,69	2,98	2,85	2,76	2,80	2,64	3,02	2,91	2,97	3,05	3,14	2,86	2,82
Mean	3,09	3,65	3,93	4,02	3,67	3,63	3,62	3,60	3,49	2,91	2,96	2,90	3,45	3,21	3,87	3,61	3,12

HAKODATE.

1 am	4,97	3,89	4,23	3,41	2,86	2,37	2,11	2,77	3,09	2,57	4,46	4,61	3,46	4,49	3,51	2,42	3,37
2 am	4,95	3,93	4,10	3,14	2,88	2,14	1,97	2,74	2,98	2,19	4,49	4,41	3,35	4,43	3,37	2,28	3,32
3 am	5,38	3,78	3,96	3,35	2,84	2,20	2,17	2,77	2,93	2,47	4,11	4,52	3,37	4,56	3,38	2,38	3,17
4 am	5,01	3,94	3,87	3,15	2,95	2,22	2,10	2,55	2,95	2,39	3,94	4,60	3,30	4,52	3,32	2,29	3,09
5 am	5,24	3,75	3,85	3,15	3,03	2,14	2,06	2,56	2,84	2,55	3,64	4,70	3,29	4,56	3,34	2,25	3,01
6 am	5,13	3,58	3,86	3,12	2,93	2,45	2,17	2,36	2,96	2,53	3,69	4,45	3,27	4,39	3,30	2,33	3,06
7 am	5,30	3,70	3,87	3,16	3,01	2,41	2,64	2,65	3,22	2,64	3,73	4,41	3,48	4,47	3,35	2,57	3,30
8 am	5,10	3,69	3,88	3,60	3,28	3,13	2,82	2,98	3,40	2,72	3,54	4,56	3,56	4,45	3,59	2,98	3,22
9 am	5,00	3,85	4,01	4,21	4,13	3,38	3,11	3,35	3,82	3,16	3,95	4,63	3,89	4,49	4,13	3,29	3,64
10 am	5,18	4,03	4,68	4,91	4,86	3,86	3,95	4,20	4,23	3,91	4,50	5,00	4,41	4,72	4,83	4,00	4,22
11 am	5,63	4,40	5,09	5,55	5,57	4,22	4,45	4,83	4,62	4,62	5,13	5,43	4,96	5,15	5,10	4,50	4,79
Noon	6,08	5,12	5,76	6,16	6,07	4,79	4,85	5,52	5,15	5,46	5,52	5,94	5,51	5,70	6,00	5,05	5,38
1 pm	6,26	5,38	6,23	6,52	6,55	4,96	5,08	5,50	5,22	5,66	5,89	6,04	5,77	5,89	6,43	5,18	5,59

HAKODATE.

Month / Hour	Jan.	Feb.	Mar.	Apr.	May	June	July	Aug.	Sept.	Oct.	Nov.	Dec.	Year	Wint.	Spr.	Sum.	Aut.
2 pm	6,71	5,36	6,32	6,87	6,54	5,07	5,03	5,58	5,33	5,58	5,83	6,25	5,88	6,12	6,58	5,23	5,58
3 pm	6,67	5,56	6,40	6,86	6,51	5,06	4,81	5,56	5,29	5,92	6,15	6,25	5,92	6,16	6,60	5,11	5,79
4 pm	6,83	5,62	6,45	6,99	6,38	4,92	4,80	5,12	5,15	5,55	5,67	5,92	5,78	6,12	6,61	4,95	5,46
5 pm	6,09	5,56	6,16	6,15	5,82	4,73	4,40	4,93	4,68	4,90	5,26	5,71	5,37	5,79	6,01	4,69	4,98
6 pm	5,66	5,24	5,77	5,82	5,19	4,40	4,11	4,15	4,09	4,34	4,86	5,63	4,96	5,51	5,59	4,32	4,43
7 pm	5,60	4,76	5,30	4,82	4,58	3,85	3,79	3,74	3,66	3,61	4,85	5,05	4,47	5,14	4,90	3,79	4,01
8 pm	5,49	4,72	4,75	4,30	4,19	3,48	3,15	3,25	3,36	3,29	4,63	5,09	4,12	5,10	4,41	3,19	3,76
9 pm	5,06	4,36	4,74	4,02	3,69	2,87	2,74	2,84	3,31	2,97	4,12	4,63	3,80	4,68	4,15	2,82	3,57
10 pm	5,18	4,27	4,95	4,13	3,59	2,86	2,37	3,01	3,39	2,91	4,18	4,84	3,81	4,76	4,22	2,75	3,49
11 pm	4,89	4,15	4,43	3,71	3,49	2,69	2,16	2,67	3,21	2,62	4,17	4,91	3,59	4,65	3,88	2,51	3,33
M. N.	4,97	4,05	4,24	3,37	2,98	2,50	2,23	2,67	3,23	2,68	4,16	4,91	3,50	4,65	3,53	2,47	3,36
Mean	5,52	4,44	4,88	4,61	4,32	3,45	3,30	3,69	3,84	3,65	4,61	5,10	4,28	5,02	4,60	3,48	4,03

SAPPORO.

Month / Hour	Jan.	Feb.	Mar.	Apr.	May	June	July	Aug.	Sept.	Oct.	Nov.	Dec.	Year	Wint.	Spr.	Sum.	Aut.
1 am	2,91	2,65	3,77	3,36	3,60	2,68	2,52	2,38	2,65	2,08	3,36	3,74	2,98	3,11	3,58	2,53	2,70
2 am	3,09	2,69	3,11	3,47	3,73	2,95	2,55	2,16	2,65	2,00	3,40	3,62	2,98	3,13	3,55	2,55	2,68
3 am	2,88	2,60	3,63	3,52	3,93	2,69	2,33	2,36	2,75	1,87	3,30	3,60	2,95	3,03	3,69	2,46	2,61
4 am	3,02	2,56	3,54	3,56	3,65	2,75	2,41	2,35	2,65	2,17	3,59	3,54	2,98	3,04	3,58	2,51	2,80
5 am	2,68	2,72	3,56	3,58	3,61	2,66	2,16	2,25	2,58	2,00	3,40	3,46	2,91	2,95	3,58	2,16	2,66
6 am	3,05	2,61	3,52	3,56	3,88	2,89	2,59	2,20	2,57	2,17	3,30	3,47	2,98	3,01	3,65	2,56	2,68
7 am	2,65	2,64	3,85	4,29	4,77	3,35	2,93	2,76	2,89	2,36	3,38	3,60	3,29	2,96	4,30	3,01	2,88
8 am	2,64	2,57	3,62	4,70	5,38	3,79	3,67	3,31	3,30	2,63	3,38	3,64	3,55	2,95	4,57	3,60	3,10
9 am	3,07	2,96	4,68	5,48	6,24	4,45	4,30	4,02	3,60	3,20	3,88	3,58	4,13	3,20	5,47	4,26	3,59
10 am	3,20	3,63	5,30	5,81	6,65	4,77	4,72	4,17	4,39	4,26	4,11	4,20	4,60	3,68	5,92	4,55	4,25
11 am	3,50	3,67	5,33	6,32	6,91	5,15	5,05	4,48	4,63	4,84	4,90	4,60	4,96	3,95	6,19	4,89	4,79
Noon	3,68	4,14	5,95	7,25	7,12	5,51	5,33	4,95	5,08	5,13	5,12	4,83	5,39	4,22	6,77	5,27	5,31
1 pm	4,20	4,31	5,96	7,13	7,42	5,59	5,48	4,91	5,23	5,40	5,64	4,76	5,60	4,42	6,84	5,34	5,42
2 pm	4,35	4,67	5,77	6,85	7,60	7,00	5,77	5,28	5,26	5,15	5,60	4,71	5,67	4,59	6,74	6,02	5,34
3 pm	4,23	5,02	6,07	7,08	7,88	6,20	5,61	5,89	5,12	4,99	5,17	4,43	5,64	4,56	7,01	5,90	5,09
4 pm	3,51	4,69	5,39	6,74	7,22	5,93	5,23	5,12	4,73	3,91	4,38	4,01	5,07	4,07	6,45	5,43	4,35
5 pm	2,98	4,29	5,17	6,34	6,56	5,62	4,67	4,68	3,74	3,15	3,87	3,91	4,58	3,73	6,02	4,99	3,58
6 pm	2,75	3,64	4,52	5,31	5,62	4,80	4,21	3,88	2,94	2,19	3,49	3,75	3,95	3,38	5,15	4,30	2,97
7 pm	2,83	3,31	4,10	4,50	4,17	3,80	3,36	3,02	2,74	2,35	3,55	3,60	3,44	3,25	4,26	3,39	2,88
8 pm	2,84	3,12	3,92	4,16	3,93	3,10	2,91	2,59	2,70	2,38	3,43	3,51	3,22	3,16	4,00	2,87	2,81
9 pm	2,99	3,11	3,83	4,26	3,76	2,93	2,79	2,14	2,61	2,38	3,30	3,24	3,14	3,11	3,95	2,72	2,76
10 pm	2,86	2,89	3,85	4,03	3,98	2,95	2,78	2,17	2,83	2,47	3,27	3,60	3,16	3,12	3,95	2,73	2,86
11 pm	2,92	2,89	3,75	3,87	3,77	2,97	2,79	2,23	2,68	2,24	3,11	3,82	3,12	3,21	3,80	2,70	2,78
M. N.	3,02	2,85	3,81	3,48	3,64	2,85	2,89	2,32	2,91	2,33	3,41	3,88	3,12	3,25	3,64	2,69	2,89
Mean	3,17	3,34	4,43	4,95	5,21	4,03	3,72	3,43	3,47	3,09	3,92	3,88	3,89	3,46	4,86	3,73	3,49

NEMURO.

Month / Hour	Jan.	Feb.	Mar.	Apr.	May	June	July	Aug.	Sept.	Oct.	Nov.	Dec.	Year	Wint.	Spr.	Sum.	Aut.
1 am	6,12	5,05	5,51	5,18	4,85	3,32	3,20	3,48	4,06	5,05	5,74	6,10	4,83	5,76	5,28	3,33	4,95
2 am	6,12	4,60	5,26	5,42	4,87	3,15	3,13	3,50	4,44	5,07	5,76	5,89	4,75	5,54	5,22	3,26	4,98
3 am	6,27	4,76	5,21	5,20	4,87	3,30	3,19	3,47	3,83	4,90	5,66	5,92	4,71	5,65	5,10	3,32	4,80
4 am	6,29	5,01	5,28	5,33	4,81	3,47	3,12	3,33	4,02	4,84	5,98	6,17	4,81	5,83	5,14	3,31	4,95
5 am	6,44	5,01	5,13	5,23	5,08	3,63	3,21	3,41	3,96	5,00	5,92	6,24	4,86	5,90	5,15	3,43	4,96
6 am	6,09	4,91	5,00	5,27	5,38	3,88	3,40	3,48	3,91	4,51	5,85	6,11	4,82	5,70	5,22	3,59	4,76
7 am	6,19	5,13	5,00	5,67	5,64	4,06	3,73	3,79	4,11	4,57	5,91	6,02	4,99	5,78	5,44	3,86	4,87
8 am	6,17	5,17	5,12	5,99	5,71	4,55	3,87	4,12	4,46	5,02	6,02	6,00	5,20	5,78	5,61	4,18	5,17
9 am	6,28	5,23	5,80	6,22	6,14	4,72	3,96	4,51	4,73	5,61	6,10	6,17	5,45	5,89	6,05	4,40	5,48
10 am	6,46	5,55	6,17	6,67	6,41	5,03	4,18	4,70	5,15	5,82	6,62	6,71	5,79	6,24	6,42	4,64	5,86
11 am	6,51	5,98	6,17	7,29	6,62	5,25	4,32	4,88	5,34	6,17	6,65	6,91	6,01	6,47	6,69	4,82	6,05
Noon	6,65	6,15	6,41	7,50	6,98	5,45	4,63	5,19	5,44	6,87	6,67	6,85	6,18	6,55	6,96	5,09	6,13
1 pm	6,56	6,27	6,61	7,60	7,10	5,76	4,59	5,39	5,60	6,31	6,81	6,81	6,29	6,55	7,10	5,25	6,24
2 pm	6,59	6,08	7,25	7,75	7,11	5,44	4,77	5,32	5,67	6,17	6,91	6,57	6,30	6,41	7,37	5,18	6,25

NEMURO.

Hour \ Month	Jan.	Feb.	Mar.	Apr.	May	June	July	Aug.	Sept.	Oct.	Nov.	Dec.	Year	Wint.	Spr.	Sum.	Aut.
3 pm	6,41	6,05	6,76	7,44	6,80	5,21	4,86	5,31	5,61	5,97	6,59	6,30	6,11	6,25	7,00	5,13	6,07
4 pm	6,05	5,99	6,58	6,81	6,58	5,28	4,91	5,07	5,43	5,38	6,27	6,18	5,88	6,07	6,66	5,09	5,69
5 pm	5,63	5,86	5,91	6,47	6,15	4,84	4,67	4,71	4,84	5,26	6,18	6,27	5,56	5,92	6,18	4,74	5,42
6 pm	5,53	5,70	5,50	6,13	5,59	4,18	4,22	4,25	4,27	4,96	6,25	6,42	5,27	5,88	5,74	4,32	5,16
7 pm	5,84	5,52	5,43	6,07	5,22	4,05	4,13	3,76	4,10	4,61	6,34	6,27	5,14	5,88	5,57	3,98	5,02
8 pm	5,87	5,57	5,02	5,72	4,90	3,87	3,84	3,75	4,11	4,67	6,17	6,33	5,04	5,92	5,44	3,82	4,99
9 pm	5,78	5,30	5,60	5,74	4,94	3,70	3,84	3,95	4,13	4,51	6,18	6,16	4,99	5,75	5,43	3,82	4,95
10 pm	5,94	5,27	5,44	6,02	4,94	3,55	3,69	3,69	4,12	4,75	6,08	5,96	4,96	5,72	5,47	3,64	4,98
11 pm	6,16	5,12	5,16	5,44	5,02	3,30	3,58	3,59	4,22	4,74	6,00	5,98	4,85	5,75	5,20	3,49	4,98
M. N.	6,44	5,17	5,21	5,10	5,02	3,10	3,39	3,74	4,07	4,90	5,94	6,11	4,83	5,81	5,21	3,31	4,97
Mean	6,16	5,44	5,72	6,16	5,71	4,27	3,98	4,17	4,56	5,21	6,19	6,27	5,31	5,96	5,86	4,12	5,32

The above table shows us that the variation of the velocity is independent of the locality, whether south or north of this country. The variation is as follows throughout the year. It begins to increase gradually from about sunrise and reaches the maximum value at 1 o'clock pm or thereabout; then it decreases suddenly and there is almost entirely no variation during the night. Thus the velocity of wind presents one maximum and one minimum in the course of a day, though the minimum is not very marked. However, at Hiroshima and Nagano, it seems that there are two maxima and two minima of velocity in a day. (See the Plate IV).

B. ANNUAL VARIATION OF THE VELOCITY OF WIND.

In our country, the velocity of wind is everywhere great in winter and spring, and small in summer and autumn; and we have the strongest wind generally in December or January and the weakest generally from August till October. Notwithstanding the frequent attacks of violent gales in late summer and early autumn, we see that at these seasons, the mean velocity of wind is the minimum of the year. From this fact, it will be evident that at these seasons winds are not very strong generally.

It is beyond question that the velocity of wind should differ considerably between sea coasts and inland regions. If we compare the mean velocities observed at those stations lying on the coast with those at inland stations, we see the following differences:

	Winter.	Spring.	Summer.	Autumn.	Year.
Coast	3,6	3,5	2,9	3,1	3,3
Inland	2,0	2,7	2,1	2,0	2,2

A glance at this table shows clearly enough the fact that wind is stronger in the coast than in the inland districts. But even among maritime stations themselves, there are great differences, the velocities in those places, projecting far into the ocean, like Choshi, Erimo, Soya, etc., are very much greater than in other places. Thus:

	Winter.	Spring.	Summer.	Autumn.	Year.
Tokio	3,0	3,6	3,3	2,9	3,2
Choshi	5,0	5,4	4,4	5,4	5,0
Hakodate	5,3	4,5	3,2	4,2	4,3
Erimo	12,9	10,2	7,0	12,9	10,7
Soya	7,1	7,4	6,5	8,0	7,2

If different localities are compared we see that wind blows somewhat stronger on the back Nippon than on the front Nippon. The place where the strongest wind most prevails in this country is Hokkaido, as is shown in the next table.

	Winter.	Spring.	Summer.	Autumn.	Year.
Front Nippon	2.7	2.7	2.3	2.3	2.5
Back Nippon	3.4	3.3	2.5	2.7	3.0
Hokkaido	5.4	5.4	4.3	5.1	5.0

Such are the general outlines concerning the velocity and its variation of wind in this country. If minute details are required, see the following table.

Mean wind velocity m.p.s.

Locality.	Jan.	Feb.	Mar.	Apr.	May	June	July	Aug.	Sept.	Oct.	Nov.	Dec.	Year	Wint.	Spr.	Sum.	Aut.
Naha	5.7	5.8	5.1	4.2	2.8	3.7	2.9	3.0	5.0	5.2	5.0	5.2	4.5	5.6	4.0	3.2	5.1
Kagoshima	1.6	1.6	1.9	1.7	1.7	1.7	1.7	1.9	2.4	2.1	1.8	1.6	1.8	1.6	1.8	1.8	2.0
Miyazaki	3.3	3.1	3.0	2.9	2.8	2.7	2.5	3.0	2.6	2.1	2.4	2.9	2.8	3.1	2.9	2.7	2.4
Kochi	3.0	3.4	3.3	2.9	2.5	2.2	2.1	2.4	2.1	2.5	2.6	2.6	2.7	3.0	2.9	2.2	2.5
Wakayama	2.6	2.1	2.1	2.3	2.0	1.9	2.0	2.3	2.0	1.9	2.2	2.6	2.2	2.5	2.2	2.1	2.0
Oita	2.8	2.6	2.4	2.1	1.9	1.9	1.8	2.0	2.0	2.0	2.3	2.7	2.2	2.7	2.3	1.9	2.1
Yamaguchi	2.0	2.2	2.6	2.6	2.6	2.3	2.1	2.5	2.5	2.2	1.8	2.0	2.3	2.1	2.6	2.3	2.2
Hiroshima	2.2	2.2	2.5	2.3	2.3	2.0	2.1	2.4	2.6	2.8	2.6	2.3	2.4	2.2	2.4	2.2	2.7
Matsuyama	2.8	2.5	2.4	2.2	1.9	2.2	1.9	1.8	1.6	1.7	2.0	2.6	2.1	2.6	2.2	2.0	1.8
Okayama	2.5	2.2	2.7	2.3	2.1	2.5	1.6	1.9	2.5	1.9	2.2	2.1	2.3	2.4	2.5	2.0	2.2
Ozaka	4.1	4.1	4.2	4.1	3.8	3.8	3.4	4.0	3.3	3.4	3.4	3.9	3.8	4.0	4.0	3.7	3.3
Kioto	1.7	1.8	2.1	1.9	1.9	1.7	1.6	1.8	1.5	1.3	1.3	1.5	1.7	1.7	2.0	1.7	1.4
Kumamoto	1.4	1.5	2.0	2.4	1.7	1.7	1.8	1.5	1.8	1.7	1.1	1.4	1.7	1.4	1.9	1.7	1.6
Saga	2.1	1.7	2.0	2.1	2.4	2.1	2.4	1.8	2.1	2.1	2.1	2.3	2.2	2.2	2.3	2.1	2.3
Nagasaki	2.2	2.2	2.5	2.7	2.2	2.3	2.3	2.3	2.1	2.0	1.9	2.4	2.2	2.2	2.5	2.3	2.0
Fukuoka	3.2	2.1	2.8	2.5	2.1	2.4	2.1	1.8	1.8	1.6	2.2	3.5	2.4	3.0	2.6	2.2	1.9
Itsugahara	1.5	1.3	1.5	1.5	1.5	1.5	1.5	1.2	1.1	1.3	1.3	1.5	1.4	1.4	1.5	1.1	1.2
Akamagaseki	4.2	3.3	3.9	3.6	3.3	3.4	3.2	3.0	3.0	2.6	3.1	4.7	3.4	4.1	3.6	3.2	2.9
Sakai	3.5	2.8	3.4	3.1	2.7	2.5	2.3	2.4	2.5	2.3	2.6	2.8	2.7	3.0	3.0	2.4	2.5
Tsu	3.0	2.8	3.1	3.3	2.7	3.1	2.1	3.4	3.7	2.5	2.6	2.7	2.9	2.8	3.0	3.0	2.9
Nagoya	2.6	3.2	3.3	2.9	2.8	2.8	2.1	2.2	2.1	2.1	1.9	2.0	2.5	2.6	3.0	2.1	2.1
Gifu	1.6	2.0	2.1	2.0	1.6	1.4	1.3	1.5	1.1	1.2	1.2	1.4	1.6	1.7	1.9	1.4	1.3
Hamamatsu	3.5	3.6	3.8	3.1	2.6	2.5	1.9	2.1	2.0	2.1	2.6	3.3	2.7	3.5	3.2	2.2	2.2
Numazu	3.6	3.3	3.5	3.6	3.4	3.1	3.2	3.3	2.9	2.9	3.2	3.8	3.3	3.6	3.5	3.2	3.0
Tokio	2.9	3.3	3.6	3.8	3.5	3.2	3.4	3.3	3.1	2.8	2.8	2.8	3.2	3.0	3.6	3.3	2.9
Utsunomiya	1.7	2.0	2.6	2.2	2.2	2.3	1.6	0.6	2.0	1.9	1.7	1.6	1.9	1.8	2.3	1.5	1.9
Choshi	5.1	5.6	5.4	5.6	5.2	4.4	4.3	4.6	4.8	5.5	5.4	4.2	5.0	5.0	5.4	4.4	5.1
Kanazawa	2.8	2.5	2.7	2.5	2.3	1.9	1.7	1.8	1.7	2.0	2.6	3.4	2.3	2.9	2.5	1.8	2.1
Fushiki	3.2	3.1	3.1	3.7	3.4	3.2	2.7	2.9	3.2	3.2	3.3	2.8	3.1	3.0	3.4	2.9	3.2
Nagano	1.6	2.6	3.6	4.4	3.8	3.4	2.7	3.4	3.7	2.8	2.6	2.5	3.1	2.2	3.9	3.2	3.0
Niigata	4.6	4.0	3.8	3.4	3.3	2.8	2.5	2.8	2.8	3.0	3.8	5.0	3.5	4.5	3.5	2.7	3.2
Yamagata	1.6	1.7	2.5	2.7	2.9	2.0	2.1	2.2	1.9	2.1	1.9	2.0	2.1	1.8	2.7	2.1	2.0
Akita	5.4	4.5	4.9	5.0	4.4	3.4	2.9	3.2	3.1	3.2	3.9	5.1	4.1	5.0	4.8	3.2	3.5
Fukushima	3.3	3.0	3.3	4.2	3.0	2.5	2.1	2.1	2.2	2.1	3.0	3.2	2.8	3.2	3.5	2.2	2.4
Ishinomaki	5.5	5.5	5.1	5.3	4.5	3.5	3.2	3.9	4.4	4.2	4.6	5.3	4.6	5.4	5.1	3.5	4.3
Miyako	2.4	2.2	2.4	2.1	2.0	1.2	1.0	1.2	1.4	1.8	2.2	2.6	1.9	2.4	2.3	1.1	1.8
Aomori	4.7	4.4	4.3	3.7	3.4	2.6	2.3	2.1	2.5	2.7	3.6	4.6	3.4	4.6	3.8	2.3	2.9
Hakodate	5.2	4.8	4.8	4.5	4.1	3.3	3.2	3.1	3.8	3.9	4.9	5.8	4.3	5.3	4.5	3.2	4.2
Suttsu	10.4	10.0	10.2	9.9	9.9	9.9	9.8	9.3	8.8	7.7	9.1	10.3	9.6	10.2	10.0	9.7	8.5
Sapporo	2.8	3.4	4.1	4.6	5.0	3.9	3.6	3.4	3.4	3.0	3.2	3.1	3.6	3.1	4.6	3.6	3.1
Kamikawa	0.9	1.2	1.8	2.1	2.2	1.4	1.1	1.0	0.9	1.2	1.5	1.6	1.4	1.2	2.0	1.2	1.2
Soya	7.5	6.5	7.1	7.8	7.1	6.7	6.7	6.1	7.5	8.3	8.2	7.2	7.2	7.1	7.4	6.5	8.0
Abashiri	2.8	2.2	3.0	2.6	3.7	2.1	2.2	2.2	2.3	2.8	3.2	3.1	2.7	2.7	3.1	2.2	2.8
Nemuro	5.0	4.4	4.4	3.7	3.9	3.4	2.8	2.9	3.5	4.1	5.0	5.0	4.0	4.8	3.9	2.9	4.2
Kushiro	1.3	1.5	2.0	2.6	3.2	2.4	1.8	1.7	1.7	1.6	2.1	2.3	2.0	1.7	2.6	2.0	1.8
Erimo	9.3	12.1	..	10.6	9.8	..	7.6	6.3	9.5	14.3	11.9	17.3	..	12.0	10.2	7.0	12.9

C. THE MAXIMUM VELOCITY OF WIND.

In our country, strong gales blow generally in late summer and early autumn. At these times, it is no uncommon occurrences that the velocity is more than 30 metres per second. The maximum velocities observed at several stations, since their establishments, are given in the next table.

ABSOLUTE MAXIMUM OF WIND VELOCITY.

Locality.	m.p.s.	Dir.	Day	Month	Year	Locality.	m.p.s.	Dir.	Day	Month	Year
Naha	31.2	NW	21	IX	1890	Numazu	22.2	SSW	30	IX	1891
Kagoshima	29.0	S	23;14	IX	1891	Tokio	38.1	SSE	11	X	1877
Miyazaki	35.1	SSE	14	IX	1891	Utsunomiya	23.1	S	30	IX	1891
Kochi	33.7	N	8	X	1887	Choshi	30.2	N	3	VIII	1890
Wakayama	32.0	WSW	16	VIII	1891	Kanazawa	25.1	W	9	I	1886
Oita	22.9	SE	14	IX	1891	Fushiki	28.0	NE	30	IX	1891
Yamaguchi	75.0	SE	14	IX	1891	Nagano	29.0	WSW	5	II	1891
Hiroshima	33.7	S	14	IX	1891	Niigata	37.5	SW	4	XII	1891
Matsuyama	26.8	S	14	IX	1891	Yamagata	21.4	SSW	13	III	1890
Okayama	20.0	SW	14	IX	1891	Akita	32.2	SW	15	IX	1891
Ozaka	33.3	W	1	VII	1885	Fukushima	25.9	SSW	26	XI	1889
Kioto	20.1	E	23	VII	1888	Ishinomaki	28.4	S	12	IX	1889
Kumamoto	21.9	S	14	IX	1891	Miyako	20.3	SSW	12	IX	1889
Saga	22.7	WNW	14	IX	1891	Aomori	29.1	NW	30	XII	1883
Nagasaki	29.0	....	14	IX	1891	Hakodate	31.2	....	9	X	1883
Fukuoka	25.3	W	14	IX	1891	Suttsu	43.2	SSE	10	X	1891
Itsugahara	30.8	S	3	VIII	1891	Sapporo	26.8	SSE;SE	13;1	V;XI	1880;82
Akamagaseki	40.7	E	14	IX	1891	Kamikawa	16.0	W	6	V	1891
Sakai	22.7	SW	14	IX	1891	Soya	11.2	W	25	XII	1889
Tsu	38.4	ESE	10	IX	1889	Abashiri	21.9	NE	12	IX	1889
Nagoya	17.1	SSE	16	VIII	1891	Nemuro	31.1	N	29	I	1890
Gifu	21.4	SSE	31	VIII	1888	Kushiro	27.2	SSW	27	IV	1890
Hamamatsu	20.1	E	11	IX	1889	Erimo	43.0	ENE	5	XII	1890

The velocities given in the above table are generally the means of twenty minutes, except the 75 metres observed at Yamaguchi on the 14th September, 1891, which is the mean of only three minutes. If we remember the fact that at Yamaguchi the mean velocity of one hour was only 19 metres per second at that day, we can easily see that in those cases when the hourly mean velocity is thirty or fourty metres, the velocity at certain instant must have reached, with great probability, some hundred metres per second.

D. FREQUENCY OF WIND.

There is clear distinction in the direction of winds between summer and winter: viz. during the three months, June, July and August, the southerly wind predominates, and during the remaining nine months the northerly wind prevails. The transitions, moreover, both from south to north and from north to south are very sudden. Now, if we divide the whole country into two parts, one lying to east of Kioto, and the other to its west, and calculate the mean directions of wind in each month for them separately, we get

MEAN DIRECTION OF WIND.

	Jan.	Feb.	March	April	May	June	July	August	Sept.	Oc.	Nov.	Dec.
West of Kioto	N10°W	N28°W	N26°W	N31°W	N12°W	S32°W	S28°E	S16°W	N10 E	N57E	N20°W	N49 W
East of Kioto	N57W	N45W	N10W	N77W	N37W	S23W	S3W	S31E	S22E	N18W	N11W	N82W

Owing to the manner of distribution of atmospheric pressure (See the Chapt. II), there is some difference in the mean direction of wind between the two parts, yet the fact that there exists a clear distinction between summer and winter, is common to both. Though in the west of Kioto, the south wind continues to blow for the three months from June till August, yet in the east of Kioto, it continues for the four months from June till September.

The observed number of winds in various directions and also the mean directions of winds are given in the following table in detail.

Frequency of wind.

WINTER.

Locality.	N	NNE	NE	ENE	E	ESE	SE	SSE	S	SSW	SW	WSW	W	WNW	NW	NNW	Calm
Naha	22	10	10	9	6	4	6	4	6	3	3	0	2	1	3	8	3
Kagoshima	20	2	7	5	5	0	1	1	1	0	1	0	1	1	14	8	33
Miyazaki	1	5	1	3	1	5	1	1	0	1	0	15	13	32	6	7	7
Kochi	10	1	7	1	5	0	3	1	3	0	17	2	18	3	20	2	7
Wakayama	15	5	11	13	4	0	0	0	1	1	1	0	5	4	18	5	12
Oita	5	1	8	1	2	0	2	1	11	3	11	2	9	1	25	5	7
Yamaguchi	20	6	7	0	0	0	1	1	3	3	7	2	2	1	10	5	32
Hiroshima	11	8	4	1	1	0	1	0	3	1	3	1	7	1	5	4	19
Matsuyama	7	1	5	2	5	3	9	5	6	2	4	4	8	7	15	5	9
Okayama	8	3	15	2	2	0	1	0	3	2	10	11	17	2	9	2	13
Ozaka	14	1	22	1	3	0	1	0	1	0	6	1	21	1	14	1	12
Kioto	11	2	5	1	2	0	2	0	6	3	9	4	7	5	16	6	21
Kumamoto	21	3	4	1	2	1	1	0	1	1	6	2	5	3	17	7	25
Saga	7	3	6	1	0	0	0	0	1	1	1	2	7	7	9	5	50
Nagasaki	29	5	9	1	3	1	3	0	1	0	3	1	4	1	11	8	20
Fukuoka	8	1	2	1	4	2	17	3	1	0	3	2	15	2	11	2	23
Itsugahara	11	1	4	0	1	0	1	0	3	0	1	1	5	1	24	2	40
Akamagaseki	4	1	5	3	21	1	2	0	1	1	5	2	13	4	24	3	10
Sakai	3	4	6	3	2	0	0	0	1	2	8	15	16	5	6	2	27
Tsu	10	2	1	2	3	1	1	0	1	1	3	4	21	9	25	7	8
Nagoya	15	4	5	1	1	0	1	0	0	0	2	1	6	5	37	11	11
Gifu	11	3	6	0	1	0	1	0	1	1	3	2	7	3	14	5	39
Hamamatsu	2	1	6	2	2	0	0	0	0	0	1	5	30	21	14	2	8
Numazu	2	2	20	20	9	4	3	1	2	3	12	7	4	3	2	1	5
Tokio	20	6	3	2	1	2	3	8	2	2	2	1	2	4	17	25	2
Utsunomiya	10	10	11	3	2	2	2	4	4	2	3	2	2	3	8	9	23
Choshi	11	3	10	1	2	0	1	0	4	2	9	1	9	5	27	2	13
Kanazawa	3	0	3	1	8	3	16	1	8	2	12	1	7	1	9	2	20
Fushiki	4	2	1	1	1	0	1	1	12	4	35	1	8	2	6	1	17
Nagano	12	1	9	4	14	3	4	1	2	1	8	3	9	2	7	3	17
Niigata	7	1	2	0	2	1	15	1	8	4	8	2	26	3	12	2	6
Yamagata	10	3	3	1	2	1	2	5	9	11	4	2	3	2	5	6	31
Akita	7	0	2	1	13	1	5	0	1	0	1	1	22	5	22	2	15
Fukushima	6	2	3	0	2	0	2	2	7	2	5	2	11	9	19	2	26
Ishinomaki	23	2	1	0	1	0	1	0	2	1	3	1	9	13	22	12	9
Miyako	3	1	2	0	2	0	0	0	2	2	15	16	38	2	3	1	13
Aomori	1	0	1	0	2	1	3	1	8	5	18	4	22	6	16	1	8
Hakodate	14	3	2	2	3	2	1	1	1	1	1	5	15	16	14	11	8
Suttsu	6	3	2	0	0	0	1	5	14	4	3	2	14	14	21	7	4
Sapporo	5	1	2	1	4	2	9	5	13	3	5	1	8	3	16	4	18
Kamikawa	6	0	1	0	1	1	3	2	7	1	5	1	13	1	5	2	51
Soya	11	3	5	2	8	3	5	1	1	2	6	3	19	5	11	6	7
Abashiri	9	1	2	0	2	0	2	1	7	1	12	1	17	2	15	2	26
Nemuro	5	6	3	3	2	2	3	2	3	3	7	9	10	10	13	7	12
Kushiro	11	4	3	1	1	0	1	0	3	2	1	1	3	1	5	8	52
Erimo	1	1	16	5	6	0	1	0	1	0	2	1	36	13	11	1	5

SPRING.

Locality.	N	NNE	NE	ENE	E	ESE	SE	SSE	S	SSW	SW	WSW	W	WNW	NW	NNW	Calm
Naha	11	7	10	9	7	10	7	5	4	6	6	2	2	2	2	1	6
Kagoshima	6	1	8	8	7	1	3	2	6	1	3	1	6	1	13	3	30
Miyazaki	2	9	3	9	4	10	3	4	1	6	1	13	7	11	4	5	10
Kochi	8	1	6	1	7	1	11	2	8	1	11	1	12	2	16	1	8
Wakayama	14	6	14	8	2	0	1	1	6	3	5	2	5	1	9	4	21
Oita	6	4	18	2	2	1	4	2	15	2	6	1	3	1	17	5	11
Yamaguchi	17	12	11	1	2	1	5	3	4	5	8	2	1	1	4	3	20
Hiroshima	37	6	3	1	2	0	1	1	10	4	8	2	4	1	2	3	15
Matsuyama	4	3	3	2	3	2	6	1	5	2	6	7	6	4	17	5	21
Okayama	6	7	26	6	3	2	6	1	4	3	9	7	6	0	7	2	5
Ozaka	17	1	29	1	6	0	1	0	1	0	8	1	15	0	8	0	8
Kioto	11	3	8	2	6	1	3	2	6	4	5	2	3	3	12	9	20
Kumamoto	18	4	4	1	2	1	0	0	3	5	20	6	4	3	8	7	13
Saga	10	3	7	1	1	0	1	1	5	3	6	3	6	3	4	4	42
Nagasaki	18	3	8	1	5	2	5	1	4	2	16	4	5	1	5	3	17
Fukuoka	13	2	3	1	4	2	12	3	2	0	2	1	7	2	12	7	27
Itsugahara	9	1	7	0	3	0	3	1	8	1	7	1	6	1	14	2	36
Akamagaseki	4	1	5	4	30	1	2	0	1	0	3	2	11	4	16	2	11
Sakai	5	7	13	7	4	1	1	1	2	1	5	7	7	4	6	4	25
Tsu	10	2	3	3	5	8	7	3	3	1	2	2	12	6	19	5	8
Nagoya	16	3	4	1	2	1	8	3	9	2	4	1	4	6	23	10	3
Gifu	11	3	5	1	1	1	4	2	5	3	5	3	6	2	13	4	31
Hamamatsu	2	2	10	7	9	3	4	1	4	2	6	7	16	10	7	0	10
Numazu	1	2	13	15	12	7	5	1	2	3	13	8	6	4	2	0	6
Tokio	12	6	6	5	4	4	6	9	10	7	2	1	2	2	8	14	1
Utsunomiya	9	8	6	3	3	4	4	11	10	5	4	2	2	2	6	5	16
Choshi	14	6	17	2	6	1	3	1	19	6	6	0	2	1	6	1	9
Kanazawa	5	1	4	1	11	4	12	1	4	2	10	2	8	2	10	3	21
Fushiki	9	4	13	2	3	0	1	1	8	3	19	1	7	2	6	1	20
Nagano	16	1	6	2	13	2	5	1	3	2	14	5	11	2	6	4	7
Niigata	13	2	5	1	2	2	16	1	5	3	13	4	14	1	5	2	11
Yamagata	11	6	3	1	1	3	3	5	6	11	6	3	2	2	7	11	19
Akita	3	0	2	1	16	6	7	0	3	1	9	3	21	4	9	1	14
Fukushima	8	5	11	1	2	1	3	2	11	3	4	1	5	5	14	2	24
Ishinomaki	12	2	4	3	4	2	7	7	9	3	2	0	4	7	15	7	12
Miyako	5	2	8	2	6	0	1	1	7	2	12	10	26	1	2	1	14
Aomori	8	4	4	1	5	1	4	1	7	4	12	3	13	5	13	2	13
Hakodate	6	1	1	2	6	6	6	4	3	4	8	7	11	9	7	7	12
Suttsu	7	2	1	0	1	0	2	13	24	4	3	3	10	7	7	8	8
Sapporo	7	1	2	1	3	3	14	10	12	2	3	1	3	3	16	7	12
Kamikawa	8	1	1	0	1	0	3	1	9	1	7	2	17	2	8	1	39
Soya	4	2	4	2	12	4	5	2	11	6	12	4	10	2	8	3	9
Abashiri	9	2	5	1	3	1	6	3	17	2	12	1	5	1	9	1	22
Nemuro	5	5	4	4	3	4	5	6	9	12	9	6	4	3	5	5	11
Kushiro	8	3	6	2	2	1	2	3	6	7	13	2	2	1	2	4	36
Erimo	1	1	13	4	10	1	2	1	1	1	5	4	45	4	2	0	5

SUMMER.

Naha	2	2	3	5	10	11	10	6	11	9	9	4	2	2	5	2	7
Kagoshima	2	0	5	6	10	1	6	3	10	1	4	2	7	1	8	1	33
Miyazaki	1	7	2	9	3	17	5	7	2	4	1	16	4	5	2	3	11
Kochi	3	0	4	1	9	1	17	3	12	0	12	1	11	1	10	0	13
Wakayama	8	2	10	7	3	0	1	1	12	7	11	3	5	1	4	2	22
Oita	6	4	17	1	2	0	8	3	17	3	9	1	2	0	10	3	14
Yamaguchi	21	11	7	0	1	1	6	5	7	5	10	1	1	0	3	3	18
Hiroshima	25	4	3	0	1	0	1	2	18	7	10	3	4	0	1	2	19

SUMMER.

Locality.	N	NNE	NE	ENE	E	ESE	SE	SSE	S	SSW	SW	WSW	W	WNW	NW	NNW	Calm
Matsuyama	2	1	2	2	3	4	7	4	5	3	6	9	5	5	17	4	21
Okayama	6	2	20	5	10	2	6	1	4	2	11	5	4	0	2	1	19
Ozaka	9	1	23	1	6	0	1	0	1	0	13	2	21	1	7	0	10
Kioto	9	2	8	2	8	2	5	2	10	6	8	2	3	1	6	4	22
Kumamoto	8	1	2	1	3	1	1	1	4	8	21	8	7	1	4	3	21
Saga	7	1	5	1	1	0	1	1	12	7	7	2	3	2	4	2	44
Nagasaki	6	2	5	1	5	3	7	2	7	5	27	6	2	0	1	1	20
Fukuoka	7	1	1	1	4	3	20	2	4	1	3	1	4	2	12	6	27
Itsugahara	6	1	5	0	3	1	4	2	15	3	7	1	4	1	4	2	41
Akamagaseki	2	1	1	5	35	4	3	0	1	1	4	1	7	3	14	2	13
Sakai	5	8	13	5	4	1	1	1	2	1	5	7	4	2	6	4	31
Tsu	6	1	5	2	7	5	17	3	5	1	3	3	11	3	13	5	9
Nagoya	9	2	4	1	3	2	12	5	14	3	6	1	4	4	11	5	14
Gifu	4	2	4	1	3	2	10	3	8	4	7	3	7	2	4	1	36
Hamamatsu	1	1	9	6	9	3	7	2	6	5	11	10	8	3	1	0	18
Numazu	1	1	12	11	8	5	3	1	2	3	17	11	10	5	2	0	8
Tokio	5	4	5	6	6	6	6	14	18	11	4	1	1	1	2	4	3
Utsunomiya	4	6	7	4	5	8	4	10	12	7	4	1	1	1	3	2	21
Choshi	11	5	13	2	5	0	4	1	23	12	10	0	0	0	2	1	11
Kanazawa	4	1	3	1	9	4	12	1	2	1	9	3	7	3	9	3	28
Fushiki	9	5	20	3	4	0	1	0	7	2	19	1	5	1	5	1	17
Nagano	12	1	4	2	9	3	5	1	3	2	14	6	11	2	6	4	15
Niigata	17	2	6	1	2	2	18	1	4	3	14	3	7	0	4	2	14
Yamagata	14	5	3	2	3	4	7	5	9	6	4	2	1	1	5	7	22
Akita	3	0	2	1	20	6	8	1	2	1	11	3	14	2	7	1	18
Fukushima	7	8	15	1	2	0	4	2	13	2	5	1	3	1	4	1	31
Ishinomaki	7	2	8	6	6	4	9	9	14	4	2	0	1	2	6	3	17
Miyako	4	2	8	2	7	1	1	1	7	3	8	7	13	1	1	1	33
Aomori	18	7	8	0	8	2	2	1	4	3	7	1	6	3	8	3	18
Hakodate	3	1	1	2	10	13	10	8	6	4	7	4	5	4	4	3	17
Suttsu	7	2	2	1	0	0	2	22	20	2	1	2	5	2	5	6	12
Sapporo	6	1	2	1	4	7	26	10	8	1	1	1	1	1	11	7	12
Kamikawa	8	1	1	0	1	0	2	1	4	1	3	1	18	2	7	2	48
Soya	2	2	3	2	19	7	10	1	7	9	12	4	6	2	2	2	10
Abashiri	15	2	8	1	6	1	6	2	22	2	8	0	2	0	4	1	20
Nemuro	3	5	6	6	4	7	9	10	12	11	8	3	1	1	1	2	11
Kushiro	5	7	9	4	4	2	2	3	7	7	7	3	1	1	2	4	32
Erimo	1	1	15	9	18	1	2	0	2	0	4	5	28	5	2	0	7

AUTUMN.

Naha	12	11	17	12	12	7	5	3	2	1	1	1	1	1	3	5	6
Kagoshima	9	2	12	10	12	1	3	1	2	0	1	0	2	1	10	5	29
Miyazaki	2	15	3	10	2	9	3	3	1	2	1	9	6	12	4	9	9
Kochi	8	1	5	1	5	0	7	1	6	1	19	1	17	2	16	1	11
Wakayama	14	6	17	16	12	1	1	0	3	1	3	1	2	1	6	3	13
Oita	5	3	14	1	1	0	5	1	20	4	10	1	3	1	16	3	12
Yamaguchi	24	11	8	2	1	1	4	2	4	3	7	1	1	1	4	4	22
Hiroshima	48	9	6	1	1	0	1	0	5	2	4	1	3	1	3	3	12
Matsuyama	5	4	5	2	3	4	9	4	3	2	3	4	5	5	11	5	25
Okayama	14	5	30	3	8	1	4	2	3	1	4	4	7	0	5	2	7
Ozaka	18	2	35	1	7	0	2	0	1	0	4	1	10	1	7	0	11
Kioto	16	3	9	2	5	1	3	1	4	3	5	1	4	2	11	7	23
Kumamoto	26	5	7	1	2	1	1	0	1	2	7	5	7	2	11	8	14
Saga	14	5	19	2	1	0	1	1	1	1	3	1	3	2	4	4	38
Nagasaki	20	5	16	2	5	1	4	1	2	1	7	2	3	1	4	3	23
Fukuoka	10	2	5	1	8	2	13	2	1	0	1	1	5	1	10	3	35

AUTUMUN.

Locality.	N	NNE	NE	ENE	E	ESE	SE	SSE	S	SSW	SW	WSW	W	WNW	NW	NNW	Calm
Itsugahara	17	1	8	1	2	0	2	1	5	1	3	0	4	1	12	3	39
Akamagaseki	5	2	7	6	27	2	8	0	1	0	3	2	7	2	14	3	16
Sakai	4	8	12	5	3	1	1	1	2	1	4	5	6	2	5	3	37
Tsu	8	2	3	1	7	2	8	2	2	1	3	4	18	6	22	6	5
Nagoya	18	3	4	0	2	1	5	2	3	2	4	1	3	3	23	8	18
Gifu	9	2	6	1	1	1	4	1	3	2	4	2	6	2	9	3	44
Hamamatsu	2	4	15	7	8	2	3	1	2	1	3	4	15	10	6	1	15
Numazu	2	2	18	21	12	7	6	1	2	3	9	4	3	2	2	0	6
Tokio	18	8	5	5	4	3	5	5	5	4	2	1	1	2	9	20	2
Utsunomiya	12	12	10	4	3	2	3	4	8	4	3	1	2	1	6	7	18
Choshi	16	6	22	2	1	1	2	1	9	1	6	0	3	1	8	2	13
Kanazawa	1	1	4	2	12	6	21	1	4	2	7	1	4	1	6	2	23
Fushiki	4	2	15	2	4	0	1	0	11	2	29	1	7	1	6	0	15
Nagano	13	1	7	3	10	3	6	1	4	2	13	5	10	2	8	4	8
Niigata	10	2	5	1	3	3	22	3	9	3	8	2	9	1	6	2	11
Yamagata	11	7	5	2	3	3	5	8	9	7	4	2	2	1	4	7	20
Akita	6	1	3	2	23	6	8	1	1	0	2	1	13	3	12	2	16
Fukushima	6	4	6	0	3	1	4	2	10	2	6	2	9	3	11	1	30
Ishinomaki	20	3	5	3	4	2	3	3	8	3	2	1	5	6	14	11	7
Miyako	3	2	3	1	6	1	1	0	5	2	11	11	27	2	2	1	19
Aomori	4	3	5	1	6	2	5	2	9	4	13	2	13	5	11	2	14
Hakodate	10	2	3	3	5	5	5	3	2	3	3	5	10	10	10	9	12
Suttsu	6	1	1	0	0	0	2	14	23	3	2	2	10	9	13	5	9
Sapporo	4	1	2	1	4	4	15	8	12	3	4	1	6	3	13	4	15
Kamikawa	3	0	1	1	1	1	3	2	6	2	5	2	11	2	5	2	53
Soya	6	2	4	2	8	2	8	2	6	4	10	4	18	6	9	3	6
Abashiri	9	1	3	1	3	1	2	2	16	5	15	2	9	1	10	2	18
Nemuro	4	4	3	4	1	5	5	6	8	11	8	7	6	6	6	5	8
Kushiro	8	4	3	2	2	1	1	2	5	4	4	2	2	1	7	6	46
Erimo	1	1	11	6	13	1	2	1	5	1	4	3	36	7	3	0	5

YEAR.

Naha	12	7	10	8	9	8	7	4	6	5	5	2	2	1	3	5	6
Kagoshima	9	1	8	7	9	1	3	2	5	1	2	1	4	1	11	4	31
Miyazaki	2	9	2	7	2	10	3	4	1	2	1	13	8	15	4	6	9
Kochi	7	1	5	1	6	1	10	2	7	0	16	1	15	1	15	1	10
Wakayama	13	5	13	11	5	0	1	1	5	3	5	2	4	2	9	3	17
Oita	6	4	14	1	1	0	5	2	17	3	9	1	4	1	17	4	11
Yamaguchi	21	10	8	1	1	1	1	3	5	1	8	1	1	1	5	4	22
Hiroshima	38	6	4	1	1	0	1	1	9	4	6	1	5	1	3	3	16
Matsuyama	5	3	4	2	3	3	8	4	4	2	5	6	6	6	15	5	19
Okayama	8	4	23	4	5	1	4	1	3	2	9	7	9	1	6	2	11
Ozaka	14	1	27	1	6	0	2	0	1	0	8	1	17	1	9	1	10
Kioto	12	3	8	2	5	1	3	1	7	4	7	2	4	2	11	6	22
Kumamoto	18	3	4	1	2	1	1	1	2	4	14	5	6	3	10	7	18
Saga	9	3	9	1	1	1	1	1	5	3	4	2	4	4	5	4	43
Nagasaki	18	4	10	1	4	2	5	1	3	2	13	3	4	1	5	4	20
Fukuoka	10	2	3	1	5	2	15	2	2	0	2	1	8	2	12	5	28
Itsugahara	12	1	6	0	2	0	2	1	8	1	5	1	5	1	11	2	39
Akamagaseki	3	1	5	5	28	2	2	3	1	0	4	2	9	3	17	2	13
Sakai	4	6	11	5	3	1	1	1	2	1	6	9	8	8	6	3	30
Tsu	9	2	3	2	5	3	8	2	3	1	3	3	16	6	20	6	8
Nagoya	14	3	4	1	2	1	6	3	7	2	4	1	4	4	23	9	12
Gifu	10	3	5	1	1	1	5	2	4	2	5	3	6	2	10	3	38
Hamamatsu	2	2	10	5	7	2	3	1	3	2	5	7	17	12	7	1	13
Numazu	1	2	16	17	10	6	4	1	2	3	13	8	6	3	2	0	6

YEAR.

Locality.	N	NNE	NE	ENE	E	ESE	SE	SSE	S	SSW	SW	WSW	W	WNW	NW	NNW	Calm
Tokio	14	6	5	4	4	4	5	8	9	7	2	1	2	2	9	16	2
Utsunomiya	9	9	9	4	3	4	3	7	9	4	3	1	2	2	6	6	19
Choshi	13	5	16	2	4	1	2	1	13	6	8	0	3	2	11	1	11
Kanazawa	4	1	3	1	10	4	15	1	4	2	9	2	6	2	9	3	23
Fushiki	7	3	13	2	3	0	1	1	10	3	25	1	7	1	5	1	17
Nagano	13	1	7	3	11	3	5	1	3	2	12	5	10	2	7	4	11
Niigata	12	2	4	1	2	2	18	1	6	4	11	3	14	1	7	2	10
Yamagata	12	5	3	2	2	3	4	6	8	9	1	2	2	2	5	8	23
Akita	5	0	2	1	18	5	7	0	2	1	6	2	17	3	13	2	16
Fukushima	7	4	9	1	2	0	3	2	10	3	5	1	7	4	12	2	28
Ishinomaki	15	2	5	3	4	2	5	5	8	3	2	1	5	7	14	8	11
Miyako	4	2	5	1	5	0	1	1	5	2	12	12	26	1	2	1	20
Aomori	8	3	5	0	5	1	4	1	7	4	13	3	14	5	12	2	13
Hakodate	8	2	2	2	6	6	5	4	3	3	5	5	10	10	9	8	12
Suttsu	7	2	1	0	0	0	2	13	23	3	2	2	9	8	12	7	9
Sapporo	6	1	2	1	4	4	16	9	11	2	3	1	4	2	14	6	14
Kamikawa	6	1	1	0	1	0	3	1	7	1	5	2	15	1	6	2	48
Soya	6	2	4	2	12	1	7	2	7	5	10	4	13	4	7	3	8
Abashiri	10	1	4	1	3	1	4	2	15	3	12	1	8	2	10	2	21
Nemuro	1	5	4	4	3	5	5	6	8	9	8	6	5	5	7	5	11
Kushiro	8	5	5	2	2	1	1	2	5	5	7	2	2	1	4	6	42
Erimo	1	1	11	6	12	1	2	0	3	1	3	3	37	7	4	0	5

Mean direction of wind.

Locality.	Jan.	Feb.	March	April	May	June	July	August	Sept.	Oct.	Nov.	Dec.	Year
Kagoshima	N2°E	N32°W	N36 E	N33 E	N71 E	S56 E	S58 E	S50 E	N67 E	N55 E	N34 E	N15 E	N22 E
Miyazaki	N78°W	N70°W	N76 W	N90 W	N73 E	S32 W	S38 E	S43 E	N42 E	N20 E	N54 W	N78 W	N51 W
Kochi	N65 W	N31°W	N44 W	S78 W	S30 W	S6 W	S15 E	S9 E	S39 W	N69 W	N74 W	N80 W	S80 W
Wakayama	N19°W	N3°W	N1 E	N19 W	N3 W	S48 W	S20 W	S23 W	N74 E	N39 E	N25 E	N12 W	N16 E
Oita	N55 W	N52°W	N41 W	N15 W	N15 E	N71 W	N70 E	N55 E	N62 E	S68 W	N73 W	N71 W	N63 W
Hiroshima	N27 W	N16 W	N12 W	N20 W	N12 W	S71 W	S57 W	N26 W	N2 W	N2 W	N4 W	N16 W	N14 W
Matsuyama	N63°W	N19°E	N45 W	N59 W	S75 W	S56 W	N75°W	N81 W	N62 W	N3 W	N33 W	N73 W	N72 W
Ozaka	N62°W	N52°W	N8 W	N6 W	N19 E	N61 W	N68 W	N27 W	N33 E	N25 E	N2 W	N58 W	N6 W
Kioto	N72°W	N55°W	N34 W	N9 W	N5 E	S22 W	S30 E	S68 E	N34 E	N10 W	N30 W	N75 W	N25 W
Kumamoto	N33°W	N20°W	N43 W	N72 W	S90 W	S67 W	S48 W	S65 W	N4 E	N9 W	N32 W	N43 W	N54 W
Nagasaki	N13 W	N1°W	N14 W	N74°W	N47 W	S29 W	S20 W	S5 E	N9 E	N22 E	N7 E	N10 W	N11 W
Fukuoka	N66°W	N14°W	N33 W	N33 W	N2 W	S15°W	N1 E	N8 W	N7 W	N14 E	N35 W	N83 W	N3 W
Akamagaseki	N34°W	N35°W	N29 W	N43 E	N64 E	N76 E	N84 E	N80 E	N27 E	N45 E	N21 W	N28 W	N28 W
Sakai	N56°W	N28°W	N22 W	N15 W	N3 W	N31 W	N11 E	N15 E	N3 E	N27 E	N27 W	N86 W	N29 W
Nagoya	N42°W	N35°W	N46 W	N56 W	N88 W	S25 W	S27 W	S5 E	S18 E	N37 W	N35 W	N43 W	N36 W
Gifu	N46°W	N39°W	N48 W	N58 W	N88 W	S69 W	S3 E	S3 W	S89 W	N47 W	N39 W	N35 W	N48 W
Hamamatsu	N69 W	N66°W	N75 W	N81 W	S83 W	S64 W	S8 W	S44 E	N81 E	N21 W	N56 W	N67 W	N68 W
Numazu	S20°E	S31°E	S34 E	S4 W	S43 E	S39 W	S42 W	S5 E	S54 E	N81 E	N79 E	S35 W	N88 E
Tokio	N24°W	N19 W	N16 W	S50 E	S28 E	S31 E	S67 E	S22 E	S64 E	N4 W	N18 W	N29 W	N6 E
Choshi	N20 W	N5 W	N24 E	N46 E	N42 E	N88 E	S9 E	S81 E	N43 E	N14 E	N15 E	N33 W	N3 W
Kanazawa	S66°W	S68 W	S45 W	S45 W	S48 W	S63 W	S41 W	S15 W	S47 E	S48 E	S1 E	S54 W	S27 E
Fushiki	N20°W	S87 W	N26 E	N6 E	N11 E	N13 W	N24 E	N35 E	N44 E	N35 E	N62 W	S60 W	S65 W
Nagano	N28 E	N41 W	N70 W	S81 W	N88 W	N58 W	N88 W	S75 W	S76 W	N87 W	N57 W	S87 W	N39 W
Niigata	N79°W	N86 W	S68 W	S48 W	S41 W	S38 W	S35 W	S10 E	S19 E	S59 W	S77 W	S79 W	S71 W
Yamagata	N42°W	N56 W	N12 E	N62 W	S15 E	N73 W	S66 E	S62 E	S74 E	S10 W	N48 W	S30 W	N58 W
Akita	N51°W	N77 W	N89 E	S65 W	S24 E	S32 E	S27 W	S87 E	S89 E	N9 E	N2 W	N85 W	N51 W
Ishinomaki	N44°W	N31 W	N19 W	N7 W	N1 W	N33 E	N75 E	S60 E	S40 E	N11 W	N41 W	N86 W	N25 W
Miyako	S81°W	S80 W	S73 W	S58 W	S56 W	S54 W	S32 W	S31 W	S50 W	S84 W	S81 W	S77 W	S76 W
Aomori	N87°W	S88 W	N84 W	N81 W	N59 W	N21 W	N35 W	N8 E	N65 W	S86 W	S90 W	N89 W	N86 W
Hakodate	N59°W	N63 W	N72 W	S66 W	S10 W	S33 E	S12 E	S39 E	S41 E	N77 W	N66 W	N72 W	N65 W
Suttsu	N56°W	N53 W	N83 W	S25 W	S5 W	S6 E	S13 E	S8 E	S11 W	S42 W	N85 W	N70 W	S56 W
Sapporo	N77°W	N61 W	N70 W	S66 W	S38 E	S28 E	S45 E	S35 E	S21 E	S24 W	S89 W	S71 W	S10 E
Kamikawa	S77°W	N66 W	S85 W	S73 W	S88 W	S87 W	N81 W	N74 W	S87 W	S80 W	S60 W	S59 W	S87 W
Soya	N59°W	N5 W	N67 W	S70 W	S22 W	S51 E	S20 W	S10 W	S69 W	S79 W	N80 W	S82 W	S48 W
Nemuro	N43°W	N32 W	N79 W	S44 W	S16 W	S59 E	S6 E	S18 E	S82 E	S77 W	N81 W	N79 W	S37 W

CHAPTER IV.

HUMIDITY.

The tension of aqueous vapour given in this chapter is calculated by the readings of the wet and dry bulb thermometers suspended in Stevenson's double louvre-boarded box, using Angot's table given in *Annal du Bureau central météorologique de France* in their reductions. Hence, though the calculated result will be correct for warmer seasons, yet for winter, especially in northern part of this country, it does not certainly represent true conditions of humidity.

A. THE DIURNAL VARIATIONS OF THE TENSION OF AQUEOUS VAPOUR AND RELATIVE HUMIDITY.

The tension of aqueous vapour and also the relative humidity have one maximum and one minimum in a day through the year everywhere in this country.

The tension of aqueous vapour has its daily variation very much similar to that of the air temperature. The time of its minimum value exactly coincides with the time of the minimum temperature, i. e. at about sunrise. The time of the maximum tension appears a little after the time of the maximum temperature, i. e. at about two hours previous to the sunset.

The diurnal variation of the relative humidity is just opposite to that of the air temperature. Its maximum occurs at about sunrise and exactly coincides with the time of the minimum temperature, while its minimum is at 2 or 3 o'clock pm. and nearly coincides with the time of the maximum temperature. For details see the following table and also the Plate V.

Mean relative humidity.

KUMAMOTO.

Hour \ Month	Jan.	Feb.	Mar.	Apr.	May	June	July	Aug.	Sept.	Oct.	Nov.	Dec.	Year	Wint.	Spr.	Sum.	Aut.
1 am	90,00	88,60	84,04	86,16	90,01	90,85	91,00	93,71	90,08	91,61	92,46	89,35	89,82	89,32	86,74	91,85	91,38
2 am	90,97	88,68	83,85	87,88	91,04	92,40	91,93	94,06	91,33	92,01	93,13	88,71	90,50	89,45	87,59	92,80	92,16
3 am	91,13	88,61	83,80	88,56	91,03	93,15	92,42	95,21	91,62	92,18	92,81	88,93	90,79	89,57	87,80	93,59	92,20
4 am	90,03	89,36	86,32	89,06	91,21	93,05	92,68	95,63	92,35	91,90	93,37	88,84	91,15	89,41	88,86	93,79	92,54
5 am	90,10	90,23	86,27	89,65	91,76	93,43	92,67	95,51	92,67	92,48	93,48	88,93	91,43	89,75	89,23	93,87	92,88
6 am	88,94	90,96	87,30	90,36	90,58	91,82	92,46	95,22	92,70	92,79	93,75	88,96	91,32	89,62	89,41	93,17	93,08
7 am	87,97	91,66	86,71	86,30	82,95	84,06	85,45	88,27	87,90	90,59	93,23	88,85	87,83	89,49	85,32	85,93	90,57
8 am	85,68	84,93	77,21	77,73	73,86	76,61	79,69	78,72	77,86	80,52	80,65	87,56	80,59	86,06	76,27	78,34	81,68
9 am	76,29	76,53	69,48	70,55	65,94	71,31	73,87	71,37	70,70	70,72	75,05	77,74	72,42	76,85	68,66	72,02	72,16
10 am	64,23	66,68	64,72	65,88	60,02	67,08	70,11	66,11	64,42	62,92	64,77	70,76	65,61	67,22	63,54	67,77	64,04
11 am	62,94	63,71	59,61	68,71	57,25	63,96	67,42	63,51	61,83	57,84	57,52	63,60	61,91	63,42	60,19	64,96	59,06
Noon	59,68	60,28	56,43	61,68	55,74	61,96	64,32	62,06	60,23	55,40	54,30	59,87	59,33	59,94	57,95	62,78	56,64
1 pm	54,65	58,84	55,51	60,86	55,14	59,38	63,54	59,95	59,22	53,95	52,74	58,68	57,70	57,39	57,17	60,96	55,30
2 pm	52,97	58,78	56,00	59,70	53,92	58,85	61,87	57,50	59,45	52,79	52,00	57,04	56,74	56,26	56,54	59,41	54,75
3 pm	55,16	58,69	55,03	58,96	54,26	58,55	61,13	58,50	59,15	53,53	52,58	57,43	56,97	57,09	56,28	59,42	55,09
4 pm	58,39	60,00	56,48	60,87	55,13	59,53	62,16	60,55	62,08	56,87	55,22	60,13	58,95	59,51	57,49	60,75	58,06
5 pm	62,94	62,57	58,56	62,82	56,84	62,06	64,11	65,48	66,88	63,31	64,45	68,19	63,18	64,57	59,41	63,88	64,88
6 pm	71,32	70,25	63,68	68,12	61,01	65,78	68,87	71,14	73,11	73,50	74,52	75,16	69,70	72,24	64,27	68,60	73,71
7 pm	74,97	76,78	71,97	74,13	70,50	71,74	75,16	78,93	77,90	80,29	80,91	79,32	76,05	77,02	72,20	75,28	79,70
8 pm	78,26	79,50	75,61	78,78	77,83	78,42	81,19	84,66	81,06	82,53	85,20	82,14	80,43	79,97	77,41	81,42	82,93
9 pm	80,87	80,89	77,76	82,80	81,61	81,35	83,40	87,93	83,33	85,05	87,28	84,64	83,08	82,13	80,72	84,23	85,22
10 pm	84,03	83,62	81,39	84,30	85,79	85,05	85,93	90,22	84,95	87,56	89,40	86,95	85,77	84,87	83,83	87,07	87,30
11 pm	86,03	85,59	82,58	85,63	87,85	87,62	88,01	91,35	86,58	89,68	90,73	88,50	87,56	86,71	85,35	89,20	89,00
M. N.	89,65	86,73	83,58	86,83	88,39	96,45	90,17	92,95	88,61	90,47	91,90	88,55	89,02	88,31	86,27	91,19	90,33
Mean	76,12	76,76	72,68	75,88	73,73	76,60	78,32	79,10	77,33	76,68	77,80	77,87	76,57	76,92	74,10	78,01	77,27

MATSUYAMA.

Hour \ Month	Jan.	Feb.	Mar.	Apr.	May	June	July	Aug.	Sept.	Oct.	Nov.	Dec.	Year	Wint.	Spr.	Sum.	Aut.
1 am	77.8	81.3	84.1	87.6	88.4	88.6	91.1	91.9	93.0	89.3	86.4	82.7	86.8	80.6	86.7	90.5	89.6
2 am	77.1	81.0	85.4	88.4	88.5	89.7	91.7	92.1	93.5	90.0	87.4	82.4	87.2	80.2	87.4	91.2	90.3
3 am	78.0	81.1	84.8	88.6	89.6	90.0	92.2	92.8	91.0	90.7	87.8	81.6	87.6	80.2	87.7	91.7	90.8
4 am	77.5	82.5	84.8	88.8	90.5	90.7	92.1	93.3	91.2	90.9	87.7	81.1	87.8	80.4	88.0	92.0	90.9
5 am	78.1	81.9	85.1	89.4	89.6	90.6	93.1	91.2	94.7	91.1	88.2	82.8	88.3	80.9	88.0	92.6	91.3
6 am	78.0	83.7	85.3	89.3	89.9	88.3	92.3	94.1	91.4	91.5	88.6	81.0	88.0	80.9	88.2	91.6	91.5
7 am	77.6	83.4	84.3	86.1	82.0	79.6	85.6	86.0	91.9	90.4	88.8	81.3	84.8	80.8	84.1	83.7	90.4
8 am	74.5	79.1	77.6	77.2	72.2	74.8	80.7	78.0	84.3	79.8	81.3	80.8	78.4	78.1	75.7	77.8	81.8
9 am	69.7	72.2	71.0	71.2	67.1	71.8	77.6	74.3	77.7	70.9	69.8	72.8	72.1	71.6	69.8	74.0	72.8
10 am	64.0	66.8	66.0	68.3	63.8	70.1	75.9	72.0	74.0	61.8	62.2	66.6	67.8	65.8	66.0	72.7	67.0
11 am	59.4	63.1	63.7	66.1	62.4	69.3	73.3	70.6	72.3	61.7	57.8	61.7	65.2	61.4	64.2	71.1	68.9
Noon	57.9	60.7	63.5	65.9	62.0	67.7	72.8	68.6	71.6	61.2	56.9	60.4	64.1	59.7	63.8	69.7	63.2
1 pm	56.8	61.1	62.3	64.6	60.6	65.7	71.0	66.6	70.2	60.2	56.4	60.5	63.0	59.6	62.5	67.8	62.3
2 pm	59.1	60.4	62.4	65.0	60.3	65.6	71.4	66.0	70.2	60.3	55.6	60.4	63.0	60.0	62.6	67.7	62.0
3 pm	56.4	59.7	62.2	64.1	60.1	66.4	70.8	66.0	70.9	60.4	56.6	61.2	62.9	59.1	62.1	67.7	62.6
4 pm	58.5	60.2	63.0	65.7	62.6	67.1	71.8	66.7	72.4	62.4	60.1	62.6	64.4	60.4	63.8	68.5	65.0
5 pm	62.3	63.7	64.3	67.4	63.5	68.9	72.9	69.9	77.3	72.0	70.4	68.4	68.4	64.8	65.1	70.6	73.2
6 pm	69.3	71.0	70.4	72.2	68.9	70.7	76.4	75.9	81.3	83.9	77.6	74.9	74.7	71.7	70.5	74.3	81.9
7 pm	72.1	75.3	78.8	80.2	77.7	74.9	80.2	83.4	89.4	86.4	78.7	77.0	79.5	74.8	78.9	79.5	84.8
8 pm	73.3	75.3	80.3	83.3	82.6	79.8	86.1	87.3	90.7	87.1	80.7	77.0	81.9	75.2	82.1	84.4	86.2
9 pm	75.0	78.3	81.8	84.0	83.3	82.4	87.4	89.2	91.5	87.4	81.7	78.9	83.4	77.4	83.0	86.3	86.9
10 pm	76.5	79.1	82.4	86.3	85.0	83.7	88.2	89.0	91.4	88.1	83.2	80.4	84.4	78.7	84.6	87.0	87.6
11 pm	77.1	79.9	83.1	87.3	86.1	85.9	89.7	90.4	93.1	89.6	84.8	81.6	85.7	79.5	85.5	88.7	89.2
M. N.	78.0	80.7	84.4	87.1	87.6	87.3	90.6	90.8	93.2	90.2	86.1	82.1	86.6	80.3	86.4	89.6	89.8
Mean	70.2	73.4	75.5	78.1	76.0	77.9	82.3	81.2	84.6	79.2	75.6	74.2	77.3	72.6	76.5	80.5	79.8

HIROSHIMA.

1 am	84.70	83.08	85.50	87.97	84.60	90.83	92.20	86.27	89.37	86.60	89.00	88.13	87.35	85.29	86.02	89.77	88.32
2 am	85.30	83.28	86.60	88.63	85.77	91.43	92.63	87.27	90.23	87.77	89.50	88.90	88.11	85.81	87.00	90.44	89.17
3 am	86.13	83.60	87.00	89.73	86.37	91.93	93.27	87.90	90.60	88.80	90.33	88.70	88.71	86.21	87.70	91.03	89.91
4 am	86.67	84.23	87.67	89.87	86.90	92.33	93.60	88.27	90.90	88.83	90.73	89.03	89.09	86.64	88.15	91.40	90.15
5 am	86.77	84.40	87.73	90.10	87.60	92.87	94.10	88.73	90.80	89.70	90.90	89.17	89.41	86.78	88.48	91.90	90.47
6 am	86.67	85.20	88.07	89.10	88.00	92.33	94.23	89.17	91.00	89.70	91.18	89.03	89.47	86.97	88.39	91.91	90.61
7 am	87.00	85.57	87.40	87.67	85.20	89.87	92.33	87.18	90.10	89.10	91.47	89.57	88.54	87.38	86.76	89.78	90.22
8 am	85.27	83.17	82.40	81.90	78.20	84.83	88.43	81.77	85.73	84.07	87.23	87.43	84.20	85.29	80.88	85.01	85.68
9 am	78.40	77.23	76.73	76.78	70.83	78.27	83.77	74.60	81.13	78.00	80.27	82.03	78.16	79.22	74.76	78.88	79.80
10 am	71.90	69.27	71.30	72.93	65.37	75.33	80.13	69.70	76.00	71.10	73.00	75.93	72.07	72.37	69.87	75.05	73.37
11 am	66.03	63.77	67.77	69.77	62.40	72.70	77.43	66.50	72.10	64.97	67.23	72.23	68.59	67.38	66.65	72.21	68.10
Noon	62.97	60.77	64.63	69.10	61.13	70.73	75.77	64.23	69.20	61.10	62.40	68.40	65.87	64.05	64.95	70.24	64.23
1 pm	61.50	58.50	62.53	68.63	60.70	69.50	72.27	62.70	67.67	59.87	60.37	65.10	64.28	61.70	63.95	68.82	62.64
2 pm	59.43	57.53	62.30	68.53	59.80	69.08	74.07	62.63	67.60	59.50	59.40	64.17	63.67	60.38	63.54	68.58	62.17
3 pm	59.66	58.33	62.63	69.77	59.33	68.83	74.20	62.87	68.23	60.07	61.08	64.20	64.10	60.73	63.91	68.63	63.11
4 pm	61.30	59.93	63.98	70.37	60.30	70.40	74.50	63.67	70.00	62.47	65.00	66.67	65.70	62.63	64.87	69.52	65.82
5 pm	66.70	63.00	66.53	72.97	62.10	72.27	76.33	67.50	72.97	68.50	71.90	72.80	69.46	67.50	67.20	72.03	71.12
6 pm	72.13	69.20	71.57	76.47	66.13	74.97	79.03	71.10	78.33	75.23	77.23	77.97	74.11	73.10	71.39	75.03	76.93
7 pm	76.23	73.77	76.40	81.03	71.80	79.13	83.20	75.60	82.47	77.70	80.30	80.93	78.22	76.98	76.41	79.31	80.16
8 pm	78.47	77.00	79.23	83.07	75.23	83.13	86.10	78.70	84.13	80.03	81.87	83.53	80.90	79.67	79.18	82.64	82.11
9 pm	79.80	78.63	80.73	84.27	77.47	84.57	87.53	80.33	85.40	81.67	82.43	84.47	82.36	80.97	80.82	84.14	83.50
10 pm	81.50	79.77	83.40	85.93	80.00	86.43	88.93	82.47	86.57	83.67	85.53	86.70	84.25	82.66	83.11	85.94	85.26
11 pm	82.87	81.13	84.37	87.37	81.43	88.30	90.17	84.00	87.87	84.97	87.17	87.47	85.59	83.82	84.39	87.40	86.67
M. N.	84.30	82.50	84.90	87.63	83.07	89.47	91.13	85.33	88.60	85.57	87.77	87.47	86.49	84.76	85.20	88.64	87.33
Mean	76.32	74.29	77.14	80.40	74.16	81.65	84.89	77.02	81.53	77.46	79.34	80.42	78.73	77.01	77.23	81.19	79.44

OZAKA.

Hour / Month	Jan.	Feb.	Mar.	Apr.	May	June	July	Aug.	Sep.	Oct.	Nov.	Dec.	Year	Wint.	Spr.	Sum.	Aut.
1 am	79,1	82,5	83,7	83,9	84,1	85,8	87,9	86,0	87,4	86,3	85,3	83,7	84,5	82,0	83,9	86,6	86,3
2 am	80,9	83,4	83,6	85,1	85,2	86,4	88,6	87,1	88,3	87,0	86,4	84,5	85,2	82,9	84,6	87,4	87,2
3 am	81,3	84,0	83,2	84,9	86,0	87,0	89,3	88,0	89,2	87,9	87,1	84,3	85,7	83,2	84,7	88,1	88,1
4 am	81,9	84,5	84,0	85,7	87,1	87,7	89,9	88,5	89,7	88,4	87,1	84,6	86,3	83,7	85,6	88,7	88,5
5 am	83,0	85,2	84,2	86,1	87,5	87,7	90,7	89,0	90,0	88,7	87,8	84,9	86,8	84,4	85,9	89,1	88,8
6 am	83,9	82,7	84,7	85,9	85,2	85,9	89,3	88,3	90,0	89,4	88,2	84,5	86,5	83,7	85,3	87,8	89,2
7 am	82,8	85,8	82,8	81,4	78,8	79,8	85,3	82,9	86,6	87,6	87,5	85,1	83,5	84,6	81,0	82,7	87,2
8 am	79,3	79,6	76,6	74,9	71,8	74,1	80,7	76,8	78,5	80,4	83,1	81,5	77,7	80,1	74,4	77,2	80,7
9 am	71,7	72,3	68,4	68,9	66,4	69,7	76,6	71,7	71,4	71,6	71,8	75,0	71,1	73,0	67,9	72,7	72,6
10 am	62,8	65,6	63,7	64,6	62,1	66,0	73,7	68,2	67,1	64,1	66,7	68,3	65,5	65,6	63,5	69,3	66,0
11 am	59,0	62,1	61,5	62,6	59,4	64,6	71,5	65,5	64,6	60,1	61,5	64,3	62,5	61,9	61,2	67,2	62,1
Noon	56,6	60,4	59,5	61,4	58,0	62,9	70,1	63,5	62,4	57,4	58,5	60,5	60,4	59,2	59,6	65,5	59,4
1 pm	55,8	59,8	58,3	60,8	57,0	62,4	68,8	62,2	62,0	56,0	57,2	60,0	59,5	58,5	58,7	64,5	58,4
2 pm	55,3	59,8	57,5	60,7	56,1	62,0	69,1	62,3	62,0	55,9	57,1	59,1	59,1	58,1	58,1	64,5	58,3
3 pm	55,8	59,6	58,1	61,1	57,4	61,8	69,5	62,5	63,5	57,2	58,5	60,8	60,0	58,7	58,9	64,6	59,7
4 pm	57,9	61,6	59,6	62,4	59,2	64,1	70,8	63,9	66,3	59,9	62,0	64,1	62,3	61,2	60,4	66,3	62,7
5 pm	62,5	66,3	63,3	65,2	52,5	66,9	73,0	67,9	70,6	66,3	69,3	70,6	66,6	66,4	63,7	69,3	68,7
6 pm	66,6	70,3	69,0	69,8	56,5	69,6	76,2	72,3	75,4	73,3	73,3	73,9	70,9	70,3	68,4	72,7	74,0
7 pm	69,7	73,6	74,1	73,7	52,2	74,8	80,1	77,3	78,7	77,1	76,4	76,6	75,0	73,3	73,3	77,5	77,4
8 pm	70,7	75,2	76,5	77,1	55,0	78,2	82,5	79,7	81,3	79,0	78,3	78,9	77,5	74,9	76,2	80,1	79,5
9 pm	72,5	77,8	79,2	78,1	56,7	80,3	83,9	81,4	82,2	80,9	80,0	79,0	79,1	76,4	78,0	81,9	81,0
10 pm	74,9	79,0	80,1	79,8	59,1	82,5	85,1	83,0	83,8	82,8	81,8	79,9	80,7	77,9	79,7	83,6	82,8
11 pm	76,9	81,0	81,4	81,9	51,2	83,9	86,5	84,6	85,1	84,1	84,0	81,2	82,4	79,7	81,5	85,0	84,4
M. N.	77,9	82,0	82,8	82,8	53,1	84,9	87,1	85,0	86,4	85,6	84,1	82,8	83,5	80,9	82,9	85,7	85,4
Mean	70,8	74,0	73,1	74,1	52,5	85,4	80,3	76,6	77,6	75,3	75,7	75,3	74,7	73,4	73,2	77,4	76,2

WAKAYAMA.

Hour	Jan.	Feb.	Mar.	Apr.	May	June	July	Aug.	Sep.	Oct.	Nov.	Dec.	Year	Wint.	Spr.	Sum.	Aut.
1 am	71.56	73.64	76.54	80.45	82.73	84.18	87.07	87.71	87.57	84.13	80.03	77.73	81.09	74.31	79.91	86.32	83.84
2 am	71.12	73.76	77.05	81.23	82.49	85.51	87.07	88.51	88.02	81.55	81.13	78.02	81.67	74.30	80.59	87.22	81.57
3 am	71.88	74.61	78.35	81.57	84.40	86.51	88.30	88.54	88.21	84.62	81.05	78.31	82.23	74.93	81.57	87.78	84.63
4 am	71.86	75.54	79.16	82.56	85.65	86.66	88.76	88.88	88.57	85.43	81.99	78.02	82.71	75.14	82.26	88.10	85.33
5 am	72.12	75.95	79.80	82.15	84.48	86.65	89.26	89.94	88.43	85.62	82.51	78.22	82.91	75.43	82.14	88.55	85.52
6 am	73.45	76.49	79.88	82.21	82.56	84.07	87.56	88.46	88.56	86.27	82.67	78.32	82.52	76.09	81.48	86.70	85.83
7 am	73.89	76.42	76.85	78.26	76.72	78.33	83.55	82.44	84.79	83.79	82.34	78.01	79.62	76.11	77.28	81.44	83.64
8 am	71.84	72.92	71.21	73.92	70.05	72.12	79.21	76.30	78.67	77.94	77.57	75.31	74.75	73.36	71.73	75.88	78.06
9 am	66.88	66.52	65.31	68.67	65.21	68.89	76.42	71.09	72.67	71.03	70.58	70.70	69.50	68.03	66.40	72.13	71.43
10 am	60.39	61.50	61.19	64.81	62.64	67.08	74.14	68.96	68.54	64.33	63.65	65.87	65.26	62.59	62.88	70.06	65.51
11 am	56.14	56.64	59.12	61.98	60.14	64.04	72.86	66.96	65.33	60.13	59.15	61.28	61.98	58.02	60.41	67.95	61.54
Noon	52.89	54.03	56.18	60.78	58.49	62.84	70.80	65.02	63.24	58.26	56.63	58.62	59.87	55.40	58.48	66.22	59.38
1 pm	51.42	54.22	54.98	59.07	58.18	62.94	69.98	64.60	63.03	56.86	55.42	57.75	59.04	54.46	57.41	65.84	58.44
2 pm	51.50	53.68	54.35	58.72	57.13	62.83	69.99	64.27	62.33	55.93	51.91	57.28	58.60	54.15	56.83	65.70	57.72
3 pm	51.95	53.59	53.73	58.79	57.40	63.37	69.40	65.32	63.08	56.39	55.78	58.40	58.93	54.05	56.64	66.03	58.42
4 pm	53.37	54.89	55.51	59.83	59.51	65.23	70.89	66.97	64.86	58.39	58.55	60.43	60.70	56.23	58.28	67.70	60.60
5 pm	56.29	57.86	58.08	62.22	62.27	67.17	73.24	70.28	68.54	63.26	64.36	64.64	64.02	59.60	60.86	70.23	65.39
6 pm	59.62	62.31	63.00	66.46	66.26	69.84	75.72	73.50	73.59	70.07	68.66	68.75	68.15	63.56	65.24	73.02	70.77
7 pm	62.11	65.22	67.89	70.32	71.70	74.51	79.02	77.78	78.18	73.98	72.18	71.92	72.12	66.42	69.97	77.30	74.78
8 pm	64.10	67.97	70.34	72.62	74.60	77.81	82.02	80.14	80.54	76.57	73.02	72.74	74.45	68.27	72.52	79.99	77.01
9 pm	65.54	69.80	71.82	74.48	76.08	79.39	82.87	81.47	82.00	78.10	75.30	73.04	75.87	69.06	74.13	81.24	78.47
10 pm	66.83	70.80	72.84	76.47	78.40	81.43	84.22	83.30	83.31	80.06	76.87	75.20	77.48	70.93	75.90	82.98	80.09
11 pm	68.78	72.41	74.04	77.59	80.05	82.77	85.75	85.33	84.76	82.15	78.74	76.30	79.06	72.50	77.21	84.62	81.88
M. N.	70.00	73.77	75.74	78.97	80.89	83.33	86.53	86.56	86.27	82.89	79.03	76.40	80.03	73.39	78.53	85.47	82.73
Mean	63.98	66.47	68.04	71.44	71.61	74.90	79.83	77.58	77.20	73.36	71.38	70.49	72.19	66.98	70.36	77.44	73.98

NAGANO.

Month / Hour	Jan.	Feb.	Mar.	Apr.	May	June	July	Aug.	Sept.	Oct.	Nov.	Dec.	Year	Wint.	Spr.	Sum.	Aut.
1 am	88.1	83.7	84.2	81.3	78.2	85.3	85.3	81.2	81.5	85.0	84.1	81.2	83.8	85.4	81.2	84.9	83.5
2 am	89.2	84.7	83.6	78.8	76.4	86.6	88.1	85.6	87.1	89.4	81.5	83.5	84.6	85.8	79.6	86.3	86.0
3 am	89.2	84.7	85.1	81.7	79.7	87.5	88.2	85.0	83.8	87.8	86.8	86.7	85.5	86.9	82.2	86.9	86.1
4 am	89.6	85.3	84.4	82.4	82.0	88.4	88.4	85.3	86.0	89.3	86.7	87.2	86.2	87.1	82.9	87.4	87.3
5 am	91.2	84.8	86.0	83.7	83.3	88.4	89.2	86.1	87.7	89.8	87.5	88.1	87.2	88.1	84.5	87.9	88.3
6 am	91.7	88.1	85.6	80.3	79.1	84.9	88.2	87.2	86.3	91.2	89.1	89.1	86.7	89.6	84.7	86.8	88.9
7 am	91.2	87.0	82.7	79.7	78.5	80.7	83.9	81.5	82.5	88.1	87.9	90.2	84.5	89.5	80.3	82.0	86.2
8 am	88.2	84.7	78.9	75.9	73.3	75.5	79.3	75.8	77.3	84.0	84.4	88.0	80.2	86.0	76.0	75.9	81.9
9 am	85.1	79.4	75.2	71.4	67.4	69.6	75.6	71.7	73.0	77.2	79.4	84.9	75.8	83.1	71.2	72.3	76.5
10 am	82.2	73.9	67.2	64.1	55.9	62.6	72.9	68.4	67.9	68.8	73.6	82.8	70.0	79.6	62.4	68.0	70.1
11 am	80.1	72.8	64.8	63.3	55.8	60.3	67.6	62.6	63.1	66.1	68.4	76.1	66.7	76.3	64.3	63.5	65.8
Noon	77.0	71.7	62.8	60.1	52.4	57.4	65.7	59.0	59.9	60.9	65.2	71.9	63.7	78.5	58.5	60.7	62.0
1 pm	73.2	69.6	61.7	58.7	51.4	54.9	63.8	59.1	58.3	58.6	62.1	69.8	61.7	68.3	57.2	59.3	59.7
2 pm	71.8	66.2	57.2	54.5	45.3	55.4	64.2	59.4	59.4	55.9	63.1	66.1	59.8	67.0	52.3	59.7	59.1
3 pm	72.4	68.2	60.4	59.1	53.2	59.5	65.3	61.8	60.9	60.9	60.8	68.8	62.6	66.5	57.5	62.2	60.9
4 pm	74.8	69.7	62.3	61.0	56.2	61.6	68.6	65.6	65.0	66.1	65.6	72.7	65.8	72.4	59.8	65.3	65.6
5 pm	77.5	75.2	65.5	62.7	58.4	63.8	70.8	68.5	71.6	72.1	69.9	77.2	69.4	76.6	62.2	67.7	71.2
6 pm	82.2	75.2	67.3	64.3	56.8	66.4	74.8	72.7	71.7	73.1	74.9	77.1	71.6	78.2	62.8	71.3	74.2
7 pm	80.6	78.0	73.4	70.6	65.3	73.0	77.6	75.2	78.1	77.9	75.4	77.9	75.3	78.8	69.8	75.3	77.5
8 pm	82.4	79.3	75.7	72.8	68.6	76.4	79.2	78.1	80.1	80.6	77.4	79.2	77.5	80.3	74.6	77.9	79.4
9 pm	83.6	82.0	77.3	74.0	71.3	78.0	81.4	79.3	81.0	81.9	78.4	80.0	79.0	81.9	74.2	79.6	80.4
10 pm	86.9	83.7	77.7	75.1	70.9	80.3	84.1	82.3	82.7	81.4	80.3	80.9	80.5	83.8	71.6	82.2	81.5
11 pm	85.7	83.6	81.1	77.7	76.3	82.3	84.5	81.6	81.6	83.1	79.9	83.4	81.7	84.2	78.4	82.8	81.5
M. N.	87.1	84.5	81.5	79.3	77.2	83.3	85.0	82.5	82.0	84.5	82.3	86.1	82.7	84.9	79.3	83.6	82.9
Mean	84.0	78.9	73.0	69.4	64.1	72.9	78.8	75.8	75.9	76.0	77.6	75.5	76.0	79.6	68.8	75.8	76.5

TOKIO.

Month / Hour	Jan.	Feb.	Mar.	Apr.	May	June	July	Aug.	Sept.	Oct.	Nov.	Dec.	Year	Wint.	Spr.	Sum.	Aut.
1 am	68.6	69.6	75.7	81.7	87.7	90.5	92.6	91.3	91.1	87.0	81.6	73.9	82.6	70.7	81.7	91.4	86.6
2 am	69.6	70.2	76.3	82.1	88.3	91.3	93.2	92.5	91.5	87.7	82.0	74.6	83.3	71.4	82.2	92.1	87.1
3 am	70.3	70.4	76.7	82.7	88.7	92.3	93.8	93.3	91.8	87.8	81.9	74.5	83.7	71.7	82.7	93.1	87.2
4 am	70.9	71.0	77.0	83.4	88.8	92.9	94.1	93.9	92.3	87.7	82.7	74.7	84.1	72.2	83.1	93.6	87.6
5 am	71.5	71.0	77.3	83.7	88.7	92.8	94.0	94.1	92.5	88.2	83.2	75.6	84.4	72.7	83.2	93.6	88.0
6 am	72.0	71.0	77.0	83.3	86.3	90.3	92.1	92.5	92.1	88.2	83.5	76.1	83.7	73.0	82.2	91.7	87.9
7 am	73.1	70.5	74.5	79.1	82.0	85.9	88.0	87.8	88.9	86.0	82.1	76.4	81.2	73.3	78.5	87.2	85.7
8 am	67.7	66.5	68.7	74.2	76.5	80.8	83.0	87.9	83.9	80.5	75.9	69.5	75.8	67.9	73.1	81.9	80.1
9 am	61.1	61.3	63.4	69.8	72.2	76.6	79.0	76.7	78.9	74.7	68.5	61.3	70.3	61.2	68.5	77.4	74.0
10 am	55.6	56.9	59.5	66.8	70.2	73.9	76.4	72.8	74.8	70.7	63.6	55.1	66.3	55.8	65.5	74.4	69.7
11 am	51.3	53.6	57.2	65.3	68.4	71.6	74.5	70.8	72.1	67.4	60.6	51.6	63.7	52.2	63.6	72.3	66.7
Noon	49.8	51.1	56.5	63.6	67.1	69.8	72.6	68.8	70.6	65.5	59.3	49.2	62.0	50.0	62.4	70.4	65.2
1 pm	49.2	50.2	55.8	63.2	65.8	69.0	71.4	67.8	69.8	64.8	58.7	48.7	61.2	49.1	61.6	69.4	64.4
2 pm	48.8	50.1	56.0	63.1	65.5	68.8	71.1	67.3	69.9	64.8	58.5	48.8	61.1	49.2	61.5	69.1	64.4
3 pm	49.8	50.6	56.3	63.5	65.8	70.0	72.0	67.8	70.8	65.9	59.8	50.5	61.9	50.8	61.9	69.9	65.5
4 pm	52.6	52.8	58.2	65.8	67.4	71.6	73.3	70.0	73.8	69.0	65.3	55.1	64.5	53.5	63.8	71.7	69.2
5 pm	58.6	57.6	61.3	68.9	70.8	74.1	76.0	72.6	76.6	75.2	72.3	62.6	68.9	59.6	67.0	74.2	74.7
6 pm	62.7	62.7	65.8	72.2	74.2	77.6	79.1	77.5	81.2	79.4	75.5	65.8	72.8	63.7	70.7	78.1	78.7
7 pm	64.4	64.3	68.5	75.6	78.4	81.3	83.2	82.0	83.8	82.1	77.4	68.5	75.8	65.7	71.2	82.2	81.1
8 pm	66.0	65.5	70.7	77.0	81.0	83.6	85.6	84.4	85.8	84.2	79.5	71.0	77.9	67.5	76.2	84.5	83.2
9 pm	67.0	66.5	72.0	78.2	82.7	85.4	87.6	86.3	87.4	85.5	80.3	71.0	79.1	68.2	77.6	86.4	84.4
10 pm	68.0	67.9	73.3	79.5	83.8	87.0	89.1	87.9	88.5	86.5	80.7	71.5	80.3	69.1	78.8	88.0	85.3
11 pm	68.2	67.8	74.0	80.8	85.3	87.7	90.4	89.1	89.4	86.7	80.4	72.0	81.0	69.3	80.0	89.1	85.5
M. N.	68.8	68.8	75.1	81.5	86.5	89.0	91.6	90.3	90.1	87.0	80.8	72.7	81.9	70.1	81.1	90.3	86.0
Mean	62.7	62.8	67.8	74.4	78.0	81.4	83.5	81.6	82.8	79.3	73.9	65.1	74.5	63.7	73.4	82.2	78.7

SAPPORO.

Month / Hour	Jan.	Feb.	Mar.	Apr.	May	June	July	Aug.	Sept.	Oct.	Nov.	Dec.	Year	Wint.	Spr.	Sum.	Aut.
1 am	84,2	88,1	84,4	81,7	79,4	89,0	92,3	93,0	91,6	88,9	83,1	82,8	86,5	85,1	81,8	91,1	87,8
2 am	85,0	88,3	85,2	82,7	81,3	89,7	92,9	92,7	92,0	89,2	83,5	82,8	87,1	85,4	83,1	91,8	88,2
3 am	85,8	88,2	86,5	81,4	82,0	89,2	92,9	92,7	92,2	88,9	85,6	82,7	87,3	85,6	83,3	91,6	88,9
4 am	86,0	88,6	85,5	82,0	82,9	89,9	93,3	93,0	92,5	88,6	84,7	82,7	87,5	85,8	83,5	92,1	88,6
5 am	86,0	87,7	85,0	82,2	81,7	88,4	92,6	93,5	93,0	89,0	84,8	83,7	87,3	85,8	83,0	91,5	88,9
6 am	85,0	87,2	84,8	79,6	76,1	84,5	89,0	90,5	91,6	89,1	85,3	84,8	85,4	84,7	80,2	88,0	88,7
7 am	85,2	85,8	84,8	78,8	70,1	79,1	83,9	85,3	86,5	85,0	83,6	82,5	81,9	84,5	75,2	82,9	85,0
8 am	82,8	84,1	75,6	69,3	64,7	75,2	79,4	79,7	81,8	78,7	79,5	81,2	77,7	82,7	69,9	78,1	80,0
9 am	80,3	81,7	72,9	65,5	59,6	70,9	75,9	74,7	77,0	71,8	74,3	78,2	73,6	80,0	66,0	73,7	71,4
10 am	76,3	77,0	69,1	62,3	55,7	67,1	71,1	71,9	72,9	67,7	70,2	76,0	69,8	76,5	62,5	70,1	70,3
11 am	74,2	74,7	69,1	61,0	58,8	64,5	69,7	68,3	70,0	64,2	67,7	74,0	67,6	74,3	61,3	67,5	67,3
Noon	74,2	76,0	70,2	60,3	59,2	63,7	69,2	67,4	68,7	63,8	67,0	73,2	67,2	74,5	61,2	66,8	66,5
1 pm	74,6	75,6	71,2	60,8	58,3	63,8	68,6	67,4	69,6	63,4	66,9	73,5	67,1	74,6	61,8	66,6	66,6
2 pm	74,0	75,5	70,1	59,1	54,4	64,4	69,4	68,9	71,1	64,3	69,1	75,0	68,0	74,8	61,2	67,6	68,2
3 pm	75,1	75,5	71,7	58,9	56,0	66,1	71,4	70,8	72,7	69,9	72,5	75,5	69,4	75,1	62,1	69,5	70,7
4 pm	75,4	79,0	72,9	59,8	58,6	68,2	73,8	73,4	75,8	71,0	75,7	78,2	71,8	77,5	63,8	71,8	74,2
5 pm	80,1	82,5	75,9	62,6	61,8	70,5	76,3	75,7	80,1	78,3	79,2	80,0	75,3	80,9	66,2	74,2	79,2
6 pm	82,2	84,1	78,6	65,7	66,3	74,2	80,1	81,2	85,6	83,4	81,5	82,0	78,8	82,8	70,2	78,5	83,5
7 pm	83,3	86,3	79,8	69,7	72,3	79,3	84,8	86,2	87,9	85,2	84,9	83,8	81,7	84,1	73,9	83,5	85,0
8 pm	82,8	86,7	82,1	72,3	74,7	83,4	88,0	89,1	89,5	85,1	83,1	85,8	83,6	85,1	76,1	86,8	86,0
9 pm	83,5	86,8	81,3	71,5	75,8	85,3	89,7	90,3	90,4	86,5	83,3	85,3	84,4	85,2	77,2	88,4	86,6
10 pm	83,3	88,9	82,3	75,3	76,3	86,9	90,5	91,6	89,6	86,7	83,7	84,8	85,0	85,7	78,0	89,7	86,7
11 pm	82,3	88,3	83,3	77,7	78,4	87,3	91,2	92,0	90,6	87,5	83,8	83,1	85,5	84,6	80,0	90,2	87,3
M. N.	82,7	87,7	83,0	80,4	79,6	88,3	92,0	92,5	90,8	88,0	83,1	82,7	85,9	84,1	81,0	90,9	87,3
Mean	81,6	83,5	78,5	70,8	68,7	77,9	82,1	82,6	83,5	79,6	78,9	80,5	79,0	81,7	72,6	81,0	80,7

HAKODATE. / NEMURO.

Hour	Hakodate Winter	Hakodate Spring	Hakodate Summer	Hakodate Autumn	Hakodate Year	Nemuro Winter	Nemuro Spring	Nemuro Summer	Nemuro Autumn	Nemuro Year
1 am	84,8	82,2	91,5	83,7	85,7	78,7	86,2	95,5	82,0	85,6
2 am	85,1	82,9	92,2	84,5	86,2	77,9	86,3	95,6	82,3	85,5
3 am	85,4	83,4	91,9	85,0	86,4	77,6	86,3	95,6	82,8	85,6
4 am	85,0	84,2	92,2	85,1	86,7	77,9	85,7	95,6	82,6	85,4
5 am	85,4	84,5	92,1	85,0	86,8	77,8	86,0	95,3	83,1	85,5
6 am	85,8	82,7	90,8	85,1	86,2	78,4	84,2	94,3	82,5	84,9
7 am	85,3	78,8	87,4	83,2	85,8	78,2	81,6	91,7	80,2	82,9
8 am	83,9	73,4	84,6	79,3	80,1	76,2	78,2	88,8	76,3	79,9
9 am	80,3	70,3	81,9	73,6	76,6	75,4	76,0	86,2	72,9	77,6
10 am	78,3	68,1	80,0	69,5	74,0	74,8	74,2	83,8	70,0	75,7
11 am	73,8	67,0	78,4	67,0	71,6	73,2	72,8	81,8	68,0	74,0
Noon	72,1	66,7	77,5	66,2	70,4	72,3	72,3	81,4	66,8	73,2
1 pm	73,1	64,8	76,5	66,1	70,1	72,4	72,5	81,5	66,9	73,3
2 pm	74,2	64,6	75,8	66,5	70,2	72,3	72,9	81,6	68,4	73,8
3 pm	74,9	65,7	76,1	67,1	71,0	73,6	74,7	83,0	70,9	75,5
4 pm	77,5	66,6	77,8	69,1	72,8	75,0	76,4	84,8	74,4	77,7
5 pm	78,9	68,5	79,8	73,2	75,1	75,9	79,3	87,9	76,9	80,0
6 pm	80,4	70,8	82,2	76,1	77,4	76,3	82,3	91,2	78,9	82,2
7 pm	82,3	73,8	85,2	78,7	80,0	75,9	83,1	92,8	79,8	82,9
8 pm	83,5	75,3	87,7	80,0	81,6	76,9	84,3	93,9	80,0	83,8
9 pm	83,8	78,0	89,1	81,7	83,2	76,4	84,3	94,2	80,1	83,7
10 pm	84,0	79,4	90,3	83,2	84,3	77,7	85,5	95,1	80,6	84,7
11 pm	85,0	79,7	90,9	83,2	84,8	78,6	85,8	95,2	80,7	85,1
M. N.	84,9	81,2	91,6	83,5	85,4	77,9	85,7	95,4	80,9	85,0
Mean	81,2	74,6	85,2	77,3	79,6	76,1	80,7	90,1	77,0	81,0

B. ANNUAL VARIATION OF THE TENSION OF AQUEOUS VAPOUR AND THE RELATIVE HUMIDITY.

The annual variation of the tension of the aqueous vapour, as is shown in the following tables, goes, everywhere in this country, hand in hand with the mean temperature at that locality,—that is to say, great in summer and small in winter. The maximum is in August and the minimum in January. The range is about 15 mm in the southern parts, and slowly gets smaller as we proceed northward, and in Hokkaido it is about 13 mm. Thus though the variations of the tension of aqueous vapour and the air temperature accompany each other, yet their amplitudes do not. At the place where the amplitude of temperature variation is great, that of tension, on the contrary, is small.

Mean tension of vapour in millimetres.

Locality.	Jan.	Feb.	Mar.	Apr.	May	June	July	Aug.	Sept.	Oct.	Nov.	Dec.	Year	Wint.	Spr.	Sum.	Aut.
Naha	8.9	11.2	11.1	13.8	17.9	19.2	22.3	22.0	21.0	16.3	12.3	11.1	15.6	10.1	14.3	21.2	16.5
Kagoshima	5.6	5.7	7.3	10.8	12.6	16.5	20.1	20.7	17.9	12.7	8.6	6.6	12.1	6.0	10.2	19.2	13.1
Miyazaki	5.4	5.7	7.1	11.0	12.8	16.8	20.9	21.3	18.4	13.2	8.8	6.4	12.3	5.9	10.1	19.7	13.5
Kochi	4.6	4.7	6.3	10.0	11.8	16.0	20.3	20.5	17.6	12.0	7.8	5.5	11.4	4.9	9.4	18.9	12.5
Wakayama	4.3	4.5	5.6	8.6	10.7	14.8	19.1	19.5	16.3	11.1	7.5	5.3	10.6	4.7	8.3	17.8	11.6
Oita	5.0	5.1	6.5	9.3	11.1	15.0	19.8	20.7	17.2	11.7	8.4	6.1	11.3	5.4	9.0	18.5	12.3
Yamaguchi	4.3	4.9	6.0	8.6	10.3	14.3	19.0	18.9	15.5	9.9	7.1	5.8	10.4	5.0	8.2	17.1	10.9
Hiroshima	4.5	4.7	5.7	8.6	11.0	15.0	19.4	19.7	16.2	11.0	7.5	5.4	10.7	4.9	8.1	18.0	11.6
Matsuyama	4.5	5.4	6.4	9.2	10.8	14.4	19.3	19.5	18.0	11.2	8.0	6.4	11.1	5.1	8.8	17.7	12.4
Okayama	3.7	4.5	5.5	7.1	9.5	13.1	19.0	19.9	18.5	10.3	7.3	4.9	10.3	4.4	7.1	17.3	12.0
Ozaka	4.1	4.5	5.7	8.5	10.6	14.7	19.1	19.6	16.3	11.1	7.6	5.8	10.6	4.7	8.3	17.8	11.7
Kioto	4.3	4.3	5.2	8.0	10.0	14.2	18.5	18.8	15.7	10.6	7.3	5.1	10.2	4.6	7.7	17.2	11.2
Kumamoto	4.3	5.9	6.7	10.2	11.4	15.0	19.5	19.8	17.8	11.5	8.0	6.5	11.4	5.6	9.4	18.1	12.4
Saga	4.4	5.4	5.9	8.1	10.4	13.8	20.0	20.2	17.9	11.1	8.1	6.9	11.0	5.6	8.1	18.0	12.1
Nagasaki	5.2	5.4	6.5	9.7	12.0	16.2	20.3	20.3	17.0	11.7	8.1	6.0	11.5	5.5	9.4	18.9	12.3
Fukuoka	4.9	6.0	6.6	9.3	10.8	14.5	19.1	20.2	17.8	11.0	8.0	6.5	11.2	5.8	8.9	17.9	12.3
Isogahara	4.9	5.1	6.3	8.9	10.9	14.7	19.8	21.0	16.8	11.3	8.2	6.4	11.2	5.5	8.7	18.5	12.1
Akamagaseki	4.7	4.8	5.9	8.6	10.8	14.8	19.2	20.2	16.6	11.4	7.8	5.6	10.9	5.0	8.4	18.1	11.9
Sakai	4.5	4.5	5.5	7.9	9.9	13.9	18.3	19.6	15.9	10.7	7.5	5.4	10.3	4.8	7.8	17.3	11.4
Tsu	4.1	5.1	6.4	8.6	10.9	14.4	19.7	20.8	17.8	11.1	7.5	5.8	11.0	5.0	8.5	18.3	12.1
Nagoya	3.8	4.3	5.7	7.8	10.3	13.3	19.0	19.6	18.1	10.2	6.9	5.9	10.4	4.7	7.9	17.3	11.7
Gifu	4.3	4.3	5.4	8.4	10.7	14.6	19.3	20.1	16.7	11.1	7.3	5.4	10.6	4.7	8.2	18.0	11.7
Hamamatsu	4.1	4.2	5.6	9.0	11.2	15.0	19.5	20.1	17.1	11.6	7.6	5.3	10.9	4.5	8.6	18.2	12.1
Numazu	4.3	4.5	5.9	8.8	10.8	14.6	18.7	19.4	16.5	11.2	7.7	5.4	10.6	4.7	8.5	17.6	11.8
Tokio	3.7	4.1	5.3	8.3	11.0	14.7	18.9	19.7	16.5	10.9	7.0	4.7	10.4	4.2	8.2	17.8	11.5
Utsunomiya	2.9	3.5	5.3	6.7	10.2	13.1	18.2	18.2	17.1	9.1	6.3	5.1	9.7	3.8	7.4	16.5	10.9
Choshi	4.6	5.0	6.6	9.4	11.6	14.9	18.9	20.5	17.6	12.3	8.8	6.4	11.4	5.3	9.2	18.1	12.9
Kanazawa	4.1	4.3	5.1	7.1	9.7	13.7	18.1	18.8	15.5	10.4	7.2	5.3	10.0	4.7	7.4	16.9	11.0
Fushiki	4.4	4.4	5.5	7.8	10.1	14.5	19.3	20.9	16.6	11.0	7.9	5.7	10.7	4.8	7.8	18.2	11.8
Nagano	3.3	3.7	4.7	6.5	7.8	11.9	15.9	16.5	13.7	8.2	5.6	4.6	8.5	3.9	6.3	14.8	9.2
Niiga'a	4.1	4.0	4.8	7.0	9.2	13.1	18.0	18.8	15.2	9.9	7.0	5.0	9.7	4.1	7.0	16.6	10.7
Yamagata	3.4	3.6	4.0	6.1	8.0	12.2	15.9	17.2	14.5	8.3	5.8	4.9	8.7	4.0	6.3	15.1	9.5
Akita	3.4	3.4	4.3	6.1	8.7	12.6	17.0	18.5	14.5	9.1	6.1	4.4	9.0	3.7	6.4	16.0	9.9
Fukushima	3.7	4.4	5.3	6.4	8.3	12.7	16.7	18.2	15.3	8.7	5.9	4.9	9.2	4.2	6.7	15.9	10.0
Ishinomaki	3.6	3.7	4.9	6.6	8.9	12.4	16.9	18.8	15.1	9.5	6.7	4.8	9.3	4.0	6.8	16.0	10.4
Miyako	3.0	3.3	4.0	5.7	7.8	11.0	15.1	17.0	13.8	8.9	5.7	4.0	8.3	3.4	5.8	14.4	9.5
Aomori	3.0	3.1	3.7	5.4	7.5	10.7	15.1	16.7	12.9	8.4	5.4	3.8	7.9	3.3	5.5	14.2	8.8
Hakodate	3.1	3.2	4.1	5.6	7.7	10.5	14.8	16.1	13.0	8.2	5.4	3.9	8.0	3.4	5.8	13.9	8.9
Suttsu	2.8	3.2	4.0	5.1	7.1	9.9	13.9	15.8	12.8	8.0	5.3	3.8	7.6	3.3	5.4	13.2	8.5
Sapporo	2.4	2.6	3.4	4.8	7.0	10.2	14.3	15.5	11.7	7.2	4.6	3.1	7.2	2.7	5.1	13.3	7.8
Kamikawa	2.0	2.4	3.1	4.5	6.3	9.6	13.6	14.1	10.8	6.3	4.2	3.2	6.7	2.5	4.6	12.4	7.1
Soya	2.4	2.7	3.6	5.2	6.8	8.2	13.0	14.9	11.9	7.7	5.0	3.5	7.1	2.9	5.2	12.0	8.2
Abashiri	2.2	2.6	3.8	5.0	7.1	9.2	12.9	15.0	12.2	7.2	4.8	3.7	7.1	2.8	5.3	12.4	8.1
Nemuro	2.5	2.4	3.3	4.6	6.1	8.2	11.8	14.2	11.7	7.8	5.5	3.8	6.8	2.9	4.7	11.4	8.3
Kushiro	2.0	2.6	3.8	4.7	6.7	9.1	12.3	15.1	13.3	7.0	4.4	3.4	7.0	2.7	5.1	12.2	8.2
Erimo	2.8	3.3	4.1	5.2	6.8	9.0	12.6	14.9	12.8	8.7	5.9	4.2	7.5	3.4	5.4	12.2	9.1

Since the relative humidity depends on the amount of aqueous vapour and also on the temperature, the annual variation of the relative humidity can not be equal everywhere. There are, indeed, considerable differences between the front Nippon and the back Nippon.

The annual variation of the relative humidity on the back Nippon has two maxima and minima, of which the absolute minimum during the year occurs between April and May, and the absolute maximum at July. The general feature of its variation is as follows. From November onward, the relative humidity increases gradually to January when it reaches a maximum. Then it decreases suddenly and at April or May it has the minimum value of the year. Again it increases quickly until July when it has the maximum value of the year. Then it decreases slowly, and after passing a minimum toward the end of autumn, it again increases to the maxinum at January.

On the front Nippon, we have a pair of the maximum and the minimum humidity in a year, of which the maximum is in July and the minimum in February. But on closer examinations, it seems to have secondary minima both at May and August. Details are given in the next table and the Plate VI.

RELATIVE HUMIDITY.

		Jan.	Feb.	March	April	May	June	July	August	Sept.	Oct.	Nov.	Dec.	Year
Back side	....	83	82	78	76	76	82	84	82	82	80	80	82	81
Front side	Sth coast	69	68	71	76	76	81	84	82	82	78	75	73	76
	Eth coast	75	76	74	72	73	83	86	85	85	81	77	73	79
	E. & S. coast of Hokkaido	81	81	85	82	83	89	91	90	88	82	79	80	85

Since our country is completely surrounded by seas, the relative humidity is very high everywhere; and especially it is highest in the southeastern part of Hokkaido. In this district, the mean of the year is 85 and even in the most dry month it is not lower than 79, and in July it reaches 91. This is probably due to the fact that the cold current from Kamtchatka touches the coast of this island, and thereby cold and warm atmospheres are mixed together and there a dense fog is frequently formed. The eastern coast of Nippon, though not so high as in Hokkaido, suffers the influences of Oyashio, and especially in its northern part, dense fogs are not unfrequently formed in summer, and hence the annual mean of the relative humidity is comparatively high, being about 80.

Next to Hokkaido, the place of high humidity is the back Nippon, where the yearly mean is 81. The high humidity in this district is not due, as in Hokkaido, to its being extremely high in summer, but to the fact that the humidity in winter is as high as in summer.

The place of the lowest relative humidity is the front Nippon. But even here, the mean of the year is still 76. Though in summer, its relative humidity is nearly equal to that of the back Nippon, yet in winter, it is considerably low falling below 68. This is the cause of the low relative humidity in the mean of the year.

The details for the relative humidity are given in the following table:—

Mean relative humidity.

Locality.	Jan.	Feb.	Mar.	Apr.	May	June	July	Aug.	Sept.	Oct.	Nov.	Dec.	Year	Wint.	Spr.	Sum.	Aut.
Naha	66	76	75	79	87	85	81	82	81	73	69	70	77	71	80	83	74
Kagoshima	74	72	73	78	78	82	81	80	79	75	75	76	77	74	76	81	76
Miyazaki	73	74	76	80	79	83	84	84	84	82	78	75	79	74	78	84	81
Kochi	68	66	69	76	76	84	86	83	82	76	72	71	76	68	74	84	77
Wakayama	68	69	68	72	72	77	78	74	76	75	73	70	73	69	71	76	75
Oita	71	72	75	79	78	81	85	83	83	77	73	70	77	71	77	83	78
Yamaguchi	78	76	76	78	75	80	85	80	79	78	80	77	78	77	76	82	79
Hiroshima	76	74	74	77	75	81	81	76	78	77	79	78	77	76	75	79	78
Matsuyama	71	73	75	78	76	78	82	84	85	80	75	74	77	73	76	80	80
Okayama	69	72	69	68	65	74	81	82	81	75	74	73	73	71	67	79	77
Ozaka	74	74	74	75	73	77	78	75	76	77	77	74	75	74	74	77	77
Kioto	78	76	73	74	72	77	78	75	78	79	81	79	77	78	73	77	79
Kumamoto	75	76	73	76	74	77	78	79	77	77	78	78	76	76	74	78	77
Saga	75	75	67	69	68	73	85	81	78	74	76	79	75	76	68	80	76
Nagasaki	75	74	73	77	78	81	82	78	78	74	75	75	77	75	76	81	76
Fukuoka	77	79	77	78	77	78	82	83	83	79	77	76	79	77	77	81	80
Isugahara	77	74	74	78	76	81	87	85	84	79	79	79	79	77	76	84	81
Akamagaseki	72	73	72	78	78	84	85	82	79	75	72	70	77	72	76	84	75
Sakai	76	75	73	76	75	79	81	79	79	76	76	74	77	75	75	80	77
Tsu	66	72	72	74	74	77	87	84	86	80	75	73	77	70	73	83	80
Nagoya	72	69	69	71	66	73	79	77	81	72	73	76	73	72	69	76	75
Gifu	77	72	72	76	75	79	80	79	81	79	79	80	77	76	74	79	80
Hamamatsu	64	62	67	76	78	82	86	82	82	77	72	68	75	65	74	83	77
Numazu	66	67	70	76	76	84	83	80	80	77	71	69	75	67	74	81	77
Tokio	66	67	70	77	79	83	84	82	83	80	75	69	76	67	75	83	79
Utsunomiya	69	66	72	73	73	84	86	83	83	76	76	84	76	70	73	83	78
Choshi	68	68	72	79	81	85	88	85	82	78	75	73	78	70	77	86	78
Kanazawa	82	81	75	76	76	84	82	78	81	79	80	80	79	81	76	80	80
Fushiki	86	84	80	82	84	88	89	85	85	82	83	84	84	85	81	87	83
Nagano	84	78	73	69	64	73	79	76	76	76	77	79	75	80	69	76	76
Niigata	81	79	75	76	75	84	82	76	78	77	78	80	78	80	75	80	78
Yamagata	87	83	76	68	64	76	81	81	82	80	82	81	79	85	69	79	81
Akita	84	82	78	75	78	85	86	85	85	83	84	82	82	83	77	85	83
Fukushima	79	76	75	65	67	79	83	83	84	79	76	77	77	77	69	82	80
Ishinomaki	82	80	75	74	76	85	[illegible]	87	85	[illegible]	80	79	81	80	75	87	82
Miyako	71	73	73	73	77	86	87	86	86	83	74	71	78	72	74	86	81
Aomori	81	79	76	73	75	80	83	81	81	77	77	79	78	80	75	81	78
Hakodate	82	80	79	78	80	87	88	85	84	80	78	81	82	81	79	87	81
Suttsu	82	83	80	74	77	84	85	84	82	78	80	81	81	82	77	84	80
Sapporo	77	79	78	73	74	80	83	82	83	80	80	79	79	78	75	82	81
Kamikawa	94	87	80	76	70	76	79	80	84	84	85	89	82	90	75	78	84
Soya	85	89	87	84	89	91	94	88	86	84	85	85	87	86	87	90	85
Abashiri	89	90	86	78	79	86	87	86	87	84	85	87	85	89	81	86	85
Nemuro	73	74	80	81	83	89	91	89	86	80	78	75	82	74	81	90	81
Kushiro	89	92	86	75	76	84	86	89	90	83	81	86	85	89	79	86	85
Erimo	80	85	88	88	89	94	96	93	88	82	79	79	87	81	88	94	83

We have mentioned above, that since our country is surrounded on all sides by seas, the atmosphere is always damp. But it is, by no means, rare that the atmosphere is very dry, and the relative humidity decreases below 30. Let us now give in the following table the minimum values of the relative humidity observed at several stations, since their establishments.

Extremes of relative humidity.

Locality.	Minimum	Day	Month	Year	Locality.	Minimum	Day	Month	Year
Naha	15	26;11,21	X;XI	1890	Tokio	10	5	I	1889
Kagoshima	14	15	II	1889	Utsunomiya	21	18	II	1891
Miyazaki	14	15	II	1889	Choshi	11	15	II	1887
Kochi	16	5	I	1889	Kanazawa	19	30	III	1891
Wakayama	17	1	V	1891	Fushiki	21	20	V	1891
Oita	22	26	I	1891	Nagano	13	25	IV	1889
Hiroshima	16	20	III	1888	Niigata	21	5	VI	1889
Matsuyama	22	9	II	1893	Yamagata	19	31	III	1891
Okayama	23	13	V	1891	Akita	22	22	IV	1891
Ozaka	22	14	I	1891	Fukushima	17	18;20	V	1891
Kioto	10	23	III	1888	Ishinomaki	27	28	IV	1888
Kumamoto	26	25;10	III;V	1891	Miyako	14	24	III	1888
Saga	22	25	III	1891	Aomori	18	19	IV	1887
Nagasaki	20	23	II	1888	Hakodate	8	19	I	1878
Fukuoka	27	24	V	1890	Suttsu	18	27	IV	1888
Itsugahara	21	30	III	1891	Sapporo	8	27	IV	1888
Akamagaseki	29	12;19,21	V;IV	1888;91	Kamikawa	22	8;17	V	1891
Sakai	18	13	V	1888	Soya	15	27	XII	1890
Tsu	25	17	I	1891	Abashiri	20	18	V	1891
Nagoya	24	30	V	1891	Nemuro	20	8;16	V;XII	1888;91
Gifu	24	2	V	1885	Kushiro	18	12	I	1891
Hamamatsu	15	9	II	1890	Erimo	20	29	I	1891
Numazu	18	1;1,20	I;III	1887;88					

Thus the extreme minimum value of the relative humidity ever observed in this country is 8, which we had at Hakodate and Sapporo. The maximum and the minimum tensions of aqueous vapour are as follows:—

Extremes of tension of vapour.

Locality.	Maximum	Day	Month	Year	Minimum	Day	Month	Year	Range
Naha	25,9	5	VII	1890	5,2	9;11	I	1891	20,7
Kagoshima	31,3	17	VIII	1888	1,0	15	II	1889	30,3
Miyazaki	32,0	31	VII	1888	1,0	15	II	1889	31,0
Kochi	26,3	11	VIII	1891	1,3	5	I	1889	25,0
Wakayama	26,9	17	VIII	1889	1,1	16	I	1891	25,8
Oita	31,6	25	VI	1889	1,2	16	I	1891	30,4
Hiroshima	29,1	27	VIII	1891	1,4	30	XI	1882	27,7
Matsuyama	29,5	2	VIII	1891	1,4	10	II	1891	28,1
Okayama	23,7	2	IX	1891	1,5	24	I	1891	22,2
Ozaka	26,1	2	VIII	1889	1,3	14;16	I	1891	24,8
Kioto	24,9	29	VII	1888	1,6	12	I	1886	23,3
Kumamoto	27,2	11	IX	1891	1,8	11	II	1891	25,4
Saga	30,3	3	IX	1890	1,4	10	I	1891	28,9
Nagasaki	29,0	7	VII	1889	2,2	18	II	1885	26,8
Fukuoka	26,0	17	VIII	1891	2,3	17	II	1891	23,7
Itsugahara	31,6	13	VII	1888	1,7	10	II	1891	29,9
Akamagaseki	25,9	5;19	VIII;VII	1884,87	1,4	14	I	1891	24,5
Sakai	28,2	4	VIII	1883	1,6	9;12,5	I;I,II	1884;86	26,6
Tsu	28,2	23	VIII	1889	1,5	14;17	I	1891	26,7
Nagoya	25,8	12	VIII	1890	1,3	10	II	1890	24,5
Gifu	26,5	2	VIII	1889	1,4	1;10	I;II	1884;91	25,1
Hamamatsu	25,8	3	VIII	1889	0,9	10	II	1891	24,9
Numazu	24,6	22	VIII	1880	1,0	19	II	1885	23,6

EXTREMES OF TENSION OF VAPOUR.

Locality.	Maximum	Day	Month	Year	Minimum	Day	Month	Year	Range
Tokio	28,9	24	VII	1879	0,8	25,26;10	I	1881;84	28,1
Utsunomiya	23,9	2	VIII	1891	1,0	10	II	1891	22,9
Choshi	30,0	16	VII	1890	1,0	15	II	1887	29,0
Kanazawa	34,2	5	VIII	1886	1,9	12	II	1887	32,3
Fushiki	30,2	8	VIII	1889	1,5	12	I	1886	28,7
Nagano	22,5	9	VIII	1889	1,0	27	II	1891	20,8
Niigata	31,5	14	IX	1891	1,7	18	II	1885	29,8
Yamagata	23,6	21	VIII	1889	1,2	29	I	1891	22,4
Akita	23,2	13	VIII	1884	0,7	5;11	II;I	1888;89	22,5
Fukushima	28,8	16	VII	1889	0,9	3	II	1891	27,9
Ishinomaki	25,8	19	VII	1890	1,6	6;10	II	1888;91	24,2
Miyako	26,6	14	V	1888	1,1	19	I	1886	25,5
Aomori	24,0	5	VIII	1882	1,1	7;18	II;I	1883;84	22,9
Hakodate	23,8	30	VII	1883	0,5	16;2	XII	1884;86	23,3
Suttsu	25,6	31	VII	1889	0,7	5	II	1885	24.9
Sapporo	24,8	20	VII	1883	0,5	17	I	1891	24,3
Kamikawa	25,7	28	VII	1888	0,1	24	II	1890	25,6
Soya	29,3	5	IX	1890	0,8	15	I	1890	28,5
Abashiri	25,9	30	VIII	1890	0,8	15,10;28,29	I	1890;91	25.1
Nemuro	25,5	30	VII	1888	0,6	19;30	II;I	1888;89	24,9
Kushiro	25,8	9	IX	1890	0,2	13;12	I	1890;91	25,6
Erimo	24.4	20	VIII	1890	0,4	29	I	1891	24,0

CHAPTER V.

THE AMOUNT OF CLOUD.

The amount of cloud given here is expressed in a proportional number from 0 to 10. When the whole sky is clear and no cloud is seen, it is taken as zero, and when the whole sky is obscured by cloud it is taken as 10.

A. DIURNAL VARIATION OF THE AMOUNT OF CLOUD.

Throughout the country, the amount of cloud is always great in daytime, and small by night without any exceptions among different seasons. There is sudden increase in its amount from about two hours before the sunrise, and it reaches its maximum value at the sunrise. Then after some increment and decrement, it has again another maximum just before the sunset; and during about three hours after sunset, the rate of its decrease is very quick.

The diurnal variation of the amount of cloud has generally two maxima and minima. The maxima occur constantly for all seasons at the sunrise and the sunset, but the times of the minimum amounts vary greatly at different seasons and at different places so that they seem to have no fixed rule for their occurrences. For the details, see the following tables and the Plate VII.

HOURLY MEANS OF CLOUD AMOUNT.

KUMAMOTO.

Hour \ Month	Jan.	Feb.	March	April	May	June	July	August	Sept.	Oct.	Nov.	Dec.	Year
1 am	5,29	4,75	6,58	6,52	5,49	6,88	5,54	4,44	5,67	4,21	3,36	5,55	5,35
2 am	5,37	4,65	6,63	6,31	5,83	7,20	5,77	4,70	5,69	4,79	3,41	5,50	5,49
3 am	5,88	5,26	6,76	6,41	5,58	7,10	6,40	5,13	5,65	4,47	3,71	5,27	5,63
4 am	6,06	5,38	6,72	6,95	5,95	7,53	6,71	5,20	5,59	4,74	4,40	5,26	5,85
5 am	6,27	5,72	7,00	7,03	6,47	7,86	7,27	5,84	5,98	4,43	4,27	4,97	6,07
6 am	6,02	5,57	7,51	7,35	7,03	7,59	6,96	5,95	7,06	4,41	4,67	5,29	6,29
7 am	6,82	6,37	7,15	7,43	6,80	7,32	6,94	5,02	6,12	4,58	4,63	6,61	6,32
8 am	6,71	6,39	7,14	7,01	6,95	7,50	6,60	4,24	5,97	4,77	4,60	6,19	6,17
9 am	7,01	6,70	7,33	6,82	6,43	7,70	6,56	4,55	6,25	4,56	4,22	6,30	6,20
10 am	5,82	6,64	7,27	6,80	6,14	7,58	6,58	4,80	6,16	4,74	4,43	6,06	6,08
11 am	5,82	6,71	6,99	6,70	6,28	7,52	6,76	5,15	6,36	5,17	4,23	5,98	6,09
Noon	6,10	7,01	6,90	6,80	6,29	7,87	6,40	4,71	6,82	5,18	4,36	5,93	6,23
1 pm	6,95	7,07	6,80	6,94	6,25	7,89	6,46	5,01	6,19	5,05	4,48	5,63	6,25
2 pm	6,85	7,13	7,07	6,81	6,25	7,81	6,51	5,01	6,49	5,32	4,13	5,69	6,31
3 pm	6,74	7,17	7,34	6,70	6,12	7,64	6,81	5,15	6,61	5,19	4,28	5,82	6,35
4 pm	6,21	6,86	6,90	6,61	6,62	7,68	7,30	5,72	7,26	4,69	3,90	5,95	6,31
5 pm	5,79	6,65	7,25	6,87	6,36	7,41	7,60	6,03	7,35	4,90	4,27	5,69	6,35
6 pm	5,80	6,92	7,25	6,91	6.78	7,61	7,91	5,98	7,56	5,29	4,46	5,54	6,50
7 pm	5,03	5,71	6,96	7,06	7,23	8,05	8,39	6,57	7,71	4,77	3,53	5,19	6,35
8 pm	4,67	4,76	6,00	6,35	6,52	7,84	8,08	5,55	6,35	3,59	3,12	5,42	5,71
9 pm	4,83	5,16	6,06	6,35	5,50	6,60	6,02	4,75	6,10	4,04	3,15	5,76	5,36
10 pm	4,79	5,17	6,38	6,30	5,48	6,69	6,33	4,19	6,49	4,29	3,40	5,82	5,44
11 pm	4,76	4,86	6,06	6,47	5,85	6,38	5,89	3,99	6,31	4,61	3,25	5,88	5,36
M. N.	4,84	4,97	6,37	6,62	5,55	6,40	5,86	4,23	5,94	4,37	3,46	5,75	5,36
Mean	5,86	5,99	6,85	6,76	6,25	7,40	6,71	5,09	6,42	4,66	4,00	5,69	5,98

HIROSHIMA.

Hour \ Month	Jan.	Feb.	March	April	May	June	July	August	Sept.	Oct.	Nov.	Dec.	Year
1 am	5,40	5,13	5,13	5,97	4,50	5,60	6,63	4,00	6,43	4,93	3,43	4,83	5,14
2 am	5,53	5,67	5,17	6,07	4,60	5,63	6,63	3,83	5,67	4,83	3,77	5,20	5,22
3 am	5,73	5,80	5,20	6,07	4,93	5,73	6,73	4,20	5,87	5,07	3,70	5,60	5,39
4 am	5,67	5,60	5,10	6,20	5,00	6,53	7,13	4,00	6,43	4,80	3,80	5,70	5,50
5 am	5,67	5,17	5,70	6,70	5,50	6,90	8,23	5,47	6,87	4,37	3,43	5,53	5,79
6 am	5,63	5,87	6,37	7,60	6,37	7,93	8,27	5,70	7,27	5,27	4,33	5,83	6,31
7 am	6,83	7,10	6,67	7,57	6,07	7,67	8,00	6,13	7,37	5,20	4,63	6,37	6,66
8 am	6,37	6,90	6,63	6,97	6,23	7,63	7,77	5,90	7,10	5,00	4,67	6,47	6,47
9 am	7,00	7,17	6,67	6,73	6,13	7,67	7,50	6,00	7,23	4,57	4,63	6,40	6,47
10 am	7,40	7,13	6,93	6,73	5,90	7,43	7,33	5,83	6,90	4,37	4,50	6,13	6,38
11 am	7,83	7,43	7,13	6,83	6,50	7,33	7,53	5,97	6,90	4,97	5,03	6,10	6,65
Noon	8,27	7,73	7,00	6,97	6,53	7,53	7,40	5,93	6,86	5,37	4,97	6,63	6,77
1 pm	8,40	7,73	6,93	7,23	6,70	7,37	7,47	5,77	6,73	5,77	4,93	6,83	6,82
2 pm	7,83	7,47	6,73	7,10	6,57	7,47	7,53	5,70	6,63	6,13	5,17	6,57	6,74
3 pm	7,67	7,33	6,40	7,10	6,70	7,27	7,67	5,63	7,10	5,70	5,40	6,63	6,69
4 pm	7,37	7,17	5,80	6,67	6,57	7,27	7,73	5,73	7,10	5,73	5,50	6,20	6,57
5 pm	7,00	7,23	6,00	6,70	6,40	7,33	7,73	6,40	7,13	5,83	5,00	5,83	6,52
6 pm	6,73	6,47	5,70	6,80	6,43	7,67	7,70	6,20	7,10	6,07	4,03	5,03	6,33
7 pm	5,03	5,43	5,37	6,67	6,53	7,70	7,90	6,27	7,00	5,20	3,33	4,37	5,90
8 pm	5,60	5,03	4,97	5,77	5,77	7,23	7,40	4,60	6,20	4,97	3,47	5,27	5,52
9 pm	5,30	5,13	5,13	5,47	5,17	6,37	6,17	4,23	5,63	5,17	3,27	4,67	5,11
10 pm	5,90	5,13	5,33	5,73	5,00	6,10	6,00	4,00	5,70	5,37	2,90	4,47	5,14
11 pm	5,50	5,30	5,47	5,93	4,83	5,70	6,07	4,00	5,70	4,93	3,17	4,78	5,11
M. N.	5,80	5,40	5,10	6,03	4,93	5,77	6,37	4,00	6,13	4,90	3,57	5,40	5,28
Mean	6,48	6,31	5,94	6,57	5,83	6,95	7,29	5,22	6,61	5,19	4,18	5,71	6,02

MATSUYAMA.

1 am	4,90	5,70	5,35	5,55	5,05	6,25	5,85	3,95	5,70	4,70	2,50	5,55	5,09
2 am	4,35	5,50	5,70	6,00	5,05	5,40	6,05	4,95	5,75	4,75	3,00	5,75	5,19
3 am	4,95	5,65	5,50	6,00	4,90	5,60	6,30	4,95	6,00	4,80	3,40	6,05	5,34
4 am	5,30	5,10	4,85	6,35	5,05	6,70	6,70	5,00	6,80	4,65	3,50	6,10	5,51
5 am	5,15	5,00	5,60	6,40	5,45	6,80	7,25	5,35	7,25	3,95	3,50	5,90	5,63
6 am	5,20	5,00	6,35	6,70	5,85	7,00	7,45	5,40	7,55	4,55	4,00	5,65	5,94
7 am	6,40	6,05	6,25	7,15	5,65	7,25	7,10	5,35	7,30	5,05	4,35	6,25	6,18
8 am	6,20	5,85	6,35	6,55	6,05	7,30	7,05	5,05	7,05	4,60	4,90	6,20	6,09
9 am	6,30	6,75	6,00	6,35	6,20	7,40	7,05	5,10	7,09	5,00	4,75	6,30	6,18
10 am	6,65	6,50	6,20	6,55	6,09	7,50	7,05	5,30	6,75	4,80	4,50	6,45	6,19
11 am	6,90	6,65	6,40	6,55	6,15	6,80	7,00	5,30	6,65	4,85	4,30	6,65	6,18
Noon	6,65	6,70	6,85	6,05	6,00	7,35	7,15	5,40	6,35	5,10	4,20	6,50	6,19
1 pm	7,05	6,70	6,80	6,10	6,00	7,40	6,75	6,00	6,45	5,20	4,55	5,85	6,24
2 pm	7,15	6,65	6,65	6,65	6,20	7,50	6,95	5,85	6,85	5,20	4,20	6,00	6,32
3 pm	7,15	6,15	6,50	6,85	6,55	7,25	6,80	6,30	6,25	5,50	4,60	6,15	6,34
4 pm	6,90	6,50	6,20	6,35	6,20	7,55	7,40	6,20	7,15	5,35	4,80	6,10	6,39
5 pm	6,15	6,50	6,15	6,55	5,80	7,35	7,35	6,80	7,30	5,50	4,70	5,55	6,31
6 pm	6,25	6,35	6,05	6,70	5,80	7,35	7,15	7,00	7,30	5,15	4,15	4,80	6,18
7 pm	5,15	5,80	6,10	6,35	5,80	7,35	6,90	6,45	6,65	4,25	4,15	4,35	5,77
8 pm	4,60	5,20	5,35	6,00	5,75	6,65	6,50	5,40	6,00	4,20	3,50	4,85	5,33
9 pm	4,50	5,05	5,05	5,80	5,10	5,55	5,40	4,60	5,80	4,30	3,90	4,75	4,98
10 pm	4,45	5,10	5,45	5,70	4,75	5,90	5,65	4,45	5,35	4,20	3,15	5,20	4,93
11 pm	4,80	5,15	5,30	5,85	5,49	5,65	5,90	4,35	5,50	4,20	3,05	5,15	5,02
M. N.	5,05	5,25	5,30	5,80	5,15	6,15	6,15	4,15	5,99	4,50	2,85	5,35	5,13
Mean	5,76	5,87	5,93	6,29	5,66	6,79	6,70	5,36	6,53	4,76	3,96	5,73	5,78

OZAKA.

Hour \ Month	Jan.	Feb.	March	April	May	June	July	August	Sept.	Oct.	Nov.	Dec.	Year
1 am	4.50	5.27	5.97	6.17	5.22	6.15	6.35	4.50	5.50	5.10	4.42	4.45	5.30
2 am	4.43	5.77	5.73	6.17	5.02	5.87	6.70	4.87	5.60	5.12	3.92	4.57	5.31
3 am	4.80	5.40	5.50	6.42	4.97	5.97	7.02	4.90	5.15	4.87	4.30	4.65	5.33
4 am	4.53	5.63	5.07	6.30	5.80	6.70	7.97	5.10	5.25	4.67	4.47	4.47	5.50
5 am	4.47	5.63	5.20	7.25	6.77	7.25	8.67	6.77	6.07	5.00	4.17	4.35	5.97
6 am	4.93	5.97	5.70	7.75	7.07	7.25	8.85	7.10	6.62	5.85	4.95	5.00	6.42
7 am	6.37	6.40	6.07	7.30	6.87	7.27	8.65	7.07	6.55	5.87	5.52	5.87	6.65
8 am	6.37	6.40	6.00	7.60	6.42	7.60	8.45	6.85	6.15	5.62	5.37	5.82	6.55
9 am	6.53	6.57	6.67	7.27	6.52	7.77	8.22	6.62	6.45	5.40	5.32	5.80	6.59
10 am	6.67	6.90	6.70	7.22	6.52	8.07	7.92	6.47	6.40	5.75	5.02	5.62	6.60
11 am	6.67	7.80	7.03	7.62	6.82	7.95	7.87	6.22	6.90	5.85	5.40	6.05	6.85
Noon	6.97	7.67	7.27	7.72	7.25	8.07	7.55	5.95	6.65	6.27	5.75	6.17	6.94
1 pm	7.40	8.23	7.30	7.65	7.27	8.02	7.37	6.12	6.57	6.15	5.52	6.12	6.98
2 pm	7.63	7.83	7.27	7.72	7.20	7.70	7.45	6.17	6.75	6.00	6.20	6.35	7.02
3 pm	7.50	8.17	7.67	7.60	7.15	7.60	7.60	6.20	6.85	5.87	5.95	6.15	7.03
4 pm	7.47	7.97	7.30	7.67	7.40	7.90	7.67	6.45	7.02	5.67	5.95	5.95	7.03
5 pm	6.83	7.53	7.13	7.67	7.27	7.97	7.92	6.85	7.37	5.50	5.95	5.57	6.96
6 pm	5.57	6.90	6.50	7.50	7.02	7.77	7.82	6.90	7.17	5.47	4.17	4.17	6.41
7 pm	4.23	5.90	5.03	6.72	6.77	7.42	7.92	6.60	6.55	4.52	3.72	3.95	5.78
8 pm	4.30	5.83	5.07	6.02	5.95	7.55	7.40	4.77	5.77	4.60	3.75	4.17	5.43
9 pm	4.07	5.73	4.93	6.00	5.47	6.50	6.07	4.62	5.32	5.27	3.65	4.25	5.16
10 pm	3.73	5.30	5.20	6.00	5.10	6.00	6.37	4.72	5.62	5.05	3.65	3.90	5.05
11 pm	3.90	6.10	5.27	6.07	5.07	6.10	6.60	4.95	5.57	5.07	4.10	4.17	5.25
M. N.	4.13	5.40	5.23	6.12	4.95	6.75	6.72	4.40	5.60	4.97	4.42	4.45	5.26
Mean	5.58	6.51	6.12	6.98	6.33	7.22	7.55	5.88	6.23	5.40	4.83	5.00	6.14

WAKAYAMA.

1 am	5.36	5.94	5.74	6.26	5.72	6.39	6.85	4.75	5.83	5.54	4.66	5.46	5.71
2 am	5.18	5.88	5.74	6.95	5.62	6.15	6.72	5.18	5.74	5.51	4.68	5.36	5.73
3 am	5.32	5.84	5.79	6.84	5.48	6.45	6.85	4.98	5.67	5.72	4.78	5.37	5.76
4 am	5.81	5.90	5.67	7.21	6.26	7.48	7.64	5.66	5.93	5.91	5.00	5.11	6.13
5 am	5.50	6.30	5.82	7.58	7.33	7.90	8.42	6.76	6.87	5.92	5.12	5.13	6.55
6 am	5.81	6.76	6.23	8.39	7.29	7.55	8.40	7.05	7.13	6.88	5.71	5.83	6.92
7 am	7.08	7.58	6.84	8.09	7.02	7.16	8.20	6.99	6.98	6.75	6.01	6.28	7.11
8 am	7.17	7.35	6.57	7.88	7.00	7.82	8.30	6.63	6.62	6.32	6.11	6.60	7.03
9 am	7.40	7.52	6.79	7.72	6.86	8.19	8.27	6.91	6.51	6.45	6.14	6.50	7.10
10 am	7.24	7.52	6.75	7.40	6.88	8.38	8.17	6.92	7.03	6.26	5.98	6.99	7.13
11 am	7.07	7.39	6.97	7.61	7.24	8.44	7.95	6.85	6.91	6.42	6.06	7.08	7.17
Noon	6.93	7.39	7.10	7.71	7.45	8.42	7.87	6.61	6.90	6.14	5.88	7.37	7.15
1 pm	7.31	7.63	7.08	7.88	7.59	8.37	7.55	6.46	6.70	6.09	6.04	7.12	7.15
2 pm	7.01	7.38	7.19	7.71	7.43	8.14	7.62	6.61	6.85	5.94	5.93	7.08	7.07
3 pm	6.94	7.02	6.78	7.15	7.33	7.87	7.68	6.51	6.58	5.74	6.04	7.31	6.91
4 pm	6.52	7.18	6.85	7.42	7.56	8.03	7.76	6.69	6.99	5.74	6.17	7.01	6.99
5 pm	6.45	7.27	6.52	7.76	7.56	8.43	8.17	7.13	7.38	5.98	6.50	6.90	7.17
6 pm	6.72	6.93	6.64	7.55	7.39	8.47	8.14	7.22	7.83	5.95	6.17	5.58	7.05
7 pm	5.59	5.96	5.67	7.32	7.20	8.13	8.23	7.19	7.08	5.40	5.17	5.18	6.51
8 pm	5.82	6.22	5.53	6.32	6.37	7.52	7.06	5.80	6.07	5.47	4.65	5.09	6.04
9 pm	5.98	6.15	5.31	6.58	5.67	6.58	6.19	4.82	5.98	5.44	4.52	5.18	5.70
10 pm	5.99	6.12	5.10	6.53	5.60	6.67	6.60	4.85	6.22	5.40	4.71	5.23	5.75
11 pm	5.37	5.83	5.57	6.35	5.61	6.46	6.41	5.08	6.04	5.54	4.84	5.22	5.69
M. N.	5.35	5.57	5.52	6.29	5.81	6.51	6.47	5.07	5.93	5.55	4.85	5.32	5.69
Mean	6.29	6.69	6.24	7.27	6.72	7.58	7.59	6.20	6.57	5.92	5.49	6.06	6.55

NAGANO.

Hour / Month	Jan.	Feb.	March	April	May	June	July	August	Sept.	Oct.	Nov.	Dec.	Year
1 am	6,10	6,65	7,35	6,60	6,00	7,35	7,15	6,10	7,55	5,75	5,30	4,95	6,32
2 am	6,35	6,25	5,90	6,90	6,00	6,90	7,35	6,25	7,15	5,70	5,30	4,70	6,23
3 am	6,30	6,40	5,85	6,45	6,30	7,10	7,05	6,05	6,70	5,85	5,50	4,70	6,19
4 am	6,10	6,25	6,00	6,65	6,70	7,70	8,15	6,80	6,95	6,00	5,55	5,70	6,55
5 am	5,15	5,75	6,35	6,90	7,30	7,90	8,30	7,15	7,70	6,10	5,90	5,55	6,70
6 am	5,90	6,65	6,65	7,05	6,90	7,10	8,30	7,00	7,85	7,45	6,55	6,50	7,02
7 am	7,00	7,05	6,60	7,10	6,90	7,15	8,25	7,10	7,50	7,25	5,75	6,95	7,07
8 am	7,45	7,10	6,85	7,35	7,30	7,25	8,15	7,15	7,05	7,30	5,55	7,15	7,16
9 am	6,95	6,95	7,00	7,50	7,20	7,10	7,70	5,90	7,10	6,85	5,45	6,95	6,91
10 am	6,75	6,85	7,30	7,70	7,20	7,90	7,60	5,70	6,55	6,80	6,05	6,60	6,92
11 am	6,35	6,95	7,00	7,10	7,25	7,80	7,25	5,70	6,45	6,10	5,85	6,10	6,71
Noon	6,85	7,00	7,05	7,25	7,30	7,90	7,80	6,10	6,50	6,40	6,20	6,35	6,89
1 pm	7,10	7,60	7,05	7,30	7,50	7,90	7,65	6,05	6,55	5,95	6,20	7,00	6,90
2 pm	7,50	7,60	6,95	7,30	7,35	7,90	7,75	6,30	6,50	6,55	5,55	7,10	7,03
3 pm	7,60	7,60	6,80	7,70	7,30	8,00	7,80	6,65	6,70	5,90	5,40	6,65	7,01
4 pm	7,20	7,40	6,55	7,70	7,80	7,90	8,30	6,85	6,85	6,30	4,95	6,45	7,02
5 pm	7,55	7,50	6,60	7,50	7,60	7,95	8,40	7,70	7,35	6,35	4,95	6,10	7,13
6 pm	6,50	7,30	6,05	8,15	8,00	7,90	8,90	8,15	8,15	6,20	3,50	5,30	7,00
7 pm	6,05	6,10	5,50	7,55	7,60	7,80	9,10	7,95	6,95	5,20	2,95	5,75	6,54
8 pm	5,60	6,40	5,65	6,90	6,55	8,00	8,50	6,50	6,60	5,60	4,20	5,80	6,44
9 pm	5,80	6,40	6,10	7,00	6,30	6,75	7,70	6,40	6,15	5,90	4,30	6,10	6,24
10 pm	6,05	6,30	5,95	6,85	6,10	6,55	7,85	6,10	6,55	5,90	4,25	6,30	6,25
11 pm	6,05	6,15	6,45	6,95	5,85	6,60	7,15	6,55	6,95	5,80	5,15	5,70	6,30
M. N.	6,15	6,30	6,55	6,85	5,90	6,90	7,25	6,25	7,15	5,50	5,25	5,60	6,30
Mean	6,53	6,77	6,15	7,19	6,97	7,51	7,90	6,63	6,98	6,20	5,23	6,09	6,70

TOKIO.

Hour / Month	Jan.	Feb.	March	April	May	June	July	August	Sept.	Oct.	Nov.	Dec.	Year
1 am	4,19	4,71	5,64	6,83	6,97	7,37	6,54	5,11	6,53	6,16	4,86	3,11	5,67
2 am	3,96	4,94	5,61	6,89	6,97	7,64	6,73	5,50	6,70	6,33	5,01	3,07	5,78
3 am	3,84	4,81	5,51	7,01	6,83	7,70	7,03	5,71	6,93	6,30	5,16	3,10	5,83
4 am	4,17	4,91	5,51	7,33	7,29	8,04	7,97	6,16	6,83	6,26	5,11	3,11	6,06
5 am	4,19	4,91	5,83	7,70	7,56	8,33	8,30	7,03	7,57	6,87	5,00	3,14	6,37
6 am	4,40	5,36	6,01	7,80	7,49	8,49	8,24	7,11	7,89	7,11	5,64	3,46	6,58
7 am	4,47	5,49	6,11	7,76	7,24	8,39	8,14	7,09	7,73	7,24	5,59	3,67	6,58
8 am	4,56	5,30	6,19	7,53	7,07	8,13	8,17	7,23	8,16	6,84	5,56	3,73	6,54
9 am	4,37	5,14	6,19	7,34	7,13	8,00	8,03	7,27	7,97	6,74	5,26	3,74	6,43
10 am	4,49	5,07	6,14	7,20	7,24	7,66	7,81	7,09	7,86	6,57	5,26	3,64	6,33
11 am	4,26	4,93	6,09	7,26	7,29	7,56	7,41	6,70	7,80	6,69	5,16	3,63	6,23
Noon	4,13	5,03	6,16	7,19	7,33	7,51	7,11	6,17	7,53	6,83	4,91	3,46	6,11
1 pm	4,16	5,11	6,36	7,14	7,23	7,53	7,21	5,91	7,39	6,96	4,81	3,61	6,12
2 pm	4,20	5,18	6,50	7,21	7,34	7,51	7,26	5,71	6,93	6,89	4,79	3,73	6,10
3 pm	4,53	5,23	6,43	7,34	7,31	7,74	7,31	5,79	6,97	6,70	5,11	3,69	6,18
4 pm	4,50	5,47	6,56	7,47	7,27	7,80	7,36	5,90	6,96	6,73	5,23	3,73	6,26
5 pm	4,46	5,69	6,67	7,43	7,59	8,01	7,54	5,99	6,93	6,69	5,11	3,73	6,35
6 pm	4,31	5,54	6,39	7,49	7,79	8,21	7,76	6,06	7,17	6,14	4,63	3,19	6,23
7 pm	3,93	5,26	5,67	7,01	7,59	8,17	7,76	6,13	6,57	5,90	4,47	3,29	5,98
8 pm	4,14	4,91	5,51	6,74	6,74	7,94	6,64	5,06	6,61	5,86	4,47	3,23	5,65
9 pm	4,30	4,54	5,36	6,91	6,76	7,21	6,21	5,07	6,67	6,27	4,39	3,23	5,58
10 pm	4,03	4,86	5,61	6,71	6,74	7,19	6,33	5,34	6,57	6,17	4,51	3,49	5,63
11 pm	4,27	4,67	5,50	6,73	6,79	7,19	6,40	5,23	6,79	6,06	4,59	3,30	5,63
M. N.	4,16	4,79	5,59	6,70	6,90	7,11	6,27	5,19	6,64	6,31	4,57	3,10	5,65
Mean	4,25	5,08	5,97	7,20	7,18	7,78	7,31	6,08	7,16	6,53	4,98	3,42	6,08

HAKODATE.

Hour \ Month	Jan.	Feb.	March	April	May	June	July	August	Sept.	Oct.	Nov.	Dec.	Year
1 am	7,00	7,03	6,03	5,70	5,23	7,50	7,57	6,30	5,80	4,95	5,95	7,07	6,34
2 am	7,20	6,90	6,33	6,00	5,00	7,43	7,70	6,42	5,90	4,77	5,95	6,92	6,38
3 am	6,87	7,03	6,53	5,73	5,17	7,80	7,97	6,60	5,92	4,55	6,10	6,82	6,45
4 am	7,03	6,60	5,97	6,70	6,40	8,00	8,57	7,55	6,37	4,10	6,03	7,00	6,72
5 am	6,47	7,07	6,73	6,80	6,87	8,37	8,60	7,60	7,17	5,35	6,22	7,02	7,02
6 am	7,20	7,73	7,43	7,13	6,77	8,60	8,50	7,77	7,32	5,97	6,97	7,80	7,43
7 am	7,67	7,70	7,30	6,87	6,47	8,57	8,42	7,87	7,32	5,62	6,65	7,82	7,36
8 am	7,40	7,80	7,43	6,83	6,33	8,30	8,25	7,75	7,25	5,62	6,52	8,12	7,30
9 am	7,30	7,43	7,23	6,80	6,50	8,23	8,27	7,20	7,62	5,87	7,02	8,10	7,30
10 am	7,70	7,83	7,60	7,03	6,10	8,10	7,70	7,00	7,50	6,15	7,25	7,97	7,33
11 am	7,57	7,77	7,63	6,77	6,13	7,87	7,32	7,00	7,35	6,15	7,20	8,27	7,25
Noon	7,60	8,27	7,53	6,77	6,27	7,87	7,22	6,87	7,50	6,22	7,27	8,37	7,31
1 pm	7,70	7,97	7,10	6,53	6,03	7,73	7,22	6,70	7,47	6,45	7,35	8,50	7,23
2 pm	7,73	7,63	7,20	6,77	5,93	7,43	7,20	6,62	7,15	6,42	7,20	8,40	7,16
3 pm	7,73	7,77	7,43	7,30	6,17	7,20	7,10	6,62	7,35	6,27	7,20	8,45	7,19
4 pm	7,67	7,13	7,33	7,27	6,40	7,13	7,22	6,37	7,27	6,27	7,02	8,10	7,12
5 pm	8,10	7,73	7,07	7,13	6,70	7,10	7,12	6,40	7,12	6,20	6,92	7,95	7,13
6 pm	6,73	6,90	7,47	7,07	6,70	7,60	7,19	6,62	7,22	5,47	6,35	7,00	6,88
7 pm	6,53	6,17	6,43	6,47	6,63	7,19	7,92	6,47	6,45	4,80	6,72	6,95	6,55
8 pm	6,23	5,37	6,03	5,50	5,53	7,43	8,02	5,75	5,82	4,87	6,47	7,15	6,18
9 pm	6,23	5,57	5,63	5,53	5,07	7,00	7,52	5,62	5,60	4,57	6,05	7,17	5,96
10 pm	6,37	5,93	6,27	5,63	5,13	6,93	7,70	5,67	5,50	4,42	6,15	7,37	6,09
11 pm	6,73	6,47	6,23	5,43	5,10	7,30	7,50	5,92	5,47	4,50	6,30	7,45	6,20
M. N.	6,77	6,60	6,47	5,50	5,20	7,33	7,55	6,05	5,95	4,67	6,15	7,50	6,31
Mean	7,15	7,11	6,84	6,47	5,99	7,68	7,73	6,70	6,74	5,14	6,64	7,62	6,81

SAPPORO.

Hour \ Month	Jan.	Feb.	March	April	May	June	July	August	Sept.	Oct.	Nov.	Dec.	Year
1 am	6,49	7,81	7,09	6,07	5,98	6,84	6,57	5,39	6,24	5,37	6,33	7,60	6,46
2 am	6,22	7,65	6,82	6,11	5,96	7,08	6,58	5,54	6,17	5,38	6,16	7,55	6,46
3 am	6,35	7,57	6,56	6,26	6,13	7,83	7,51	5,82	6,16	5,61	6,18	7,41	6,64
4 am	6,20	7,63	6,79	6,61	6,87	7,89	7,86	6,87	6,49	5,77	6,31	7,53	6,90
5 am	6,02	7,66	6,90	6,92	7,22	8,18	7,85	6,94	6,74	6,54	6,50	7,70	7,10
6 am	6,74	7,87	7,44	7,25	6,67	8,21	7,68	7,01	6,95	6,94	8,95	8,28	7,50
7 am	6,27	7,80	7,04	6,88	6,81	8,11	7,33	6,99	6,68	6,70	6,99	8,41	7,25
8 am	7,21	8,31	6,51	6,56	6,46	8,22	7,17	6,69	6,90	6,25	7,10	8,27	7,14
9 am	7,38	8,20	6,87	6,98	6,29	7,87	7,32	6,59	6,88	6,40	7,07	8,27	7,18
10 am	7,22	8,15	6,62	7,27	6,49	7,82	7,08	6,27	7,19	6,76	7,53	8,47	7,24
11 am	7,42	8,18	7,02	7,11	6,67	7,77	7,05	5,76	7,39	6,63	7,27	8,43	7,22
Noon	7,38	8,18	7,02	7,22	6,54	7,60	7,08	5,80	7,49	6,70	7,02	8,47	7,21
1 pm	7,74	7,85	7,54	6,94	6,56	7,62	6,85	5,84	7,83	6,73	7,58	8,53	7,30
2 pm	7,60	8,26	7,30	7,00	6,35	7,71	6,78	5,95	7,76	7,01	7,55	8,55	7,32
3 pm	7,62	8,00	7,51	6,99	6,32	7,89	7,08	5,71	7,72	7,24	7,32	8,48	7,32
4 pm	7,36	8,09	7,45	6,96	6,26	7,69	6,91	5,82	7,83	6,82	7,10	8,41	7,23
5 pm	7,38	7,99	7,46	6,91	6,22	7,59	6,83	5,66	7,71	6,45	6,79	7,88	7,07
6 pm	5,91	7,36	7,44	7,05	6,37	7,50	6,88	5,94	7,71	5,97	6,01	7,06	6,77
7 pm	5,82	7,18	6,61	6,81	6,39	7,75	7,21	6,13	6,87	5,27	6,29	7,23	6,63
8 pm	5,84	6,81	6,36	5,94	6,19	7,94	6,86	5,18	6,45	5,46	6,61	7,57	6,43
9 pm	6,14	7,17	6,55	5,28	5,58	6,92	6,47	4,85	6,35	5,60	6,27	7,84	6,26
10 pm	6,37	7,11	6,77	5,25	5,91	7,04	6,29	4,58	6,43	5,48	6,44	7,32	6,25
11 pm	6,34	7,35	6,80	5,20	6,05	7,09	6,03	4,64	6,22	5,58	6,70	7,36	6,33
M. N.	6,03	7,30	7,10	5,82	6,06	6,72	6,57	4,78	6,44	5,03	6,55	7,37	6,29
Mean	6,74	7,73	6,98	6,56	6,36	7,62	7,02	5,87	6,92	6,15	6,79	7,92	6,89

NEMURO.

Hour \ Month	Jan.	Feb.	March	April	May	June	July	August	Sept.	Oct.	Nov.	Dec.	Year
1 am	4,51	5,81	5,47	6,17	7,02	8,10	7,91	7,30	7,01	4,91	5,46	4,38	6,17
2 am	4,51	5,53	5,19	6,01	6,80	8,33	8,06	7,34	7,17	4,78	5,61	4,48	6,18
3 am	4,64	5,60	5,61	5,91	7,27	8,52	8,65	7,86	7,37	4,78	5,41	4,33	6,33
4 am	4,58	5,72	5,63	6,49	7,70	8,67	8,71	8,57	7,56	5,30	5,25	4,53	6,56
5 am	4,60	5,62	6,54	6,76	7,46	8,66	8,69	8,65	8,03	5,81	5,08	4,63	6,71
6 am	5,56	6,56	6,62	6,66	7,50	8,37	8,63	8,49	8,01	5,79	5,76	5,39	6,91
7 am	5,73	6,50	6,70	6,59	7,23	8,23	8,54	8,30	7,86	6,10	5,79	5,72	6,91
8 am	5,68	6,21	6,30	6,46	7,11	8,01	8,28	7,91	7,84	6,19	5,74	5,32	6,76
9 am	5,60	6,52	6,21	6,28	6,88	7,90	7,80	7,61	7,69	6,00	5,54	5,30	6,61
10 am	5,35	6,56	6,15	6,72	6,95	7,51	7,37	7,55	7,37	6,02	5,57	5,28	6,54
11 am	5,76	6,68	6,27	6,60	7,10	7,33	7,28	7,26	7,26	5,85	5,80	5,43	6,55
Noon	5,82	6,81	6,07	6,60	6,95	7,31	7,38	7,22	7,45	5,98	5,75	5,55	6,57
1 pm	6,01	6,65	6,18	6,55	6,77	7,28	7,13	7,10	7,25	5,96	5,93	5,78	6,55
2 pm	6,23	6,53	6,13	6,69	6,61	7,23	7,25	7,03	7,35	5,88	6,21	5,96	6,59
3 pm	6,15	6,39	6,23	6,94	7,01	7,50	7,55	7,35	7,41	6,07	6,33	5,82	6,73
4 pm	6,20	6,45	6,76	6,91	7,07	7,52	7,67	7,30	7,52	6,21	6,52	6,01	6,84
5 pm	5,92	6,68	6,89	7,10	7,59	8,08	7,72	7,52	7,69	6,12	6,09	5,55	6,91
6 pm	5,45	6,04	6,88	7,54	7,59	8,22	8,20	7,89	7,60	5,62	6,15	5,08	6,83
7 pm	5,24	5,91	5,84	6,74	7,10	8,31	8,29	7,74	6,51	5,11	5,66	4,78	6,46
8 pm	5,21	5,98	5,78	6,03	6,53	8,32	8,19	7,29	6,15	5,08	5,30	4,62	6,24
9 pm	5,04	5,90	5,77	6,22	6,46	7,82	7,91	7,46	6,17	4,96	5,40	4,59	6,14
10 pm	4,95	6,04	5,65	5,94	6,64	7,97	7,91	7,39	6,49	5,28	4,86	4,93	6,17
11 pm	5,21	6,11	5,56	6,05	6,72	8,24	7,75	7,60	6,99	4,94	5,15	4,91	6,27
M. N.	4,69	5,84	5,85	5,93	6,69	8,18	7,87	7,62	7,05	4,87	5,30	4,69	6,21
Mean	5,35	6,19	6,11	6,49	7,04	7,98	7,95	7,64	7,28	5,57	5,65	5,13	6,53

The above tables or the Plate VII show us that the variations at Tokio and Nemuro, are so regular that the times of the maximum and the minimum amounts can be determined very precisely. Though this must partly depend upon their comparatively long periods of observations, yet it must also due to their positions being favorable for the determination of the amount of cloud. In mountainous region, as most part of our country is, it is difficult to look round at whole heavens in a single view, hence when a great mass of cumulus appears on the sky, the field of view is greatly obstructed and it seems to the observer as if the whole sky is covered completely by cloud. Consequently the determination of the amount of cloud is not unfrequently very inaccurate. Now Tokio and Nemuro have no high mountains for considerable distances around them, and the field of vision is very wide so that they are most appropriate places for the observations on the amount of cloud. Hence we may safely say that the variation of the amount of cloud observed at these two stations represents the variation in our country.

B. ANNUAL VARIATION OF THE AMOUNT OF CLOUD.

The variation of the amount of cloud, as that of the amount of precipitation, is remarkably different on the two sides of Nippon. They are almost opposite to each other.

On the front Nippon, the amount of cloud is small in winter and great in summer. Thus during the six months from April to September, it is always beyond the mean of the year, while from October to March, it is always below the mean. The manner of variation is as follows. If we start from the minimum value 5.0 at December, it increase gradually up to 6.7 at April. In May, we have slight decrease to 6.5. Then it suddenly increases to 7.7 in June, and this is the maximum amount in the year. After that month, it decreases gradually and in August it is 6.5. There is again an increase to 7.0 in September. Then it keeps on decreasing and reaches to the minimum at December.

On the back Nippon, the amount is great in winter and small in summer. Thus from March to October, it is generally below the mean of the year, and from November to February it is far above the mean. The general outlines of the annual variation are as follows. The December, we have the maximum of the year equal to 8.6. It decreases until May and then begins to increase slightly in June, but in August it attains the value 6.1, which is the minimum value in the year. There is sudden increase to 6.9 in September and again a sudden decrease to 6.4 in October. After that month, it increases steadily and at last it reaches to the maximum in January.

Of the front side, the manners of variation at the western coast of Kiushu and Setouchi are slightly different from the variation elsewhere on that side. Here the minimum of the year is in November instead of in December.

The noteworthy facts on the annual variation of the amount of cloud common to the whole country are that:—

i) The amount is comparatively small in May and August.

ii) The amount is comparatively great in June and September.

With respect to the amounts during several seasons, we notice the following facts:—

1.) On the back Nippon, the amount of cloud remains nearly constant during spring, summer, and autumn, and it is far smaller than the amount during winter.

2.) On the front Nippon from the southern extremity of Kiushu, to the eastern extremity of Hokkaido, it is greatest in summer and least in winter. During spring and autumn, the amount are nearly equal and lie between those during summer and winter.

3.) In the western part of Kiushu, the amounts in winter, spring, and summer are nearly equal. During autumn, it has very small value.

4.) On Setouchi, it is great in spring and summer, and small in autumn and winter. The maximum is in summer and the minimum in autumn.

Thus the amount of cloud is always great on the back Nippon. There, the yearly mean is 7.2. On the front Nippon, it is only 6.4. Details are given in the following table and the Plate VIII.

Mean amount of cloud.

Locality.	Jan.	Feb.	Mar.	Apr.	May	June	July	Aug.	Sept.	Oct.	Nov.	Dec.	Year	Wint.	Spr.	Sum.	Aut.
Naha	6.7	8.3	8.8	7.9	8.5	8.8	6.9	5.7	7.0	5.1	6.2	6.9	7.2	7.3	7.7	7.2	6.1
Kagoshima	5.8	6.1	6.3	7.1	6.6	7.8	6.6	5.6	6.0	5.6	4.8	5.0	6.1	5.6	6.7	6.7	5.5
Miyazaki	4.1	5.1	5.6	6.6	6.4	7.5	6.4	5.7	6.8	6.4	4.1	3.2	5.6	4.1	6.2	6.4	5.8
Kochi	4.4	4.9	5.3	6.6	6.2	7.6	6.7	5.9	6.6	5.8	4.4	3.6	5.7	4.3	6.0	6.7	5.7
Wakayama	6.4	6.6	5.9	7.0	6.5	7.6	6.4	5.5	6.5	6.2	5.8	5.9	6.3	6.2	6.5	6.5	6.2
Oita	5.5	5.4	5.4	6.4	5.7	6.7	6.6	4.9	6.2	5.5	4.4	4.8	5.5	5.2	5.6	6.1	5.3
Yamaguchi	6.3	6.8	6.2	6.7	6.4	7.0	7.2	5.1	6.3	4.9	5.2	5.8	6.2	6.3	6.3	6.5	5.5
Hiroshima	5.9	6.1	5.5	6.3	5.9	7.1	6.2	5.2	6.2	5.5	4.9	5.3	5.8	5.8	5.9	6.2	5.5
Matsuyama	5.6	5.8	6.1	6.4	5.6	6.7	6.7	5.5	6.6	4.7	3.9	5.6	5.8	5.7	6.0	6.3	5.1
Okayama	5.2	6.5	6.0	5.9	4.6	6.6	8.2	6.0	7.2	4.4	4.0	4.6	5.8	5.4	5.5	6.9	5.2
Ozaka	5.5	6.2	6.1	6.9	6.6	7.5	6.8	6.0	6.1	6.2	5.3	4.7	6.2	5.5	6.5	6.7	6.0
Kioto	6.4	7.2	6.5	7.0	6.8	7.8	7.3	6.4	6.9	6.3	5.9	5.6	6.7	6.4	6.8	7.2	6.4
Kumamoto	5.9	4.2	6.8	7.0	6.3	6.9	6.7	5.2	6.4	4.8	4.4	5.6	5.8	5.2	6.7	6.3	5.4
Saga	6.1	6.3	5.2	5.7	4.9	5.3	7.1	5.2	5.9	4.4	4.0	5.7	5.5	6.0	5.3	6.0	4.8
Nagasaki	6.3	6.1	5.7	6.5	6.1	7.4	6.5	5.2	5.7	5.3	5.3	5.9	6.0	6.1	6.1	6.4	5.4
Fukuoka	7.7	7.6	7.1	7.1	6.3	7.3	7.3	6.1	7.3	5.6	5.7	7.1	6.8	7.5	6.8	6.9	6.2
Itsugahara	5.4	5.3	5.0	6.0	5.5	6.5	6.9	5.2	6.1	5.2	4.0	4.8	5.5	5.2	5.5	6.2	5.1
Akamagaseki	7.2	7.0	6.1	6.6	6.1	6.9	6.1	5.5	6.3	5.7	5.9	6.7	6.4	7.0	6.3	6.3	6.0
Sakai	8.6	8.5	7.2	6.8	6.4	7.3	6.8	5.6	7.0	6.8	7.1	8.0	7.2	8.4	6.8	6.6	7.1
Tsu	4.9	6.0	5.8	6.7	6.2	7.1	7.5	6.4	7.0	5.5	4.3	4.7	6.0	5.2	6.2	7.0	5.6
Nagoya	4.6	4.7	5.0	5.9	4.7	7.0	7.2	6.0	6.6	4.8	4.2	5.1	5.5	4.8	5.2	6.7	5.2
Gifu	5.5	5.6	5.7	6.6	6.5	7.4	7.5	6.3	6.9	5.7	5.0	5.2	6.2	5.4	6.3	7.1	5.9
Hamamatsu	4.0	4.6	5.1	6.6	6.8	7.5	7.2	5.9	6.7	6.0	4.6	3.6	5.7	4.1	6.1	6.9	5.8
Numazu	4.6	5.7	6.3	7.7	7.5	8.3	8.3	7.0	7.7	6.8	5.2	3.8	6.6	4.7	7.2	7.9	6.6

Mean amount of cloud.

Locality.	Jan.	Feb.	Mar.	Apr.	May	June	July	Aug.	Sept.	Oct.	Nov.	Dec.	Year	Wint.	Spr.	Sum.	Aut.
Tokio	3,9	5,3	5,6	6,7	6,8	7,7	7,0	6,1	7,3	6,3	4,8	3,6	5,9	4,3	6,4	6,9	6,1
Utsunomiya	2,4	4,1	5,7	6,5	4,9	8,2	7,9	7,1	7,0	5,3	4,0	3,5	5,5	3,2	5,7	7,7	5,4
Choshi	4,9	5,8	6,5	7,1	7,2	7,5	7,5	6,0	6,9	6,5	5,7	4,5	6,4	5,1	7,0	7,0	6,4
Kanazawa	8,8	8,6	7,5	7,2	7,0	7,8	7,6	6,3	7,5	6,9	7,6	8,5	7,6	8,6	7,2	7,2	7,3
Fushiki	9,0	8,6	7,3	7,4	6,8	7,7	7,8	6,3	7,4	7,1	7,3	8,0	7,5	8,5	7,2	7,3	7,4
Nagano	6,7	6,6	6,4	6,7	6,1	7,1	8,0	6,8	7,1	5,6	5,4	5,5	6,5	6,3	6,5	7,4	6,0
Niigata	8,6	8,4	7,1	7,0	6,7	7,2	7,1	5,3	7,0	6,1	7,5	8,3	7,2	8,1	7,0	6,5	7,0
Yamagata	8,0	6,9	6,6	6,7	6,0	7,1	7,7	6,6	7,0	5,7	6,6	7,0	6,8	7,3	6,4	7,2	6,4
Akita	9,2	8,5	8,1	7,2	7,0	7,6	7,6	6,4	7,1	6,4	7,9	8,8	7,7	8,8	7,1	7,2	7,2
Fukushima	6,0	6,0	6,5	6,8	6,1	7,8	8,0	7,3	7,5	5,9	5,5	5,8	6,6	5,9	6,6	7,7	6,3
Ishinomaki	4,9	5,3	5,7	7,0	6,6	7,7	7,9	7,0	7,5	5,7	5,5	5,4	6,3	5,2	6,4	7,5	6,2
Miyako	4,4	4,7	5,5	6,3	6,5	7,1	7,2	6,5	7,3	5,6	5,4	4,3	5,9	4,5	6,1	7,0	6,1
Aomori	8,2	7,8	7,1	6,4	6,1	6,9	7,1	5,7	6,5	5,6	7,6	8,1	6,9	8,0	6,4	6,6	6,6
Hakodate	7,3	6,8	6,5	6,0	6,1	6,9	7,1	6,3	6,3	5,6	6,8	7,5	6,6	7,2	6,3	6,8	6,2
Suttsu	9,3	8,5	7,6	6,2	6,5	7,3	6,8	6,3	6,8	6,4	8,4	8,9	7,5	8,9	6,8	6,8	7,1
Sapporo	7,0	7,0	6,8	6,4	6,1	7,0	6,9	6,2	6,2	5,9	7,4	7,2	6,6	7,1	6,3	6,7	6,4
Kamikawa	7,5	7,6	7,1	7,1	6,9	7,9	7,5	7,2	7,8	7,5	7,9	8,7	7,6	7,9	7,0	7,5	7,7
Soya	8,7	7,9	6,5	6,6	7,3	8,2	7,2	6,4	6,5	6,6	8,4	8,9	7,1	8,5	6,8	7,3	7,1
Abashiri	7,1	7,0	5,8	6,2	6,7	7,4	7,2	5,2	6,7	5,8	6,4	6,4	6,5	6,9	6,2	6,6	6,3
Nemuro	5,4	5,5	5,8	6,1	6,7	7,7	7,3	6,9	6,6	5,8	5,7	5,2	6,2	5,4	6,2	7,3	6,0
Kushiro	4,2	5,7	5,9	6,3	6,5	8,5	7,9	8,6	8,0	5,9	4,8	4,7	6,4	4,9	6,2	8,3	6,2
Erimo	6,4	5,6	5,8	6,0	6,5	8,2	8,5	7,1	6,5	5,4	6,3	6,6	6,6	6,2	6,2	7,9	6,1
Back side	8,57	8,05	7,18	6,72	6,62	7,43	7,28	6,42	6,94	6,38	7,49	8,20	7,24	8,27	6,84	6,94	6,94
Front side	4,95	5,48	5,78	6,67	6,50	7,76	7,34	6,46	6,96	5,86	5,08	4,72	6,12	5,05	6,32	7,15	5,97
Coast of Inland sea	5,77	6,29	5,93	6,47	5,90	7,06	7,00	5,63	6,54	5,36	4,76	5,20	6,00	5,75	6,10	6,57	5,55
West coast of Kiusiu	6,50	6,22	6,20	6,67	6,05	6,93	6,82	5,47	6,27	5,23	4,97	6,00	6,10	6,24	6,31	6,41	5,49

C. The numbers of clear and cloudy days.

When the part of the sky obscured by clouds be less than $\frac{2}{10}$ of the whole sky on the average of the day, then that day is said to be clear; while on the other hand, if the part obscured be more than $\frac{8}{10}$ of the whole then it is said to be cloudy. The numbers of these clear and cloudy days for several stations are:—

Mean number of clear days.

Locality.	Jan.	Feb.	Mar.	Apr.	May	June	July	Aug.	Sept.	Oct.	Nov.	Dec.	Year	Wint.	Spr.	Sum.	Aut.
Naha	1,0	0,0	0,0	1,0	0,0	0,0	1,0	1,5	0,0	4,5	3,5	1,0	13,5	2,0	1,0	2,5	8,0
Kagoshima	5,0	3,9	4,7	2,7	4,2	2,0	3,3	3,4	4,3	6,0	6,9	6,2	52,6	15,1	11,6	8,7	17,2
Miyazaki	12,4	7,1	7,0	4,2	4,0	2,4	4,4	4,0	2,4	5,6	10,2	15,1	78,8	34,6	15,2	10,8	18,2
Kochi	8,4	5,9	6,2	3,7	5,4	1,6	3,9	4,4	2,6	5,3	9,3	12,2	68,9	26,5	15,3	9,9	17,2
Wakayama	2,9	1,8	5,2	2,8	3,7	1,8	4,1	5,5	3,6	4,1	4,3	4,0	43,8	8,7	11,7	11,4	12,0
Oita	6,2	3,0	7,0	4,1	6,2	2,4	3,2	4,6	3,4	5,2	9,8	7,0	62,4	16,2	17,6	10,2	18,4
Yamaguchi	2,0	1,2	4,5	2,0	2,5	1,2	2,5	3,0	3,0	6,2	4,7	4,5	38,3	7,7	10,0	6,7	13,9
Hiroshima	2,8	2,4	5,0	4,0	6,4	2,2	3,8	5,3	3,5	5,3	5,8	4,7	50,9	9,9	15,1	11,3	14,6
Matsuyama	3,0	3,0	4,0	3,0	7,0	2,5	5,0	3,0	5,0	7,0	10,0	3,5	56,0	9,5	14,0	10,5	22,0
Okayama	3,0	0,0	2,0	4,0	9,0	2,0	2,0	3,0	1,0	4,0	9,0	7,0	46,0	10,0	15,0	7,0	14,0
Ozaka	4,3	1,3	3,4	2,9	3,4	1,9	2,3	3,6	3,9	4,8	5,9	6,3	44,0	11,9	9,7	7,8	14,6
Kioto	1,2	0,8	2,9	2,6	3,1	1,3	1,7	1,5	3,0	3,7	3,7	3,8	29,3	5,8	8,6	4,5	10,4
Kumamoto	5,0	4,5	2,0	3,0	5,0	3,0	2,5	4,5	1,5	7,5	9,5	4,5	52,5	14,0	10,0	10,0	18,5
Saga	3,0	2,0	6,0	5,0	5,0	1,0	2,0	5,5	3,5	5,5	9,0	3,5	54,0	8,5	16,0	11,5	18,0

Mean number of clear days.

Locality.	Jan.	Feb.	Mar.	April	May	June	July	Aug.	Sept.	Oct.	Nov.	Dec.	Year	Wint.	Spr.	Sum.	Aut.
Nagasaki	3,0	3,2	5,8	3,5	4,9	2,0	2,9	4,4	4,9	5,7	6,1	4,0	50,1	10,2	14,2	9,3	16,7
Fukuoka	0,5	1,5	1,5	2,0	3,5	3,0	2,5	1,0	2,5	3,0	5,0	1,5	27,5	3,5	7,0	6,5	10,5
Itsugahara	5,1	4,6	8,8	5,2	7,4	3,2	3,6	5,0	3,2	5,2	8,3	6,3	64,2	16,3	19,4	11,8	16,7
Akamagaseki	0,8	1,7	3,9	2,8	3,9	3,0	3,3	2,7	2,9	4,2	3,1	1,8	34,1	4,3	10,6	9,0	10,2
Sakai	0	0,6	2,3	3,4	3,9	2,1	3,3	3,8	1,9	2,0	1,3	0,8	25,4	1,4	9,6	9,2	5,2
Tsu	6,0	1,0	3,0	2,0	6,0	2,5	0,7	0,7	2,7	4,7	9,7	6,3	45,3	13,3	11,0	3,9	17,1
Nagoya	4,0	6,0	4,0	3,0	8,0	3,0	1,0	2,0	3,5	9,0	8,5	5,0	57,0	15,0	15,0	6,0	21,0
Gifu	3,4	3,1	4,4	3,4	2,2	1,1	1,3	2,4	2,0	5,0	6,3	4,6	40,5	11,1	11,0	5,1	13,3
Hamamatsu	9,7	7,4	7,8	2,6	3,3	1,0	1,1	2,0	2,2	2,8	7,9	12,3	61,4	29,4	13,7	4,4	13,9
Numazu	9,1	5,7	4,8	1,3	2,2	1,1	0,8	1,4	1,1	2,3	7,1	12,3	49,2	27,1	8,3	2,3	10,5
Tokio	12,0	6,4	6,1	3,5	3,2	1,4	2,1	2,2	0,9	3,7	7,7	12,9	62,2	31,3	12,8	5,7	12,3
Utsunomiya	17,0	8,0	2,0	2,0	4,0	1,0	1,0	0,0	0,5	3,5	9,0	13,5	61,5	38,5	8,0	2,0	13,0
Choshi	9,2	4,0	3,4	1,6	1,4	1,4	0,6	2,8	1,2	3,8	4,1	9,4	43,2	22,6	6,4	4,8	9,4
Kanazawa	0,2	0,6	2,1	2,6	2,8	1,3	2,0	4,1	1,4	2,8	1,7	0,6	22,2	1,4	7,5	7,4	5,9
Fushiki	0,2	0,6	2,6	1,9	3,1	1,6	1,3	2,0	1,3	3,4	2,3	1,1	22,1	1,9	7,6	5,0	7,0
Nagano	1,3	2,0	3,0	3,3	2,3	1,0	0,3	2,3	2,3	4,3	3,0	5,0	30,1	8,3	8,6	3,6	9,6
Niigata	0,0	0,3	1,5	1,8	3,2	1,2	2,4	6,3	1,5	3,5	1,3	0,5	23,5	0,8	6,5	9,9	6,3
Yamagata	0,5	1,0	2,5	1,5	2,5	0,0	1,0	1,7	1,0	3,0	3,0	2,0	19,7	3,5	6,5	2,7	7,0
Akita	0,1	0,8	0,8	1,1	2,2	1,2	1,9	3,8	0,7	2,6	0,8	0,6	16,6	1,5	4,1	6,9	4,1
Fukushima	2,0	3,0	2,5	2,5	3,3	0,3	0,7	0,7	1,3	2,7	2,7	3,3	26,0	8,3	8,3	1,7	7,7
Ishinomaki	3,0	1,7	4,2	1,0	2,2	0,7	0,2	1,7	0,8	3,8	3,2	2,0	25,5	7,7	7,4	2,6	7,8
Miyako	7,7	6,0	4,4	2,3	3,2	2,0	1,9	3,2	1,2	4,8	5,2	7,4	49,3	21,1	9,9	7,1	11,2
Aomori	0,3	0,6	1,6	2,8	3,5	2,0	1,2	3,6	2,3	4,7	1,0	0,6	24,2	1,5	7,9	6,8	8,0
Hakodate	0,5	1,5	1,9	3,9	3,2	2,8	2,1	2,5	2,9	4,4	1,6	0,5	27,8	2,5	9,0	7,4	8,9
Suttsu	0,0	0,6	1,4	3,8	2,9	1,9	2,0	2,9	2,4	3,5	0,5	0,1	22,0	0,7	8,1	6,8	6,4
Sapporo	1,4	0,9	2,1	3,7	3,2	1,7	3,1	4,1	2,8	4,2	1,1	0,7	29,0	3,0	9,0	8,9	8,1
Kamikawa	1,3	1,0	3,0	1,3	1,7	0,3	1,2	0,5	0,0	0,5	0,7	0,0	11,5	2,3	6,0	2,0	1,2
Soya	0,2	0,4	1,4	2,4	1,8	0,6	2,2	2,0	1,6	2,0	0,6	0,2	15,4	0,8	5,6	4,8	4,2
Abashiri	2,0	0,0	5,5	3,5	1,5	1,5	0,0	7,0	2,0	2,7	2,0	2,3	30,0	4,3	10,5	8,5	6,7
Nemuro	4,7	4,8	3,5	3,9	1,7	1,6	2,3	1,9	2,8	4,8	2,7	4,8	40,5	14,3	9,1	5,8	11,3
Kushiro	8,5	2,0	4,0	2,5	1,5	0,5	0,5	1,0	0,5	4,0	6,5	9,0	40,5	19,5	8,0	2,0	11,0
Erimo	2,0	2,4	4,0	3,4	3,4	0,8	0,2	1,0	1,7	5,5	1,3	1,3	27,0	5,7	10,8	2,0	8,5

Mean number of cloudy days.

Naha	12,0	18,0	26,0	16,0	20,0	23,0	12,0	7,0	19,0	6,0	12,0	13,5	178,5	43,5	62,0	42,0	31,0
Kagoshima	9,9	10,3	12,9	15,8	14,6	19,7	12,8	8,9	10,7	10,8	7,1	6,3	141,1	26,5	43,3	42,4	28,9
Miyazaki	6,2	7,2	11,1	11,0	12,0	17,2	11,9	8,8	12,8	11,9	6,8	3,9	124,8	17,3	38,1	37,9	31,5
Kochi	5,4	6,9	8,9	13,7	12,0	17,8	15,0	10,7	13,7	10,9	6,5	3,6	125,1	15,9	34,6	43,5	31,1
Wakayama	8,2	10,9	11,0	14,6	12,7	17,7	13,2	9,2	12,5	11,6	8,7	7,1	137,4	26,2	38,3	40,1	32,8
Oita	5,2	6,4	9,0	11,8	9,2	12,4	11,0	6,0	11,2	9,0	4,4	6,8	105,4	18,4	30,0	32,4	24,6
Yamaguchi	9,7	10,5	10,2	12,7	10,5	15,5	16,0	7,0	10,2	6,5	7,2	11,0	127,0	31,2	33,4	38,5	23,9
Hiroshima	7,9	8,5	8,9	12,0	10,8	16,1	11,7	7,2	11,2	9,1	5,9	6,5	115,8	22,9	31,7	35,0	26,2
Matsuyama	7,5	7,5	11,5	10,5	12,0	14,5	16,5	8,0	15,5	6,5	4,0	5,5	119,5	20,5	34,0	39,0	26,0
Okayama	4,0	8,0	10,0	11,0	6,0	13,0	23,0	12,0	15,0	2,0	3,0	6,0	113,0	18,0	27,0	48,0	20,0
Ozaka	7,4	8,7	10,1	14,9	13,1	17,6	14,3	10,1	12,3	11,7	7,3	5,5	133,0	21,6	38,1	42,0	31,3
Kioto	10,0	12,9	11,5	15,1	14,2	17,8	15,9	10,8	14,2	11,6	8,6	7,3	149,9	30,2	40,8	44,5	34,4
Kumamoto	11,0	9,0	14,5	13,0	13,5	15,5	15,5	6,0	9,5	6,5	4,5	7,5	126,0	27,5	41,0	37,0	20,5
Saga	11,0	11,0	8,0	10,0	5,0	7,0	19,0	5,5	9,5	2,5	3,5	6,5	98,5	28,5	23,0	31,5	15,5
Nagasaki	11,7	10,1	10,8	12,8	12,9	16,8	12,3	7,0	9,1	8,9	7,9	10,9	131,2	32,7	36,5	36,1	25,9
Fukuoka	18,5	16,0	15,0	13,5	12,5	18,0	18,0	9,5	15,5	7,0	10,0	14,0	167,5	48,5	41,0	45,5	32,5
Itsugahara	7,6	7,8	9,4	11,6	8,6	13,6	15,8	7,4	10,2	7,0	4,8	6,3	110,1	21,7	29,6	36,8	22,0
Akamagaseki	13,7	12,2	11,3	12,6	10,4	14,6	12,1	6,7	11,0	8,4	8,0	12,6	133,6	38,5	34,3	33,4	27,4
Sakai	23,7	21,2	16,1	13,9	12,8	16,2	14,2	8,1	14,3	12,9	15,7	19,6	188,7	64,5	42,8	38,5	42,9
Tsu	4,0	8,5	9,5	13,5	13,5	15,0	19,3	9,3	14,7	9,7	5,3	6,3	128,6	18,8	36,5	43,6	29,7
Nagoya	3,0	6,0	7,0	9,0	5,0	16,0	17,0	9,0	13,5	9,0	4,5	6,5	105,5	15,5	21,0	42,0	27,0
Gifu	7,0	7,6	9,1	12,6	13,2	15,7	17,1	10,1	11,4	10,1	7,2	5,7	130,1	23,2	24,9	43,2	31,7

MEAN NUMBER OF CLOUDY DAYS.

Locality.	Jan.	Feb.	Mar.	April	May	June	July	Aug.	Sept.	Oct.	Nov.	Dec.	Year	Wint.	Spr.	Sum.	Aut.
Hammatsu	4,3	6,2	8,8	12,9	13,7	16,7	15,6	9,2	13,1	9,7	6,3	1,1	120,6	14,6	35,4	41,5	29,1
Numazu	7,1	9,7	12,7	17,8	19,0	21,1	21,2	13,2	17,0	13,1	9,9	4,9	167,0	22,0	49,5	55,5	40,0
Tokio	5,6	8,1	9,6	13,9	14,2	16,8	14,3	9,1	15,1	12,2	7,6	4,6	131,2	18,3	37,7	40,2	34,9
Utsunomiya	1,0	1,0	10,0	11,0	7,0	20,0	17,0	11,0	11,5	7,0	5,5	5,5	110,5	10,5	28,0	48,0	24,0
Choshi	8,6	9,0	12,2	17,0	14,6	16,8	16,0	9,0	13,8	13,2	8,6	6,0	144,8	23,6	43,8	41,8	35,6
Kanazawa	24,0	20,8	17,4	15,3	14,1	18,5	18,6	11,5	16,7	14,3	17,6	12,4	211,5	67,2	47,1	48,6	48,6
Fushiki	27,0	21,7	17,1	16,0	13,9	17,4	19,4	11,9	16,7	12,9	16,3	20,1	210,1	68,8	47,0	48,7	45,9
Nagano	10,7	9,7	13,0	13,0	11,3	15,7	20,3	12,7	15,3	6,7	7,3	9,3	145,0	29,7	37,3	48,7	29,3
Niigata	22,5	19,1	16,3	14,5	13,1	15,8	14,5	9,0	13,9	12,8	16,8	20,9	189,2	62,5	43,9	39,3	43,5
Yamagata	20,0	12,0	13,5	12,0	8,5	16,0	17,3	11,7	13,7	8,0	11,3	15,3	159,3	47,3	34,0	45,0	33,0
Akita	27,0	20,8	20,1	15,2	14,3	17,3	18,1	12,7	14,7	12,4	17,6	24,7	215,2	72,5	49,6	48,4	44,7
Fukushima	8,5	8,5	12,5	12,5	12,3	17,3	19,3	15,0	15,7	7,0	7,7	7,3	143,6	24,3	37,3	51,6	30,4
Ishinomaki	4,0	6,0	9,0	13,2	13,2	17,0	18,0	14,5	16,1	9,4	8,0	5,8	134,5	15,8	35,4	49,5	33,8
Miyako	3,7	5,0	7,8	10,9	12,6	16,4	15,7	12,3	15,1	8,4	7,0	4,3	119,2	13,0	31,3	44,4	30,5
Aomori	20,4	16,2	15,1	9,9	10,2	13,9	15,7	8,9	10,6	9,2	16,6	20,2	166,9	56,8	35,2	38,5	36,4
Hakodate	14,6	14,4	12,1	9,3	11,9	15,2	15,9	10,9	11,1	7,9	12,4	14,8	147,5	40,8	33,3	42,0	31,4
Suttsu	27,9	21,1	18,1	9,8	13,4	16,0	14,1	13,7	12,7	12,4	21,5	26,1	206,8	75,4	41,3	43,8	46,6
Sapporo	13,1	12,1	13,0	10,5	11,7	14,4	14,5	11,0	10,7	9,4	12,9	13,5	146,5	38,7	35,2	39,6	33,0
Kamikawa	17,3	15,0	16,0	14,7	12,7	18,0	16,0	14,0	13,5	14,2	19,0	23,5	193,9	55,8	43,4	48,0	46,7
Soya	23,2	15,0	14,2	11,8	16,0	20,2	15,4	12,4	9,8	12,4	20,0	24,8	195,2	63,0	42,0	48,0	42,2
Abashiri	16,0	10,5	9,5	9,0	11,0	14,5	14,5	8,7	12,3	9,3	8,0	12,7	136,0	39,2	29,5	37,7	29,6
Nemuro	7,7	6,6	8,6	10,7	12,3	17,6	16,5	12,5	11,8	9,1	8,6	6,5	128,5	20,8	31,6	46,6	29,5
Kushiro	4,5	8,5	11,0	12,5	12,5	21,0	18,0	24,5	21,5	8,5	6,0	9,5	158,0	22,5	36,0	63,5	36,0
Erimo	8,4	6,0	9,2	9,6	12,6	21,0	23,2	14,0	13,5	8,2	8,8	11,0	145,5	25,4	31,4	58,2	30,5

If the average numbers of those days for the two sides of Nippon be calculated separately, we have the following numbers:—

MEAN NUMBER OF CLEAR AND CLOUDY DAYS.

Number of clear days.

	Jan.	Feb.	March	April	May	June	July	August	Sept.	Oct.	Nov.	Dec.	Year
Back side	0,1	0,6	1,7	2,4	2,8	1,4	2,2	3,7	1,5	2,8	1,2	0,6	21,1
Front side	5,4	3,6	4,5	3,0	3,9	1,6	2,5	2,4	2,3	4,7	6,3	7,5	47,6

Number of cloudy days.

	Jan.	Feb.	March	April	May	June	July	August	Sept.	Oct.	Nov.	Dec.	Year
Back side	25,0	20,0	17,0	13,8	14,0	17,3	16,4	11,4	14,1	12,9	12,9	22,7	202,4
Front side	6,0	7,9	10,0	12,6	11,9	16,8	16,2	10,5	13,6	8,7	7,0	6,6	129,8

If the ratios which those days bear to the number of days in each month be expressed in percentages, then we have:

Clear days.

	Jan.	Feb.	March	April	May	June	July	August	Sept.	Oct.	Nov.	Dec.	Year
Back side	0	2	6	8	9	5	7	12	5	9	4	2	6
Front side	18	13	15	10	13	5	8	8	8	15	21	24	13

Cloudy days.

	Jan.	Feb.	March	April	May	June	July	August	Sept.	Oct.	Nov.	Dec.	Year
Back side	81	71	55	46	45	58	53	37	47	42	43	73	56
Front side	19	28	32	42	38	56	52	34	45	28	23	21	36

First of all, for the mean of the year we observe that the clear sky may be more expected on the front Nippon than on the back Nippon, for the number of cloudy days on the back Nippon is 56 per cent of the number of days in a year, yet on the front Nippon it is only 36 per cent, and the number of clear days is 13 per cent on the front Nippon and only 6 per cent on the back Nippon.

The annual variations of the cloudy days on the two sides of Nippon are as follows:—

On the back Nippon, we have many cloudy days in winter, especially in January when the percentage of those days is 80. The minimum is in August, when it is about 37. From the maximum in January, it decreases continually to May, and in June there is a sudden increase to 58, and still increases till July. In August, as just mentioned, it has the maximum value of the year. It is increasing in September and again decreasing in October and November. In December, it increases suddenly to 73.

It is just opposite to the preceding case on the front Nippon. Here we have very few cloudy days in winter, especially in January, when the percentage is only 20; and great many cloudy days in June when the ratio is 56. The variation of the percentage ratio of the number of cloudy days is as follows. Starting from the minimum, 20, in January, it reaches 42 in April through constantly increasing stages. In May, it slightly decreases, and then suddenly increases up to the maximum of the year in June. In July, still it keeps rather great value, but in August it is diminished to 34. After passing an increasing stage in September, it decreases from October onward, until it reaches the said minimum in January.

It is a peculiar feature of this country common to both sides, that June, July and September have comparatively many cloudy days, while August has only a few.

The variations of the number of the clear days are as follows:

On the back Nippon, we have very few clear days in winter, especially in January when it is almost entirely absent. After January, the percentage ratio of those days increases to 9 in May. In June, there is a decrease to 6, but after July it increases and reaches the maximum 12 of the year in August. There is a decrease in September and an increase in October. After November, it diminishes continually to the minimum in January.

On the front Nippon, however, clear days are most numerous in winter. Indeed the maximum ratio, 24 per cent, is in December, and the minimum, 5 per cent, in June. The general feature of the variation of the percentage ratio is this. It decreases gradually from winter toward summer, (though February and April present some decrements compared to the preceding and the succeeding months), and reaches the minimum in June. From July to September it retains almost constant value of 8 per cent, but from October it begins to increase until at last it reaches to the maximum in December.

D. SUNSHINE.

It is manifest that the sunshine forms an important element in the study of climate. What we have mentioned in the preceding articles, relate solely to the amounts of cloud and no references had been made on the forms of clouds, so that nothing can be concluded from them about the duration of sunshine.

Now of our stations, there are only a few number of stations provided with sunshine recorders. Moreover, the number of observed years is so small, that we have as yet no satisfactory results. But for the sake of future references, some remarks are added here. The recorders, now used in our stations are Jourdan's form, in which the sunlight, after passing through a narrow slit, is made to fall on a blue print sensitive paper, and the sunshine is thus recorded on it.

The monthly duration of sunshine expressed in hours and the percentages of possible duration of sunshine at several stations are as follows:—

Number of hours of sunshine.

Locality.	Jan.	Feb.	Mar.	April	May	June	July	Aug.	Sept.	Oct.	Nov.	Dec.	Year	Wint.	Spr.	Sum.	Aut.
Wakayama	119,1	128,2	164,0	155,3	174,2	157,2	164,7	210,4	151,7	158,0	152,1	110,3	1844,0	349,2	464,5	477,3	454,9
Matsuyama	117,8	125,5	207,7	155,5	217,8	168,4	186,5	215,7	159,2	178,8	184,0	132,7	2168,9	376,0	580,9	570,5	518,0
Ozaka	117,7	113,3	149,7	152,9	188,5	163,2	167,6	214,7	164,0	170,0	152,9	132,6	1887,1	363,6	491,1	545,5	486,9
Kioto	116,1	110,8	161,2	174,6	218,7	163,5	179,1	224,6	165,1	183,4	153,8	131,0	2047,1	357,9	554,6	596,6	502,3
Kumamoto	118,8	101,8	173,2	172,3	238,6	157,5	164,4	217,7	163,6	171,9	158,8	110,2	1948,8	330,8	584,1	539,6	494,3
Gifu	145,1	129,2	170,8	157,0	188,3	169,0	139,3	183,8	143,6	185,7	171,6	149,8	1891,1	394,1	516,1	492,1	507,9
Tokio	202,2	175,6	149,6	157,3	181,9	157,5	151,2	183,7	178,3	164,6	179,2	168,9	2059,0	544,7	491,8	498,4	522,1
Kanazawa	43,4	71,5	180,7	207,1	302,7	180,0	101,7	216,8	164,1	214,8	105,7	65,3	1886,9	181,2	690,5	501,5	510,6
Hakodate	84,5	87,2	138,6	163,0	224,8	114,1	151,3	164,4	113,0	183,9	121,8	65,2	1617,5	238,0	522,5	455,2	421,7

Percentages of possible duration.

Wakayama	38	42	44	39	40	36	37	51	42	45	50	37	42	39	44	41	46
Matsuyama	37	41	56	40	51	40	43	53	43	51	57	43	49	40	49	45	50
Ozaka	38	37	40	39	44	38	38	52	44	49	49	43	43	39	44	43	47
Kioto	37	36	44	45	51	45	41	51	45	53	50	43	46	39	47	47	49
Kumamoto	37	33	45	44	56	36	38	52	44	49	51	35	43	35	48	42	48
Gifu	47	43	44	40	43	39	31	44	40	54	57	40	43	43	43	38	49
Tokio	65	57	40	40	42	36	35	44	48	47	58	56	47	59	44	38	51
Kanazawa	14	25	49	52	69	45	24	52	44	69	35	22	43	24	58	39	50
Hakodate	29	29	37	41	45	32	33	38	30	51	42	24	36	27	42	34	42

Thus the annual mean of the duration of sunshine is almost everywhere constant in our country; the sun shines for about $\frac{42}{100}$ of a day. The annual variations are remarkably different on the two sides of Japan. The most remarkable fact is that the duration of the sunshine in winter is very much smaller in the back Nippon than in the front Nippon. That it is great in spring and autumn, and small in winter and summer is a fact common to everywhere.

The result of seven years' observations on sunshine at Tokio, where Compbell Stolk's form of the sunshine recorder is used, is as follows:—

Number of hours of sunshine at Tokio.

Month.	4–5	5–6	6–7	7–8	8–9	9–10	10–11	11–N	N–1	1–2	2–3	3–4	4–5	5–6	6–7	Total	
	h.	h.	h.	h.	h.	h.	h.	h.	h.	h.	h.	h.	h.	h.	h.	h.	%
January	..	..	0,0	6,3	17,1	18,9	19,9	20,5	21,3	20,3	18,7	11,7	3,7	0,0	..	161,2	52
February			0,3	9,9	15,7	17,3	17,7	17,8	18,2	17,3	16,8	15,0	8,3	0,2		154,4	51
March		..	1,8	11,8	15,7	17,2	17,8	17,5	18,0	16,5	14,8	13,7	5,2	0,7	..	150,7	41
April		1,1	3,8	8,8	10,7	11,7	12,2	13,1	12,6	12,2	10,0	7,5	2,8	1,1	0,0	107,6	27
May	0,1	1,1	8,0	10,4	11,9	11,7	11,8	12,1	11,3	10,7	9,6	8,5	7,9	1,4	0,6	120,1	28
June	0,0	1,8	6,7	9,3	11,6	12,9	13,5	11,2	9,3	11,9	11,1	9,5	8,0	5,9	1,6	121,1	29
July	0,0	1,1	7,5	10,5	12,1	14,7	15,5	15,6	14,6	15,1	14,3	12,2	10,4	7,0	1,6	152,5	35
August	0,0	0,6	10,0	13,8	15,9	17,0	18,6	20,1	21,0	20,8	19,8	18,6	16,0	11,5	1,3	205,3	49
September		0,0	2,2	8,1	10,3	11,7	11,6	12,6	12,1	13,1	13,4	11,9	6,1	1,6	0,0	115,4	31
October			0,2	9,2	12,8	14,4	15,8	15,0	14,9	13,8	13,0	12,4	7,1	0,3	..	128,9	37
November			0,0	6,3	14,8	17,2	18,3	18,0	18,6	18,7	17,7	19,1	7,6	0,1		153,2	50
December	..	..	0,0	6,4	19,2	21,1	22,2	22,6	22,6	20,8	18,2	14,0	1,7	0,0	..	169,0	56
Year	0,1	5,7	40,4	110,9	168,2	185,7	194,8	196,3	195,0	191,1	177,3	154,0	85,1	32,7	5,2	1742,7	39

Thus in Tokio, the maximum sunshine lies somewhere between 10 am and 1 pm. On the mean of the year it lies between 11 o'clock am and the noon.

The days without sunshine are:—

NUMBER OF DAYS WITH NO SUNSHINE AT TOKIO.

	Jan.	Feb.	March	April	May	June	July	August	Sept.	Oct.	Nov.	Dec.	Year
Number of days	4.2	4.9	7.1	10.0	8.3	8.9	7.0	2.4	5.9	6.3	5.3	3.0	72.6
id %	13	17	23	33	27	30	23	8	20	20	18	10	20

Thus the maximum number of days without sunshine is at April, when it is about $\frac{1}{3}$ of the number of days in that month. June has next great number of days without sunshine. The minimum number of days without sunshine is in August, when it is only $\frac{8}{100}$ of the number of days in that month. December is the next to August.

On the annual average, the days without sunshine are about $\frac{20}{100}$ of the days in a year.

CHAPTER VI.

PRECIPITATION.

By precipitation, we mean all the amount of water fallen on earth from the upper regions, whether in the form of rain, snow, graupel or hail. Its amount is subject to great fluctuations, and hence it is very difficult to determine its constant for any district by observations of some fifty or sixty years only. Thus, though the mean of the total precipitations in a year observed in Tokio during the sixteen years from 1876 to 1891 is 1,485 mm, yet in some years this amount reached to the great value of 1,958 mm, while in others only to 1,221 mm. Thus the difference between the two extremes is 737 mm, which is more than 50 per cent of the total amount. If we take the monthly precipitations, the difference between the two extremes is even greater than this; namely in September the mean precipitation is 208 mm, the greatest 461 mm, and the least only 72 mm. Thus, the difference 349 mm is 168 per cent of the monthly mean.

If we calculate, the deviations of precipitation of each year from the mean of sixteen years at Tokio, we obtain the following.

		Jan.	Feb.	Mar.	April	May	June	July	Aug.	Sept.	Oc'.	Nov.	Dec.	Year
Mean amount (1876-1891) in mm.		54,2	78,5	121,8	116,1	148,4	177,0	125,5	109,7	207,8	188,0	107,9	50,1	1485,0
Mean deviation from the mean of 16 years	in mm.	±31,6	±51,2	±58,0	±33,1	±57,9	±68,4	±63,8	±66,8	±100,0	±81,8	±62,9	±37,5	±204,9
	% of the means	±54	±65	±42	±28	±39	±39	±51	±61	±48	±43	±58	±75	±14
Mean errors of the means	in mm.	±8,2	±13,2	±13,1	±8,5	±14,9	±17,6	±16,4	±17,2	±25,8	±21,1	±16,2	±9,7	±52,9
	% of the means	±15	±17	±11	±7	±10	±10	±13	±16	±12	±11	±15	±19	±4

This table shows that the deviations of the monthly precipitation of each year from the monthly means deduced by the observations of sixteen years at Tokio, are, on the average 30—75 per cent of the mean values, and the deviations of the yearly mean is about 14 per cent. The mean monthly amounts of precipitation deduced from the sixteen years' observations have, therefore, errors of 10—19 per cent; and even in the yearly mean, there still exists the mean error of about 4 per cent.

The number of years required for reducing the mean error of the mean value to within one per cent are given in the next table.

	Jan.	Feb.	March	April	May	June	July	August	Sep.	Oc'.	Nov.	Dec.	Year
No. of year required	3000	4500	1800	850	1600	1500	2600	3800	2100	2000	3400	5600	200

Thus at least eight or nine hundred years, and sometimes six thousand years are required in order to find the monthly precipitation true to one per cent of the whole, and even for the yearly amount 200 years are required.

The number of days with precipitation, viz. the days whose precipitation reached or exceeded 0.1 mm, differs greatly year to year, as the amount of precipitation does. The mean numbers of those days observed at Tokio during sixteen years are as follows:—

Mean number of days with precipitation at Tokio.

	Jan.	Feb.	March	April	May	June	July	August	Sept.	Oc'.	Nov.	Dec.	Year
No. of days with precipitation	7,2	9,4	11,6	14,9	13,2	14,2	11,1	11,9	15,3	12,5	8,8	6,1	139,5

The greatest and the least number in a year which we had during the sixteen years, were 181 and 118 days respectively, the range being 63 days. In the month of September, the greatest number was 21 days, and the least 8 days, ranging 13 days. Since the mean deviation of the total number of days with precipitation in each year from the mean of the sixteen years, is about 14 days, the yearly mean number has the mean error of about 3,5 days. Hence, if we wish to reduce this error within one day, observations of about 300 years are required.

Thus it is evident that fifty or sixty years are not sufficient to give the amount of precipitation and the number of days with precipitation so accurately as to furnish us constants for a district.

A. DIURNAL VARIATION OF PRECIPITATION.

In almost all the meteorological observations in any country which we have heard of, the amount of precipitation is not observed several times in a day but generally only once. Consequently, on the diurnal variation of precipitation, we have no discussions by any great authorities. In our country, at present there are ten stations where hourly observations of precipitation are taken. We hope that these observations may contribute some informations regarding its diurnal variation. But we are sorry that the observed number of years are very small, e. g. only seven years in Tokio and generally two or three years in others, so that we can not say that we have sufficient data for the investigation at present.

The hourly amounts of precipitation at the stations of first order are given in the following tables :—

Amount of precipitation.

KUMAMOTO.

Month / Hour	Jan.	Feb.	Mar.	Apr.	May	June	July	Aug.	Sept.	Oct.	Nov.	Dec.	Year	Wint.	Spr.	Sum.	Aut.
M.N.—1	0,1	4,8	1,3	1,8	8,3	2,7	8,9	1,2	9,1	4,2	2,0	5,6	59,7	10,9	17,5	15,9	15,4
1—2	0,9	3,1	7,5	5,5	9,8	2,5	11,0	1,9	7,9	3,1	2,7	2,5	58,9	6,8	22,8	15,4	13,7
2—3	0,8	6,7	8,5	5,5	11,4	7,1	10,2	1,6	1,6	5,7	2,3	2,1	67,2	9,6	25,5	19,3	12,6
3—4	1,0	1,5	3,5	5,2	10,8	3,0	5,1	2,1	8,4	2,7	1,1	4,9	52,8	10,1	19,6	10,5	12,3
4—5	1,9	3,5	3,7	1,3	10,3	5,8	10,7	5,1	1,9	3,1	2,3	6,1	61,8	11,5	18,3	21,6	10,3
5—6	1,2	2,7	1,6	7,9	9,5	1,3	6,5	21,2	5,1	2,2	1,4	2,2	72,2	6,1	22,1	35,1	8,8
6—7	1,0	3,2	7,2	7,3	10,2	5,1	1,4	17,7	4,1	2,1	1,6	4,3	68,7	8,5	21,7	27,2	8,2
7—8	3,3	4,0	7,0	7,0	7,0	6,1	6,1	12,1	2,8	2,4	1,2	2,1	61,1	9,4	21,0	21,1	6,4
8—9	3,5	2,8	11,0	21,1	3,7	7,5	11,1	4,1	5,3	4,1	1,9	3,0	79,5	9,3	35,8	23,0	11,3
9—10	2,5	5,2	2,9	8,8	5,8	6,2	20,5	3,5	4,7	5,1	0,5	6,2	72,1	13,9	17,5	30,3	10,6
10—11	6,6	6,9	2,7	6,5	5,3	5,0	18,3	1,9	1,1	4,6	1,9	1,4	65,4	11,9	14,5	25,2	10,7
11—N.	5,4	8,2	1,0	1,0	7,6	2,9	7,3	2,8	10,4	3,0	2,1	8,5	63,1	22,1	12,7	13,1	15,5
Noon—1	3,9	8,1	1,5	13,4	9,0	2,2	9,4	2,5	4,6	2,8	1,3	34,0	92,9	46,0	23,9	14,1	8,8
1—2	2,4	7,1	2,6	3,2	7,8	4,2	6,8	6,8	4,0	2,7	0,7	11,1	60,0	20,9	13,7	17,9	7,1
2—3	2,2	10,3	3,7	5,2	15,3	20,3	14,0	5,3	20,7	2,7	1,1	7,3	108,1	19,8	21,2	39,7	24,6
3—4	2,5	3,7	4,1	9,0	13,0	19,2	15,2	5,6	6,1	3,8	0,1	18,7	101,5	24,9	26,4	40,1	10,1
4—5	1,9	5,1	1,1	23,9	7,4	11,2	15,1	7,6	8,8	6,8	0,0	8,9	101,3	15,9	35,8	33,9	15,6
5—6	1,7	8,9	5,7	22,4	6,2	8,3	18,4	3,1	2,3	2,5	0,0	2,0	81,9	12,6	34,3	30,1	4,9
6—7	2,6	1,7	5,0	11,9	4,0	7,6	13,8	1,5	2,1	6,6	0,0	1,4	61,4	8,7	20,9	23,0	8,8
7—8	2,1	4,2	5,1	12,5	3,1	8,1	17,5	1,2	2,0	3,8	0,4	3,1	63,9	9,7	20,8	27,1	6,2
8—9	0,2	2,3	4,5	9,1	2,6	3,6	6,1	0,0	2,1	2,8	2,9	1,4	40,8	6,9	16,2	9,7	7,9
9—10	0,2	1,4	8,3	15,7	8,6	3,6	6,6	0,7	5,5	2,2	0,5	2,6	59,3	7,3	32,6	11,0	8,3
10—11	0,2	1,9	8,1	10,5	16,1	3,1	16,3	0,5	5,2	1,5	0,4	0,3	67,1	5,1	34,8	19,9	7,2
11—M.N.	0,2	6,6	7,6	5,1	7,8	2,8	10,5	4,4	7,6	2,0	4,7	0,1	56,7	7,2	20,2	17,8	11,1
Sum	48,9	126,6	125,2	230,3	200,9	153,5	271,1	121,3	142,9	83,8	30,7	144,0	1679,4	319,5	556,1	545,9	257,5

MATSUYAMA.

Month / Hour	Jan.	Feb.	Mar.	Apr.	May	June	July	Aug.	Sept.	Oct.	Nov.	Dec.	Year	Wint.	Spr.	Sum.	Aut.
M.N.—1	1,7	2,0	2,9	3,1	9,1	6,2	2,9	0,8	5,2	6,5	0,4	6,8	47,8	10,5	15,2	9,9	12,4
1—2	2,4	1,4	3,3	2,7	14,0	5,9	12,8	4,0	13,3	7,8	0,4	2,7	70,9	6,6	20,0	22,8	21,5
2—3	1,7	2,6	3,2	6,1	20,3	4,5	4,5	3,8	9,2	6,2	0,3	0,7	63,5	5,1	29,7	12,9	15,8
3—4	0,8	1,8	6,7	3,3	18,1	4,3	4,0	2,3	10,5	7,7	1,7	1,1	57,5	3,7	28,2	10,7	19,9
4—5	0,8	3,5	2,2	8,1	11,1	4,7	3,8	0,5	12,6	5,3	0,4	1,1	54,6	5,7	21,5	9,0	18,3
5—6	0,5	4,9	1,2	6,8	8,7	3,2	6,5	1,7	11,8	6,0	3,4	1,8	56,7	7,2	16,8	11,4	21,2
6—7	2,5	4,3	2,2	5,7	5,5	3,5	4,2	6,8	15,2	7,1	1,4	3,6	62,3	10,5	13,5	14,6	23,7
7—8	0,6	1,5	2,4	7,1	3,4	3,7	3,6	1,6	15,1	7,8	1,4	3,5	51,8	5,6	12,9	8,9	21,3
8—9	1,8	2,1	2,7	5,9	4,4	4,1	7,3	3,3	11,1	4,4	1,0	3,1	51,8	7,1	13,0	14,8	16,5
9—10	2,3	4,2	1,6	7,1	2,0	1,9	12,4	5,7	8,6	7,6	1,9	2,1	54,5	5,6	10,7	20,0	18,1
10—11	3,0	0,4	1,9	8,1	5,4	1,5	17,7	2,5	7,9	8,5	0,5	4,0	64,8	7,4	15,5	24,8	17,0
11—N.	3,5	1,2	0,7	9,3	5,8	7,9	7,2	3,1	1,6	7,7	0,5	4,6	56,3	9,3	15,8	18,3	12,8
Noon—1	7,5	1,7	5,0	10,9	3,1	7,8	7,3	2,1	9,8	5,0	0,7	12,7	73,8	21,9	19,1	17,2	15,5
1—2	5,3	4,4	2,7	2,8	2,0	9,9	3,4	2,5	9,4	3,3	0,0	6,6	52,6	16,3	7,5	15,9	12,8
2—3	4,2	4,2	4,2	9,4	3,4	3,5	9,6	2,1	14,3	11,0	3,0	7,5	70,6	15,9	17,0	9,2	28,3
3—4	3,5	3,7	5,4	6,1	8,0	4,6	4,3	1,4	18,4	3,8	2,0	5,0	66,5	12,2	19,6	10,4	24,2
4—5	1,4	1,5	4,0	2,6	6,0	12,9	10,9	6,4	8,0	1,9	2,6	4,1	65,4	10,0	12,6	30,2	12,5
5—6	0,8	5,7	4,6	5,0	4,5	4,1	8,3	5,8	5,0	4,9	0,2	7,9	57,2	14,5	14,2	18,3	10,1
6—7	0,2	6,1	4,1	6,4	3,0	1,8	2,8	6,2	1,9	7,2	0,6	11,5	55,0	17,8	13,5	13,8	9,8
7—8	0,3	4,6	3,4	14,7	3,7	3,3	2,5	2,0	3,1	3,0	0,8	5,2	47,2	10,1	21,9	7,9	7,3
8—9	1,1	2,2	3,1	3,1	3,8	1,1	11,8	3,2	2,3	3,6	0,2	1,8	37,5	5,2	10,0	16,1	6,1
9—10	0,8	1,7	2,5	6,8	4,1	0,8	3,4	6,4	4,6	8,5	1,1	3,4	44,2	5,9	13,4	10,6	11,2
10—11	0,7	0,2	3,2	5,3	5,8	2,4	2,1	1,6	3,0	4,5	0,9	2,4	35,7	3,3	14,3	9,5	8,5
11—M.N.	0,7	0,4	3,3	7,9	8,2	2,8	2,5	0,8	3,0	6,8	0,9	4,1	41,7	5,2	19,4	6,2	10,8
Sum	48,5	67,2	77,1	154,8	159,0	113,0	150,8	79,9	208,6	146,7	26,7	107,9	1310,6	223,6	391,0	343,8	382,1

HIROSHIMA.

Month / Hour	Jan.	Feb.	Mar.	Apr.	May	June	July	Aug.	Sept.	Oct.	Nov.	Dec.	Year	Wint.	Spr.	Sum.	Aut.
M.N.—1	0,6	2,5	4,4	6,0	11,3	13,9	20,4	4,2	5,6	4,9	1,9	2,3	77,9	5,5	21,7	38,4	12,3
1—2	1,3	2,0	5,1	8,7	10,7	9,4	22,3	6,0	13,1	6,1	4,1	2,8	91,6	6,2	24,4	37,8	23,2
2—3	1,0	2,0	4,8	6,5	10,6	12,2	11,2	4,5	9,0	5,2	5,1	2,0	74,2	5,0	21,9	27,9	19,3
3—4	1,3	2,3	3,1	9,9	7,2	8,6	20,0	2,8	10,5	6,0	7,0	4,1	82,7	7,6	20,2	31,3	23,5
4—5	1,3	3,3	2,9	11,2	6,6	7,4	28,1	1,2	7,4	9,4	4,1	3,9	86,8	8,4	20,7	36,7	20,9
5—6	0,5	2,0	2,9	13,2	5,7	14,1	21,3	4,5	10,7	8,2	4,0	3,3	101,2	5,7	21,8	50,9	22,9
6—7	0,3	1,3	3,8	18,4	6,9	16,5	28,5	3,3	9,8	8,6	4,9	5,6	107,9	7,3	29,1	48,3	23,3
7—8	0,9	0,7	4,4	13,3	6,4	12,7	31,0	3,0	12,5	5,5	5,8	4,9	101,2	6,5	24,2	46,7	23,8
8—9	1,4	1,1	3,7	9,8	5,5	14,6	30,1	5,4	9,3	10,1	6,7	7,1	104,9	9,6	19,1	50,1	26,1
9—10	2,4	2,5	5,0	14,3	5,5	22,1	15,7	5,1	10,6	6,8	3,3	4,3	97,7	9,2	24,8	42,9	20,7
10—11	2,8	2,7	3,6	7,3	6,9	17,8	14,3	6,3	9,0	4,8	2,0	8,7	86,3	14,2	17,9	38,4	15,9
11—N.	2,1	2,9	7,7	4,4	7,6	6,2	9,9	4,5	8,9	5,4	1,3	7,5	68,4	12,5	19,7	20,6	15,7
Noon—1	1,9	6,8	4,0	8,8	6,0	7,8	14,5	6,3	4,2	4,5	2,1	4,0	70,9	12,7	18,8	28,5	10,8
1—2	2,2	2,7	3,0	9,1	3,5	6,0	8,8	5,2	4,9	3,2	1,4	2,2	52,2	7,1	15,6	20,0	9,5
2—3	2,5	3,9	3,1	11,9	4,0	9,8	19,1	5,3	4,7	3,2	2,2	2,9	72,6	9,3	19,0	34,2	10,1
3—4	1,6	4,5	2,9	19,8	8,7	14,3	11,3	3,0	2,2	2,2	2,1	1,1	76,8	7,3	31,4	31,6	6,6
4—5	1,3	4,2	2,9	14,9	6,8	11,2	14,2	4,4	3,2	2,9	2,0	4,1	72,2	9,7	24,6	29,8	8,1
5—6	1,8	5,5	4,1	12,3	3,9	7,0	21,2	12,5	1,4	3,0	1,1	4,5	78,2	11,8	20,3	40,6	5,6
6—7	1,0	5,8	3,9	14,9	4,4	9,5	10,2	0,8	3,3	2,2	0,5	4,8	61,4	11,7	23,2	20,6	6,0
7—8	1,3	5,1	4,2	11,5	4,7	3,2	13,7	0,7	12,4	4,1	2,3	4,0	67,1	10,4	20,4	17,6	18,7
8—9	1,0	3,3	2,3	7,6	3,2	2,4	5,3	0,3	3,0	3,0	0,1	1,3	32,9	5,6	13,1	8,0	6,1
9—10	0,7	3,6	5,1	11,7	5,5	6,2	7,6	0,4	3,3	2,8	0,0	5,6	52,5	10,0	22,3	14,1	6,1
10—11	1,0	1,1	4,9	8,4	6,8	7,2	11,8	0,8	4,9	2,6	0,1	2,6	52,3	4,8	20,1	19,8	7,7
11—M.N.	0,9	2,7	5,1	5,9	7,5	6,9	13,8	0,8	3,7	2,6	1,0	1,8	52,7	5,4	18,5	21,5	7,3
Sum	33,2	74,8	96,7	256,0	156,0	247,0	420,2	89,3	167,7	117,4	65,2	95,5	1822,9	203,4	512,7	756,5	350,3

OZAKA.

Month / Hour	Jan.	Feb.	Mar.	Apr.	May	June	July	Aug.	Sept.	Oct.	Nov.	Dec.	Year	Wint.	Spr.	Sum.	Aut.
M.N.—1	1,7	7,3	13,8	29,9	15,4	7,6	49,8	23,5	46,4	13,0	8,8	7,1	158,1	16,1	59,1	80,9	68,2
1—2	1,3	4,3	14,7	21,9	24,7	8,1	21,4	11,7	35,7	10,4	9,5	7,1	123,7	12,7	61,3	41,2	55,6
2—3	2,3	5,6	19,5	20,1	21,4	20,2	16,4	18,6	31,8	7,6	9,9	3,0	126,4	10,9	61,0	55,2	49,3
3—4	1,2	5,3	14,8	21,1	25,9	12,1	19,3	2,3	29,7	6,4	14,6	9,7	140,3	16,2	61,8	33,7	50,7
4—5	1,6	5,8	28,3	23,5	31,3	12,9	20,4	1,9	18,5	14,6	12,8	6,9	149,4	14,3	83,1	35,2	45,9
5—6	5,8	6,3	19,5	25,2	27,2	15,5	19,9	7,6	22,8	47,1	11,8	7,1	158,4	19,2	71,9	43,0	81,7
6—7	3,3	9,9	21,5	35,9	33,1	15,3	22,9	1,4	33,3	21,8	13,5	5,6	177,3	18,8	90,5	39,6	71,6
7—8	4,7	8,4	21,0	24,5	39,7	18,9	14,2	3,2	13,0	22,9	11,4	4,8	159,2	17,9	85,2	36,3	47,3
8—9	3,0	11,6	15,8	33,8	37,0	15,9	25,6	4,7	9,5	23,4	12,1	8,0	163,7	22,6	86,6	46,2	45,0
9—10	2,2	9,7	23,9	59,9	22,2	18,6	17,1	4,6	20,8	14,0	11,4	8,4	178,1	20,3	106,0	40,3	49,2
10—11	3,4	8,4	10,6	48,3	9,1	48,1	16,7	17,3	16,6	13,0	12,4	7,9	168,3	19,7	68,0	82,4	42,0
11—N.	6,4	7,2	10,1	48,3	14,1	25,0	32,1	28,1	18,7	26,2	9,9	15,6	200,2	29,2	72,5	85,2	54,8
Noon—1	8,7	11,4	9,8	54,5	18,6	38,6	42,3	17,1	8,3	21,6	8,2	26,8	211,3	46,9	82,9	98,0	38,1
1—2	7,3	10,6	12,9	34,0	15,5	45,2	41,8	30,3	12,0	19,1	9,1	26,2	228,2	44,1	62,4	117,3	40,2
2—3	7,4	8,1	21,5	36,3	17,8	19,8	29,4	39,7	11,8	20,8	13,2	26,3	214,6	41,8	75,6	88,9	45,8
3—4	10,2	7,4	15,3	38,5	20,9	17,7	21,6	12,5	23,9	20,1	13,3	25,9	184,1	43,5	74,7	51,8	57,3
4—5	11,0	9,2	11,1	39,1	27,8	12,8	39,8	14,7	15,4	18,6	23,9	25,6	200,5	45,8	78,0	67,3	57,9
5—6	7,7	13,9	10,9	30,4	25,2	20,9	22,0	15,5	20,3	27,1	9,5	18,5	184,3	40,1	66,5	58,4	56,9
6—7	5,0	11,4	8,3	20,0	26,7	21,1	32,8	16,8	41,7	24,7	9,4	10,1	194,9	26,5	55,0	70,7	75,8
7—8	3,4	16,0	12,3	20,5	23,4	46,8	16,0	32,1	54,6	23,0	10,3	12,0	231,0	31,4	56,2	94,9	87,9
8—9	1,7	14,7	8,0	28,8	20,7	41,0	16,1	28,3	31,7	11,5	16,6	13,2	166,2	29,6	57,5	85,4	59,8
9—10	3,1	14,7	12,4	51,9	13,2	25,7	20,4	29,0	11,8	16,3	13,3	19,9	183,1	37,7	77,5	75,1	41,4
10—11	4,1	11,9	8,9	34,0	8,3	19,0	23,7	27,7	20,5	23,4	9,3	14,9	167,3	30,9	51,2	70,4	53,0
11—M.N.	3,2	9,0	13,1	18,7	7,7	16,3	29,5	15,8	31,0	16,0	15,0	11,0	134,8	23,2	39,5	61,6	62,0
Sum	109,7	228,4	358,0	793,1	526,9	513,4	611,2	404,4	579,8	465,6	292,0	321,6	4200,4	659,1	1684,0	1559,0	1337,4

WAKAYAMA.

Month / Hour	Jan.	Feb.	Mar.	Apr.	May	June	July	Aug.	Sept.	Oct.	Nov.	Dec.	Year	Wint.	Spr.	Sum.	Aut.
M.N.—1	0,4	2,0	4,5	5,1	3,8	3,3	6,1	8,1	11,5	2,3	3,5	2,1	53,4	4,6	13,5	17,9	17,1
1—2	0,6	0,8	5,0	4,6	4,9	6,1	5,1	2,3	6,8	4,3	2,4	2,1	45,2	3,5	14,5	13,6	13,6
2—3	0,9	1,4	5,2	7,5	5,3	5,2	7,8	1,1	11,2	4,0	2,8	2,1	57,6	4,5	18,0	14,1	21,0
3—4	0,6	1,6	5,9	8,8	6,2	4,6	5,8	0,7	8,2	9,0	6,3	2,6	60,5	4,9	20,9	11,2	23,6
4—5	0,5	1,5	3,0	6,4	5,7	3,2	9,1	1,9	9,0	8,6	3,6	4,0	56,7	6,0	15,2	14,3	21,2
5—6	0,4	1,2	3,3	4,3	4,6	2,7	8,5	1,0	13,9	9,6	1,2	1,8	52,6	3,5	12,2	12,3	24,7
6—7	1,1	4,3	7,2	6,1	9,3	3,8	12,7	2,2	4,5	10,7	2,4	3,5	67,9	8,9	22,9	18,8	17,3
7—8	0,8	3,3	5,1	8,1	10,3	2,1	11,3	2,9	8,3	6,5	5,5	2,8	67,0	6,9	23,5	16,4	20,3
8—9	0,7	5,9	2,3	12,1	7,0	2,5	8,8	3,8	13,2	8,0	4,9	1,7	70,9	8,4	21,4	15,1	26,1
9—10	1,9	3,5	3,2	10,3	5,2	2,7	8,8	4,2	13,1	5,9	5,3	3,5	67,7	8,9	18,7	15,8	21,2
10—11	2,5	2,3	4,0	7,2	3,5	2,2	12,7	12,4	4,3	5,0	5,9	3,3	65,3	8,1	14,7	27,3	15,2
11—N.	1,8	2,0	5,4	16,0	2,9	6,8	6,8	7,2	6,6	5,1	1,3	7,8	69,8	11,6	24,3	20,8	13,0
Noon—1	2,6	2,2	5,1	9,6	4,8	6,5	8,6	18,5	5,0	3,5	4,0	7,3	77,8	12,1	19,5	33,6	12,5
1—2	2,0	2,2	6,4	5,7	4,4	7,1	20,3	5,1	10,8	1,7	1,8	5,6	73,3	9,9	16,5	32,6	14,4
2—3	4,0	3,4	5,8	7,1	3,9	15,1	14,5	5,7	10,6	3,5	5,0	6,3	84,9	13,7	16,8	35,3	19,4
3—4	3,3	1,7	6,7	6,1	3,3	5,5	6,6	2,9	6,9	3,6	3,4	11,9	62,0	16,9	16,2	15,1	13,9
4—5	2,4	1,8	4,4	7,6	5,7	3,3	13,1	6,4	7,2	2,7	2,4	12,4	69,5	16,6	17,7	22,7	12,3
5—6	2,0	2,8	3,6	6,5	8,4	6,0	8,8	8,6	11,1	5,6	2,0	4,2	69,5	8,9	18,5	23,4	18,7
6—7	1,1	8,0	3,9	4,4	5,5	6,3	5,3	11,3	12,5	4,3	2,8	6,0	71,1	15,0	13,8	22,9	19,6
7—8	0,6	5,1	1,9	10,0	3,6	6,0	15,3	6,8	8,2	5,2	3,4	4,0	73,3	9,7	18,6	28,1	16,8
8—9	0,3	3,0	4,6	2,9	4,3	11,3	4,6	7,6	9,2	4,9	1,7	3,5	58,2	6,9	11,9	23,5	15,9
9—10	0,5	3,2	4,7	7,0	3,9	6,0	6,3	6,0	8,3	8,4	3,0	4,0	61,4	7,7	15,7	18,3	19,7
10—11	0,2	3,3	3,6	14,2	2,8	3,3	7,3	13,5	17,0	7,0	1,1	5,3	78,8	8,8	20,6	24,2	25,1
11—M.N.	0,2	2,8	3,7	10,4	2,5	4,3	5,8	6,9	12,9	3,7	2,8	2,6	58,7	5,6	16,6	17,0	19,5
Sum	31,8	69,6	111,6	188,5	122,1	126,3	220,3	147,5	233,3	133,2	78,5	110,6	1573,1	212,0	422,2	494,2	445,0

NAGNO.

Hour \ Month	Jan.	Feb.	Mar.	Apr.	May	June	July	Aug.	Sept.	Oct.	Nov.	Dec.	Year	Wint.	Spr.	Sum.	Aut.
M.N.—1	3,7	1,1	4,0	8,6	2,6	6,1	1,3	7,6	10,3	7,5	2,8	5,9	61,8	3,3	5,1	5,0	6,9
1—2	8,1	1,9	11,5	9,9	13,9	37,7	17,8	15,0	16,8	13,6	12,1	23,8	182,4	11,1	11,8	23,5	14,2
2—3	3,9	0,7	3,9	14,1	1,4	6,2	1,6	5,8	1,6	7,0	1,2	2,9	55,4	2,5	7,2	5,5	2,3
3—4	1,7	0,8	2,8	18,4	1,6	7,1	10,6	4,8	4,9	4,9	2,0	2,7	59,3	1,7	7,6	7,5	2,9
4—5	1,5	1,5	1,9	19,7	0,6	8,4	14,0	5,8	6,0	4,6	3,8	3,8	71,6	2,3	7,4	9,4	4,8
5—6	9,2	11,1	18,8	15,3	4,2	26,0	38,0	7,3	11,4	5,3	10,1	17,7	174,4	12,7	12,8	23,8	8,9
6—7	3,3	8,7	5,0	12,9	3,6	11,1	21,6	4,8	0,9	2,8	8,7	7,8	91,2	6,6	7,2	12,5	4,1
7—8	2,9	2,9	3,3	8,6	5,9	7,6	23,8	1,2	0,5	5,6	7,8	8,1	78,5	4,7	5,9	10,9	4,6
8—9	3,3	3,1	3,6	7,1	5,7	5,9	11,4	2,5	4,7	2,5	8,1	8,8	64,0	5,1	5,6	6,6	4,1
9—10	9,3	11,1	15,6	6,0	16,9	11,2	37,0	7,0	16,5	4,0	19,3	8,6	162,5	9,7	12,8	18,4	13,3
10—11	1,3	1,3	2,4	7,1	17,6	4,3	4,5	0,0	0,7	6,0	1,7	3,5	50,7	2,0	9,4	2,9	2,8
11—N.	0,8	2,8	1,7	8,2	11,3	6,8	6,9	0,4	3,9	5,7	4,7	4,4	57,6	2,7	7,1	4,7	4,8
Noon—1	1,3	3,4	5,6	10,8	6,6	3,9	6,7	1,0	11,8	4,3	6,7	1,9	64,0	2,2	7,7	3,9	7,6
1—2	10,8	9,7	19,4	19,5	2,6	15,9	67,4	29,3	11,1	9,0	20,4	2,2	217,3	7,6	13,8	37,5	23,5
2—3	3,7	3,3	9,7	6,1	2,0	4,8	9,7	4,7	5,7	17,6	8,3	1,7	77,9	2,9	6,1	6,4	10,5
3—4	2,6	4,7	6,1	10,8	3,9	7,6	9,2	9,0	9,5	13,9	5,8	3,3	86,4	3,5	6,9	8,6	9,7
4—5	4,0	2,4	3,6	6,8	3,7	4,7	34,9	12,5	11,6	15,4	3,3	5,6	108,5	4,0	4,7	17,4	10,1
5—6	19,6	7,4	10,6	22,1	28,9	18,4	79,7	37,9	18,8	15,5	13,4	11,9	283,9	10,6	20,5	45,3	15,8
6—7	4,1	3,4	3,2	10,4	2,5	17,6	15,0	31,0	7,6	10,4	1,9	5,7	112,8	4,4	5,4	31,2	66,3
7—8	2,8	4,3	5,2	8,7	8,4	12,4	16,1	27,1	3,8	8,2	5,0	4,1	106,4	3,8	7,4	18,5	5,7
8—9	2,7	5,0	4,2	6,5	2,0	13,8	29,5	9,3	4,9	14,6	3,9	3,5	96,9	3,7	4,2	17,5	6,8
9—10	26,5	16,1	14,4	22,2	8,3	38,3	46,8	24,6	13,9	5,7	11,5	22,2	250,5	21,6	15,0	36,6	10,4
10—11	3,5	3,9	3,1	15,4	8,5	3,1	6,7	6,2	9,1	6,0	2,9	3,4	71,8	3,6	9,0	5,3	6,0
11—M.N.	4,1	3,3	6,0	20,5	7,5	3,6	1,0	10,7	5,5	9,4	3,0	4,9	82,5	4,1	11,3	6,1	6,0
Sum	135,0	114,2	164,7	296,3	173,8	282,5	520,2	262,5	212,5	199,5	168,1	169,0	2398,3	139,4	211,6	355,6	193,1

TOKIO.

M.N.—1	1,2	1,6	3,6	4,6	7,1	5,7	7,6	3,2	10,2	7,9	5,4	2,5	5,1	5,1	15,7	16,5	23,5
1—2	1,3	3,5	5,7	5,5	9,3	4,6	6,5	5,7	10,7	7,9	4,3	3,8	5,7	8,6	20,4	16,8	22,9
2—3	1,2	2,4	5,1	5,6	8,3	6,8	6,6	6,0	13,9	8,6	3,8	1,4	5,8	4,8	19,0	19,3	26,4
3—4	1,2	1,9	4,7	5,7	8,5	6,2	5,7	4,3	17,9	8,1	3,9	0,9	5,7	4,0	18,9	16,4	20,9
4—5	2,3	1,6	6,7	6,5	9,8	5,5	7,0	5,0	11,4	6,4	1,1	2,2	5,7	6,1	23,0	17,5	21,6
5—6	2,4	2,4	6,1	3,5	11,8	6,8	7,9	4,0	9,1	8,3	4,6	2,1	5,8	6,9	21,4	18,7	22,0
6—7	1,6	2,1	5,4	2,9	10,5	6,7	5,6	3,5	6,8	6,1	4,3	0,5	4,7	4,2	18,8	15,8	17,3
7—8	1,6	1,7	6,1	3,6	9,6	9,7	4,7	4,8	7,0	6,0	3,2	0,8	4,9	4,1	19,6	19,2	16,2
8—9	2,0	1,8	9,9	4,1	9,9	6,5	3,6	2,4	8,1	7,3	4,2	1,3	5,1	5,1	23,8	12,5	19,6
9—10	2,1	1,5	8,5	2,9	9,4	4,6	4,0	2,3	8,1	8,2	2,6	0,9	4,6	4,5	20,8	11,0	18,9
10—11	1,7	2,1	5,9	1,4	7,0	4,6	3,4	10,6	5,0	9,3	5,4	1,0	4,8	4,8	14,3	18,7	19,7
11—N.	2,5	2,5	5,2	2,4	7,2	4,0	1,7	5,8	8,1	7,1	6,2	0,6	4,4	5,6	14,8	11,4	21,1
Noon—1	2,0	3,2	4,9	3,9	8,8	5,8	2,1	6,7	7,9	5,3	8,3	0,4	4,9	5,6	17,6	14,6	21,5
1—2	1,8	3,4	4,9	3,8	7,6	5,8	1,9	3,3	5,6	3,6	7,5	0,8	4,2	6,0	16,3	11,0	16,8
2—3	1,4	3,2	4,3	3,6	8,1	10,4	3,3	1,5	6,1	7,3	5,8	1,2	4,9	5,7	16,0	18,3	19,2
3—4	1,3	3,4	4,7	3,9	5,7	11,2	4,9	8,5	7,9	7,4	7,9	1,8	5,7	6,5	14,3	21,6	22,9
4—5	2,1	2,2	5,0	5,6	7,0	8,5	4,6	3,0	6,3	6,0	7,7	1,3	4,9	5,7	17,6	16,1	20,0
5—6	1,6	2,1	4,6	7,0	6,7	9,9	5,5	3,4	6,2	7,4	6,8	2,7	5,3	6,4	18,3	18,8	20,4
6—7	1,4	2,8	4,7	7,1	8,3	8,3	4,8	3,1	5,9	6,7	5,1	2,2	5,0	6,3	20,0	16,2	18,1
7—8	1,8	3,1	4,5	6,5	6,8	7,6	3,3	6,1	7,3	5,3	4,2	2,5	4,9	7,4	17,7	17,0	16,7
8—9	1,4	1,5	4,2	6,2	5,6	6,0	3,9	2,2	7,0	5,4	3,8	1,4	4,0	4,4	15,9	12,0	16,3
9—10	1,7	1,7	3,9	6,4	6,8	6,9	7,0	4,7	6,2	7,5	6,9	2,1	5,1	5,5	17,1	18,5	20,6
10—11	1,4	1,6	3,0	5,6	5,7	3,9	6,7	2,3	5,7	8,7	5,8	2,0	4,4	4,9	14,4	12,9	20,3
11—M.N.	1,0	1,5	3,9	5,4	5,5	6,0	6,4	3,0	7,1	9,2	5,8	2,1	4,7	4,6	14,7	15,4	22,1
Sum	40,1	54,6	125,7	113,8	191,5	162,0	118,7	108,3	195,1	170,8	128,1	38,5	120,6	133,2	430,6	389,0	494,0

HAKODATE.

Hour \ Month	Jan.	Feb.	Mar.	Apr.	May	June	July	Aug.	Sept.	Oct.	Nov.	Dec.	Year	Wint.	Spr.	Sum.	Aut.
M.N.—1	2,9	1,4	2,6	5,3	0,4	4,5	12,9	3,8	7,4	8,7	2,2	2,5	54,6	6,8	8,3	21,2	18,3
1—2	1,7	1,1	3,9	5,7	0,3	2,7	7,8	1,7	4,9	5,4	3,7	3,8	42,7	6,6	9,9	12,2	14,0
2—3	1,6	2,8	5,3	6,5	0,1	1,3	9,2	6,0	10,4	6,6	4,1	3,8	57,7	8,2	11,9	16,5	21,1
3—4	1,3	2,2	7,7	6,2	0,7	0,5	3,3	4,8	9,8	2,7	5,9	3,7	48,8	7,2	14,6	8,6	18,4
4—5	1,3	1,8	4,7	5,3	0,5	1,9	3,3	4,3	8,2	5,1	4,8	4,0	45,2	7,1	10,5	9,5	18,1
5—6	1,8	2,1	3,7	4,0	0,8	2,5	3,6	7,0	5,8	6,4	2,5	3,3	43,5	7,2	8,5	13,1	14,7
6—7	2,1	3,0	2,0	7,2	1,3	2,1	5,5	8,1	11,4	5,9	2,7	2,6	54,8	7,7	11,4	15,7	20,0
7—8	6,2	3,4	6,5	5,2	1,0	3,4	8,5	5,8	7,1	4,9	3,8	4,7	60,5	14,3	12,7	17,7	15,8
8—9	3,4	2,4	6,3	5,5	1,2	4,6	6,8	3,8	6,7	10,7	3,4	4,1	58,9	9,9	13,0	15,2	20,8
9—10	2,6	3,1	4,5	6,0	0,6	3,9	5,4	3,3	12,0	4,8	4,1	3,2	53,5	8,9	11,1	12,6	20,9
10—11	1,1	1,3	2,8	5,5	0,4	4,0	6,5	3,4	9,5	8,4	2,4	3,4	48,7	5,8	8,7	13,9	20,3
11—N.	0,9	1,2	3,6	5,7	2,1	3,6	3,4	4,3	5,3	7,2	3,1	3,8	44,2	5,9	11,4	11,3	15,6
Noon—1	1,1	1,9	0,7	6,8	2,1	4,3	5,1	3,4	8,1	5,4	3,5	5,3	47,7	8,3	9,6	12,8	17,0
1—2	1,8	1,8	0,8	7,0	3,7	2,0	1,2	3,7	8,7	3,1	5,7	4,2	43,7	7,8	11,5	6,9	17,5
2—3	1,8	2,1	1,5	8,0	6,4	3,4	3,1	4,9	13,0	3,9	3,2	5,4	56,4	9,3	15,9	11,1	20,1
3—4	0,2	1,6	2,0	7,3	2,7	4,1	3,6	3,6	7,1	6,2	5,8	3,6	47,8	5,4	12,0	11,3	19,1
4—5	0,3	0,4	2,8	6,9	3,1	3,8	3,6	3,4	5,9	6,3	3,9	4,8	45,2	5,5	12,8	10,8	16,1
5—6	0,2	1,3	2,6	5,6	2,0	2,9	2,3	2,5	7,2	6,4	5,7	6,4	45,1	7,9	10,2	7,7	19,3
6—7	1,0	0,6	2,9	4,1	1,2	4,7	7,7	2,5	9,0	6,7	6,3	5,7	52,4	7,3	8,2	14,9	22,0
7—8	0,4	1,2	3,3	4,1	1,0	6,3	3,7	2,4	7,2	5,5	5,9	4,6	45,6	6,2	8,4	12,4	18,6
8—9	0,6	1,0	4,1	3,5	1,5	6,7	5,5	2,3	7,7	5,8	6,5	3,7	48,9	5,3	9,1	14,5	20,0
9—10	1,0	1,7	5,0	3,6	1,9	5,8	16,5	6,9	13,2	5,7	5,4	4,1	70,8	6,8	10,5	29,2	21,3
10—11	2,0	2,7	2,8	4,4	1,3	5,7	15,1	2,1	6,5	3,4	2,5	3,2	51,7	7,9	8,5	22,9	12,4
11—M.N.	6,0	2,5	3,7	5,8	1,2	7,2	9,5	1,3	9,0	5,1	1,5	3,3	56,1	11,8	10,7	18,0	15,6
Sum	43,3	44,6	86,7	135,2	37,5	91,6	153,1	95,3	201,1	140,3	98,6	97,2	1224,5	185,1	259,4	340,0	440,0

SAPPORO.

Hour \ Month	Jan.	Feb.	Mar.	Apr.	May	June	July	Aug.	Sept.	Oct.	Nov.	Dec.	Year	Wint.	Spr.	Sum.	Aut.
M.N.—1	1,9	4,3	1,8	3,3	1,6	2,0	5,7	5,7	11,4	1,7	3,6	3,7	46,7	9,9	6,7	13,4	16,7
1—2	1,4	4,5	2,8	3,9	1,2	2,1	4,7	3,7	6,6	5,6	3,7	4,2	44,4	10,1	7,9	10,5	15,9
2—3	2,4	3,6	2,4	4,0	2,2	2,6	3,6	2,4	9,4	2,5	4,5	4,5	44,1	10,5	8,6	8,6	16,4
3—4	3,1	5,2	2,8	3,8	1,9	2,6	4,3	2,2	7,7	3,6	4,8	5,3	47,3	13,6	8,5	9,1	16,1
4—5	2,4	4,4	2,9	3,6	1,9	2,5	5,0	3,4	7,6	4,2	3,5	5,7	47,1	12,5	8,4	10,9	15,3
5—6	2,4	2,4	4,5	3,9	2,0	2,4	5,0	3,9	10,7	4,6	3,6	4,4	49,8	9,2	10,4	11,3	18,9
6—7	2,1	2,5	5,0	4,1	2,8	2,0	3,5	10,8	10,8	4,1	2,7	6,0	56,4	10,6	11,9	16,3	17,6
7—8	2,5	4,9	4,8	4,6	1,6	2,6	3,4	5,9	8,3	2,8	1,2	2,8	45,4	10,2	11,0	11,9	12,3
8—9	4,5	3,1	4,0	4,3	0,8	2,8	5,1	4,9	4,5	6,5	2,1	3,8	46,4	11,4	9,1	12,8	13,1
9—10	2,0	1,5	3,7	5,9	1,0	4,6	4,3	7,6	3,3	5,1	3,0	3,3	45,3	6,8	10,6	16,5	11,4
10—11	5,2	2,2	0,7	4,7	0,8	4,0	6,1	2,3	3,4	4,8	1,6	3,3	39,1	10,7	6,2	12,4	9,8
11—N.	5,9	1,3	1,2	4,6	1,9	4,0	8,4	3,0	6,8	6,7	2,4	3,3	49,5	10,5	7,7	15,4	15,9
Noon—1	2,2	1,6	2,2	3,2	2,5	2,7	3,7	8,4	13,8	3,4	2,2	5,2	51,1	9,0	7,9	11,8	19,4
1—2	1,0	2,2	1,1	4,6	0,4	3,2	7,3	6,2	11,2	6,9	7,6	5,7	57,1	8,9	6,1	16,7	25,7
2—3	1,1	3,0	1,4	4,4	0,9	2,3	3,3	4,2	9,0	7,1	6,4	6,8	49,9	10,9	6,7	9,8	22,5
3—4	1,6	3,5	1,5	4,6	1,3	3,0	2,8	6,3	6,0	4,5	2,6	8,0	45,7	13,1	7,4	12,1	13,1
4—5	2,4	4,0	1,7	4,2	1,3	4,0	3,0	3,0	3,3	2,3	6,1	5,5	40,8	11,9	7,2	10,0	11,7
5—6	1,5	3,0	2,8	3,4	1,2	2,6	2,9	5,2	9,7	5,9	5,8	4,1	48,1	8,6	7,4	10,7	21,4
6—7	1,6	4,0	1,5	3,2	0,6	2,9	6,1	3,5	9,5	6,7	2,9	3,8	46,3	9,4	5,3	12,5	19,1
7—8	1,1	5,0	1,5	3,1	0,6	3,8	5,6	3,5	4,9	5,3	3,8	2,6	41,1	8,7	5,5	12,9	14,0
8—9	1,2	2,1	1,6	2,4	1,3	3,9	5,9	2,8	3,3	2,7	3,9	3,3	34,7	6,9	5,3	12,6	9,9
9—10	2,2	6,3	2,4	3,4	2,1	4,2	5,8	6,5	4,5	3,1	3,9	3,6	48,0	12,1	7,9	16,5	11,5
10—11	3,2	5,8	2,9	2,6	2,6	2,2	5,1	5,8	6,0	2,7	3,6	4,1	46,6	13,1	8,1	13,1	12,3
11—M.N.	1,7	4,9	2,7	2,7	2,8	2,3	5,2	6,7	5,8	2,4	5,4	4,3	46,9	10,9	8,2	14,2	13,6
Sum	56,6	85,6	59,9	92,8	37,3	71,3	115,8	117,9	177,5	105,2	90,9	107,3	1118,1	249,5	190,0	305,0	373,6

NEMURO.

Month / Hour	Jan.	Feb.	Mar.	Apr.	May	June	July	Aug.	Sept.	Oct.	Nov.	Dec.	Year	Wint.	Spr.	Sum.	Aut.
M.N.—1	0,5	0,6	0,7	6,8	3,7	3,9	3,5	12,1	13,0	3,2	12,8	5,6	66,4	6,7	11,2	19,5	29,0
1—2	0,0	0,5	0,9	7,1	1,2	2,7	4,2	1,7	8,9	3,7	11,8	5,9	57,9	6,4	12,5	11,6	27,1
2—3	0,0	0,9	1,5	5,9	3,9	3,4	7,9	3,5	7,1	7,6	7,1	7,3	56,1	8,2	11,3	14,8	21,8
3—4	0,0	0,1	1,1	1,7	9,0	4,7	8,0	1,5	10,7	8,6	3,2	6,4	58,3	6,8	11,8	14,2	22,5
4—5	0,1	0,5	2,2	1,3	7,4	8,1	4,7	0,4	7,3	4,9	2,5	6,3	48,7	6,9	13,9	13,2	14,7
5—6	0,4	0,2	1,7	5,6	1,8	8,3	3,8	0,3	8,1	6,3	1,6	7,3	18,4	7,9	12,1	12,4	16,0
6—7	0,2	0,5	2,4	6,4	2,9	7,6	3,4	0,1	8,7	1,8	0,5	2,4	40,2	3,1	11,7	11,4	14,0
7—8	0,7	0,8	1,9	5,7	3,7	4,8	3,2	0,0	11,0	2,4	0,9	2,3	40,4	3,8	11,3	8,0	17,3
8—9	1,1	1,6	1,1	1,7	3,7	3,5	2,2	0,2	18,1	1,9	0,8	2,6	11,5	5,3	9,5	5,9	20,8
9—10	3,2	0,8	0,9	3,0	3,9	0,9	1,1	0,1	12,4	3,1	1,4	5,2	36,3	9,2	7,8	2,1	17,2
10—11	0,8	0,2	1,7	4,0	1,3	0,6	2,0	0,0	10,9	6,1	1,6	2,7	31,9	3,7	7,0	2,6	18,6
11—N.	0,6	0,2	1,7	2,3	0,7	1,8	2,3	2,9	3,2	5,2	2,0	3,5	26,4	1,3	4,7	7,0	10,4
Noon—1	1,9	0,5	2,6	2,1	1,2	1,3	2,3	3,4	2,0	7,1	1,9	1,8	28,1	4,2	5,9	7,0	11,0
1—2	0,9	0,6	2,5	3,7	0,5	1,6	3,9	1,6	3,3	6,2	2,8	4,4	35,0	5,9	6,7	10,1	12,3
2—3	2,0	0,9	3,0	5,5	1,0	1,4	4,0	7,1	1,0	5,0	3,8	5,2	42,9	8,1	9,5	12,5	12,8
3—4	1,5	1,2	3,4	5,3	1,2	2,6	4,2	2,6	2,4	4,3	2,3	4,6	35,6	7,3	9,9	9,4	9,0
4—5	1,1	2,2	2,4	4,0	1,6	5,5	5,8	2,4	4,8	3,3	1,7	6,1	40,9	9,4	8,0	13,7	9,8
5—6	0,4	1,1	0,5	2,7	3,3	1,8	3,9	7,1	6,7	6,5	3,3	10,4	48,0	12,2	6,5	12,8	16,5
6—7	0,3	1,1	1,3	4,2	2,7	0,9	4,0	2,9	6,3	3,6	4,5	7,3	39,1	8,7	8,2	7,8	11,1
7—8	0,4	0,6	1,0	3,6	4,1	1,0	2,1	1,1	4,8	4,5	3,5	6,5	33,5	7,5	8,7	4,5	12,8
8—9	0,3	0,9	0,6	5,3	3,3	0,4	1,5	2,9	7,4	3,2	3,8	5,6	35,2	6,8	9,2	4,8	11,1
9—10	1,1	1,3	1,5	7,0	6,1	3,9	2,5	5,1	12,6	3,2	2,4	11,6	58,3	11,0	14,6	11,5	18,2
10—11	0,3	1,3	1,2	6,0	5,2	5,3	1,5	7,3	11,7	2,3	2,7	5,9	53,7	7,5	12,4	11,1	19,7
11—M.N.	0,7	0,8	0,5	8,0	3,8	4,2	3,5	8,5	14,1	3,7	10,0	7,5	65,3	9,0	12,3	16,2	27,8
Sum	18,5	20,0	38,3	118,2	83,2	80,2	85,8	81,1	205,5	111,0	91,9	131,4	1068,1	172,9	230,7	247,1	408,4

As is evident, the phenomena of precipitation is very irregular, sometimes we have no precipitation at all for ten or more days, while there are abundant precipitation in a space of a few hours. Hence even the annual amounts differ considerably from one another. It is beyond question then that the hourly amount shall be subject to still greater fluctuations. Hence we can not deduce the true feature of its diurnal variation from the observations continued for a few years only. The great discrepancies in the above table are probably due to this irregularity. Notwithstanding this apparent irregularity, however, we observe some general law in this variation. Thus according to seven years' observations in Tokio, there are two maximum and minimum precipitations in a day. The first maximum is at 2 or 4 o'clock am and the second maximum at 4—6 o'clock pm. The first minimum occurs at 7—10 o'clock am and the second minimum at 8—11 o'clock pm

The first maximum is generally greater than the second throughout the year, excepting summer when the maximum in the afternoon is far greater than that of the forenoon. In the two seasons of summer and autumn, the chief minimum is in day, but in spring it is at night. Both the maxima and the minima in winter are not so marked as in other seasons.

Since the precipitation is subject to great disturbances quite accidental, the observed quantity at any time can not be taken directly as representing the true amount of precipitation at that moment. Hence in the following table, we have added together the observed quantity at any time with those of the preceding and the succeeding hours, and its mean is taken as the actual precipitation at that instant. By this means, we believe, accidental disturbances are more or less eliminated, and the true manner of the diurnal variation may be found.

Hour	Winter	Spring	Summer	Autumn	Year
1 am	6.2	16.9	16.3	22.9	62.2
2 am	6.3	18.1	17.6	21.3	66.5
3 am	5.8	19.5	17.1	26.6	69.1
4 am	5.0	20.3	17.6	25.9	68.9
5 am	5.7	21.1	17.4	21.5	68.7
6 am	5.8	21.1	17.3	20.3	64.4
7 am	5.1	20.0	17.9	18.5	61.4
8 am	4.5	20.7	15.8	17.7	58.7
9 am	4.6	21.4	14.2	18.2	58.5
10 am	4.8	19.7	14.0	19.4	57.9
11 am	5.0	16.6	13.7	20.0	55.2
Noon	5.3	15.6	14.9	20.8	56.6
1 pm	5.8	16.2	12.3	19.9	54.2
2 pm	5.8	16.6	11.6	19.1	56.2
3 pm	6.1	15.5	18.0	19.6	59.2
4 pm	6.0	16.0	19.7	20.7	62.3
5 pm	6.2	16.7	19.9	21.1	63.8
6 pm	6.2	18.6	17.0	19.5	61.3
7 pm	6.7	18.7	17.3	18.4	61.1
8 pm	6.1	17.9	15.1	17.0	56.0
9 pm	5.8	16.9	15.9	17.8	56.3
10 pm	5.0	15.8	14.5	19.0	54.2
11 pm	5.0	15.4	15.6	21.0	57.0
M. N.	5.0	14.9	14.9	22.0	56.8

B. ANNUAL VARIATION OF PRECIPITATION.

The annual variation of precipitation differs considerably on the two sides of the mountain range running through along Nippon. On the back Nippon, or that side of Nippon facing the Sea of Japan, there is more precipitation in autumn and winter, and less in spring and summer. The maximum is in December and the minimum in May. But the amplitude is so small that there is no noticeable difference between the maximum and the minimum values. We may say, therefore, that on this side the precipitation remains constant through the year.

Contrary is the case in the front Nippon, facing the broad Pacific. Here, we have plentiful precipitation in summer and autumn, and scanty in winter and spring. The range is also considerable. We may divide this side into the following five districts.

(1) Northwestern part of Kiushu.—In this district, precipitation is least in the three winter months, and increases toward summer, reaching the maximum in June. The range is about 13 per cent of the total amount of the year.

(2) A zone of land facing the south sea extending from the southeastern part of Kiushu to Choshi and Mito, through Setouchi or Inland sea.—In this zone, the precipitation is least in December, January, and February, increases toward summer, and reaches to its maximum in June, as in the last division. After that month, however, we have sudden decrease in August, and again great increase in September. The range is about 11 per cent of the yearly amount.

(3) The eastern coast of Nippon. The variation in this district is similar to that in the last division, but that the precipitation in September far exceeds that in June, and that the secondary minimum is in July, instead of in August.

(4) Central Nippon.—The variation is similar to the first division, except that the maximum is not in June but in July.

(5) Hokkaido (excepting the coast of the Sea of Japan).—Here the minimum precipitation is in February, and the maximum in September. Moreover, the precipitation in autumn is remarkably great when compared with other seasons.

What we have mentioned, are some of the general conclusions; for the detail see the following table and the Plate IX.

Annual variation of precipitation.

PERCENTAGE OF YEARLY AMOUNT.

	Jan.	Feb.	March	April	May	June	July	Aug.	Sept.	Oct.	Nov.	Dec.
Back side	10.3	7.0	6.1	6.9	5.1	7.3	8.1	6.7	10.6	9.4	10.7	11.6
Front side in general	3.7	3.6	6.2	8.2	9.1	11.8	12.0	10.0	14.0	10.1	6.2	5.0
W coast of Kiushu	3.9	3.9	5.9	10.4	10.7	16.6	16.1	8.6	10.1	5.4	3.9	4.3
S coast	3.4	4.2	7.3	10.9	10.3	14.3	11.0	7.1	12.0	10.4	5.5	3.4
E coast	4.0	4.7	6.7	6.0	6.7	13.3	9.0	11.0	18.0	10.7	5.7	4.3
Central part	3.7	3.7	6.0	8.3	8.3	11.3	18.3	11.0	11.0	8.7	5.3	4.3
SE^{rn} coast of Hokkaido	3.6	2.0	5.0	7.7	11.0	8.5	9.5	11.0	15.0	10.5	8.5	8.0

Annual variation of precipitation at each station.

PERCENTAGE OF YEARLY AMOUNT.

Locality.	Jan.	Feb.	Mar.	Apr.	May	June	July	Aug.	Sept.	Oct.	Nov.	Dec.	Wint.	Spr.	Sum.	Aut.
Naha	1	7	8	21	19	5	9	11	9	2	5	3	12	47	21	17
Kagoshima	4	1	6	12	12	18	11	10	10	5	4	4	12	30	39	19
Miyazaki	3	3	6	10	11	16	8	8	17	11	4	3	9	27	32	32
Kochi	2	3	6	12	10	14	12	10	14	9	5	3	8	28	36	28
Wakayama	3	5	7	11	9	14	10	6	13	12	6	4	12	26	30	32
Oita	2	4	7	9	11	14	14	9	13	10	4	3	9	26	38	27
Yamaguchi	4	5	8	12	10	15	15	8	8	7	5	3	12	31	37	20
Hiroshima	3	1	7	13	12	16	15	5	11	7	4	3	10	32	36	22
Matsuyama	6	4	5	11	13	13	11	6	14	11	3	3	13	30	29	28
Okayama	4	4	7	11	9	16	9	9	8	12	9	2	10	27	34	29
Ozaka	3	4	8	13	10	16	9	5	11	11	6	4	11	30	31	28
Kioto	4	4	8	12	9	15	12	8	11	8	5	4	11	29	35	25
Kumamoto	3	5	6	13	14	11	14	7	8	8	3	5	14	32	35	19
Saga	3	3	5	9	10	22	19	9	9	4	3	4	9	25	50	16
Nagasaki	4	4	6	10	11	18	13	11	10	5	4	4	12	28	40	20
Fukuoka	5	4	6	7	8	14	25	7	9	6	5	4	12	22	46	20
Itsugahara	4	3	6	11	10	15	16	8	15	4	4	4	11	27	39	23
Akamagaseki	4	4	6	11	10	15	17	8	10	6	4	5	13	26	41	20
Sakai	10	7	7	8	5	9	11	5	11	9	9	9	26	21	24	29
Tsu	2	4	7	9	9	24	9	5	13	11	5	2	8	25	38	29
Nagoya	3	4	7	10	11	12	13	8	11	11	6	4	11	28	33	28
Gifu	4	8	7	12	11	12	14	8	11	9	5	4	12	29	34	25
Hamamatsu	3	4	9	13	11	13	10	6	12	9	6	4	10	33	30	27
Numazu	3	4	9	11	10	13	11	8	11	10	6	4	11	30	32	27
Tokio	1	5	8	8	10	12	9	7	14	13	7	3	12	26	28	34
Utsunomiya	2	3	5	1	7	11	25	16	14	7	3	3	8	16	53	23
Choshi	5	8	8	9	9	8	7	5	13	16	8	4	17	26	20	37
Kanazawa	11	7	7	8	5	7	8	5	7	8	12	15	33	20	20	27
Fushiki	13	7	7	7	5	7	8	5	8	7	11	15	35	19	21	25
Nagano	5	5	6	9	7	11	16	9	8	10	8	6	16	22	37	25
Niigata	11	6	6	7	4	7	9	6	10	9	12	13	30	17	22	31
Yamagata	9	1	7	4	6	10	9	16	15	7	6	7	19	17	36	28
Akita	8	6	6	8	7	8	9	9	10	9	10	10	24	20	26	30
Fukushima	4	5	6	3	6	16	9	11	22	10	4	4	12	15	37	36
Ishinomaki	4	5	7	8	8	12	8	12	17	10	5	4	13	23	32	32
Miyako	4	4	7	7	6	12	10	10	15	12	8	5	12	21	32	35
Aomori	10	8	6	5	6	7	8	9	12	9	10	10	28	16	25	31
Hakodate	6	3	5	7	7	9	9	11	17	9	10	7	16	19	29	36
Suttsu	9	8	4	5	6	6	6	8	14	15	10	9	26	15	20	39
Sapporo	7	5	5	6	5	6	7	11	17	11	10	10	23	16	23	38
Kamikawa	8	3	4	5	9	11	8	9	14	12	9	8	20	17	28	35
Soya	6	4	3	8	9	7	7	8	16	19	12	6	16	15	23	46
Abashiri	7	3	4	5	7	8	11	10	17	13	8	7	17	16	29	38
Nemuro	2	1	5	8	11	9	9	9	17	11	10	8	11	24	27	38
Kushiro	3	2	6	8	14	8	10	11	14	8	7	9	13	28	30	29
Erimo	2	2	5	10	12	9	8	11	12	10	8	8	12	27	31	30

C. DISTRIBUTION OF PRECIPITATION OVER THE COUNTRY.

The monthly amounts of precipitations at various stations are :—

PRECIPITATION.

Locality.	Jan.	Feb.	Mar.	April	May	June	July	Aug.	Sept.	Oct.	Nov.	Dec.	Year	Wint.	Spr.	Sum.	Aut.
Naha	35.2	169.3	195.6	516.1	151.8	113.7	227.7	263.2	226.2	51.1	131.6	82.6	2170.1	287.1	1166.5	601.6	411.9
Kagoshima	79.5	86.2	131.5	243.0	242.1	368.0	211.7	207.9	208.1	111.8	83.9	79.5	2086.8	215.2	646.9	820.6	404.1
Miyazaki	79.9	85.2	148.8	265.8	294.8	415.7	225.6	223.9	440.1	291.9	109.1	74.8	2655.5	239.9	709.1	865.2	841.1
Kochi	59.9	85.0	177.3	319.1	283.3	393.1	330.4	278.7	102.2	254.5	131.5	93.6	2807.3	238.5	779.1	1002.2	787.2
Wakayama	44.3	68.9	92.4	144.9	123.5	196.2	130.4	81.3	177.6	166.2	87.8	57.2	1370.7	170.4	360.8	407.9	431.6
Oita	37.1	65.3	109.4	148.1	181.9	235.5	243.1	144.1	213.9	160.4	65.9	55.6	1660.9	158.0	439.7	623.0	440.2
Yamaguchi	74.7	96.2	159.0	239.9	195.6	287.3	289.0	146.2	156.2	127.1	94.8	68.3	1934.3	239.2	594.5	722.5	378.1
Hiroshima	43.5	63.0	100.7	197.2	182.3	245.8	217.8	83.0	166.1	102.5	67.2	44.6	1513.7	154.1	480.2	546.6	335.8
Matsuyama	68.2	46.9	58.4	138.0	158.6	150.3	134.7	68.7	169.0	135.0	38.6	10.7	1206.8	155.8	354.7	353.7	312.6
Okayama	39.1	37.8	71.0	111.3	90.1	162.2	90.9	91.8	83.0	116.2	93.2	20.7	1007.3	97.6	272.1	344.9	292.1
Ozaka	42.7	52.7	107.2	165.8	127.1	209.7	121.7	68.4	135.3	144.5	80.7	49.5	1305.6	144.9	400.1	399.8	360.5
Kioto	55.6	54.8	118.4	178.7	142.1	231.0	183.2	121.0	162.9	130.8	85.4	54.9	1524.5	165.3	438.9	541.2	379.1
Kumamoto	56.2	87.3	105.2	211.9	231.0	240.7	236.0	109.2	134.1	130.7	50.8	92.2	1685.6	235.7	458.1	585.9	315.9
Saga	55.4	57.8	115.5	202.1	213.2	469.4	424.7	196.7	194.6	91.4	70.4	81.9	2175.8	198.1	530.8	1090.8	356.1
Nagasaki	82.0	84.7	130.1	213.3	238.3	363.7	258.4	216.9	210.9	108.6	86.3	87.2	2080.1	253.9	581.7	839.0	405.8
Fukuoka	84.7	66.1	107.3	137.4	155.5	255.0	467.5	126.3	168.2	113.4	93.7	81.7	1856.8	232.5	400.2	848.8	375.3
Itsugahara	87.2	77.0	152.0	250.5	225.3	342.7	380.0	195.6	360.1	94.9	95.2	88.7	2319.5	252.9	627.8	918.3	505.5
Akamagaseki	64.9	66.6	103.8	176.9	154.3	266.3	298.1	123.2	164.3	98.1	70.8	81.6	1669.2	213.0	431.0	687.7	333.2
Sakai	185.6	124.2	135.5	150.0	97.0	161.7	191.5	100.6	205.0	163.0	169.5	167.2	1850.8	477.0	382.5	453.8	537.5
Tsu	29.2	66.3	121.0	151.8	168.1	120.0	154.4	94.9	226.7	192.9	95.1	16.7	1770.1	142.2	443.9	669.3	515.0
Nagoya	38.6	62.4	108.8	145.8	159.7	180.2	198.1	115.2	165.4	157.1	91.7	53.8	1177.1	154.8	114.3	493.8	411.5
Gifu	73.9	67.9	135.4	231.1	204.7	229.6	278.5	151.1	223.8	168.1	99.4	86.6	1952.8	228.1	570.0	662.2	491.1
Hamamatsu	50.7	66.9	164.6	232.8	117.2	166.9	113.6	44.7	211.9	158.9	109.2	63.3	1500.7	187.2	563.3	536.7	480.0
Numazu	65.8	70.1	153.9	202.8	182.1	226.9	201.6	150.5	201.4	178.4	103.2	70.7	1807.1	206.3	538.8	579.0	483.0
Tokio	54.2	78.5	121.8	116.1	148.4	177.0	125.5	109.7	207.8	188.0	107.9	50.1	1185.0	182.8	386.3	412.2	503.7
Utsunomiya	40.0	64.1	100.0	92.5	148.5	235.7	526.7	327.8	283.2	132.2	63.8	69.1	2083.6	173.2	341.0	1090.2	479.2
Choshi	82.3	122.4	129.7	135.3	140.2	133.4	114.4	76.8	201.6	216.2	133.9	69.6	1585.8	274.3	405.2	324.6	581.7
Kanazawa	290.3	174.7	169.8	193.9	131.4	186.2	197.7	123.9	176.4	206.5	296.9	371.5	2519.2	836.5	495.1	507.8	679.8
Fushiki	262.3	139.5	139.1	144.4	100.5	150.5	164.3	104.5	168.3	128.4	213.1	301.3	2018.8	706.1	384.0	418.9	509.8
Nagano	44.0	42.1	55.3	82.1	61.8	100.0	141.9	82.2	69.1	83.1	69.9	56.3	888.1	142.7	199.2	324.1	222.1
Niigata	178.3	108.1	101.7	124.3	72.5	124.7	149.1	108.0	167.1	156.4	212.3	220.2	1722.7	506.6	298.5	381.8	535.8
Yamagata	94.9	42.9	73.4	49.6	68.2	115.4	99.6	178.2	559.3	81.5	62.1	71.6	1090.7	212.4	191.2	393.2	302.9
Akita	134.6	97.8	100.8	137.8	113.7	132.3	148.7	162.8	181.5	155.8	180.6	182.0	1728.4	414.4	352.3	413.8	517.9
Fukushima	50.1	54.0	71.1	42.9	67.2	189.7	112.2	138.0	263.1	119.5	49.6	43.1	1200.5	147.2	181.2	430.9	432.2
Ishinomaki	41.7	59.1	79.1	85.3	83.0	128.8	91.1	130.4	183.7	109.2	51.8	43.1	1089.3	143.9	247.1	353.3	314.7
Miyako	47.3	52.5	94.6	93.2	81.1	154.5	135.2	130.7	198.4	153.8	100.9	60.9	1302.7	160.7	268.5	420.4	453.1
Aomori	118.5	96.5	68.6	65.5	69.1	87.1	103.0	117.3	146.4	111.7	128.0	128.6	1210.3	313.6	203.2	307.4	386.1
Hakodate	61.2	31.6	57.9	76.6	74.9	92.0	100.9	119.5	187.0	100.7	109.3	78.9	1093.5	174.7	209.4	312.4	397.0
Suttsu	107.2	95.8	55.4	65.7	70.5	69.8	70.2	106.0	179.4	181.9	127.7	119.2	1248.8	322.2	191.6	216.0	489.0
Sapporo	68.2	54.8	54.5	55.6	48.0	56.4	67.7	104.1	165.3	112.2	95.3	96.7	978.8	219.7	158.1	228.2	372.8
Kamikawa	101.8	36.6	45.1	59.3	109.0	140.3	93.8	105.6	171.4	116.3	107.1	102.1	1218.4	210.5	213.4	339.7	421.8
Soya	50.3	31.5	27.6	27.3	69.5	61.0	61.6	66.0	129.3	157.4	102.1	50.0	833.6	131.8	121.4	188.6	388.8
Abashiri	39.5	18.3	24.3	31.2	42.4	50.0	64.2	57.2	99.4	78.7	49.2	41.8	596.2	99.6	97.9	171.4	227.3
Nemuro	19.6	12.6	48.5	71.9	99.7	83.9	83.3	86.4	158.8	106.1	96.6	71.5	941.9	103.7	223.1	253.6	361.5
Kushiro	23.4	15.5	51.8	65.0	118.6	72.4	87.3	96.2	121.5	65.3	60.8	72.8	851.6	111.7	236.4	255.9	247.6
Erimo	20.3	15.8	51.6	90.2	107.0	80.4	79.3	127.3	115.3	95.6	70.8	76.2	929.8	112.3	248.8	287.0	281.7
Fusan	31.8	17.7	68.8	103.1	105.0	181.9	190.8	108.7	161.0	48.0	31.7	37.3	1091.9	89.8	276.9	484.4	240.7

The mean annual amount of precipitation is most plenty at the southeastern coast of the province of Kii, especially at Shingu, where it reaches to the great quantity of 3,100 mm; and most scanty on the northeastern part of Hokkaido, it is only 600 mm at Abashiri. Thus, in our country, the mean annual amount varies, on the average, from 3,100 mm to 600 mm according to different localities. Next to the southeastern coast of Kii, the great precipitations along the Pacific are at the southeastern coast of Kiushu and the southern coast of Shikoku; while the great precipitations along the Sea of Japan are at the two provinces, Kaga and Echizen. In all these places, the yearly amount exceeds 2,500 mm.

Among the regions of the least precipitation, the central part of Nippon, that is, Shinano, comes next to the northeastern part of Hokkaido, and then Setouchi or Inland sea follows both of them. In Shinano, the yearly amount is only 900 mm and in Setouchi only 1,000 mm. We see that these places of the least precipitation are surrounded completely by high mountains on all sides. These mountains intercept the damp wind coming from the Pacific or the Sea of Japan, and consequently the enclosed regions are kept dry. In short, these districts are, so to speak, the deserts of our country.

The amount of precipitation depends very much on the configuration of the land, so that slight differences in the configuration cause enormous change in the amount of precipitation. The form of the mountain range have powerful influences on regulating the amount of precipitation. Let us take an example. Kochi and Matsuyama in Shikoku are not very far separated from each other. They lie on the opposite sides of the Shikoku mountain range, the former on the southern side facing the Pacific, while the latter on the northern side facing the Inland Sea. Owing to this difference in their situations, there is a great difference of 1,600 mm in their precipitations. The distribution of precipitation being thus very irregular, general conclusions can not be drawn. But we may say that in our country, the precipitation decreases as we go from the southwestern extremity of this country toward the northeastern. The details are given in the Plate XV, drawn by the observations taken at all our meteorological stations and also referring to the reports issued from 150 raingauge stations distributed over the country.

D. HEAVY RAIN.

The heaviest rain that we ever observed in our country is that fallen on Tanabe in the province of Kii on the 20th of August, 1889. The precipitation in that single day reached to the enormous amount of 900 mm. The second heavy rain is that fallen on Yuasa in the same province on the same day. There the precipitation was 520 mm. A rain of 490 mm at Miyazaki, on the 24th of September, 1886, that of 350 mm at Naha on the 11th of April, 1891 and that of 345 mm at Nagasaki on the 10th of April, 1882, are the next heavy rains. The greatest quantities of rain fallen during a day, observed at several stations, since their establishments, are given in the following table:—

Maximum precipitation in 24 hours.

Locality.	in 24 h.	Day	Month	Year	Locality.	in 24 h.	Day	Month	Year
Naha	351,8	11	IV	1891	Numazu	193,3	29	VI	1885
Kagoshima	167,1	2	VII	1888	Tokio	162,2	15	IX	1878
Miyazaki	190,2	24	IX	1886	Utsunomiya	73,3	22	VIII	1890
Kochi	293,1	11	IX	1890	Choshi	146,9	5	X	1887
Tokushima	164.2	2	VIII	1891	Kanazawa	120.9	5	X	1890
Wakayama	184.2	5	X	1890	Fushiki	109,1	24	VII	1889
Oita	181,8	11	IX	1888	Nagano	69.0	20	VII	1891
Yamaguchi	173.9	24	VIII	1891	Niigata	134.5	17	VIII	1885
Hiroshima	149.8	23	VII	1889	Yamagata	87,3	30	IX	1891
Matsuyama	80,6	6	X	1890	Akita	91.5	6	VII	1890
Okayama	79.2	3	VI	1891	Fukushima	152.2	30	IX	1891
Ozaka	133,0	7	X	1887	Ishinomaki	151.5	7	X	1888
Kioto	127,0	19	VIII	1889	Miyako	172.5	7	X	1888
Kumamoto	132.2	28	XII	1890	Aomori	107.9	2	XI	1886
Saga	131,1	8	IX	1891	Hakodate	146,8	7	XI	1887
Nagasaki	345,1	11	IV	1882	Suttsu	80,5	6	VIII	1888
Fukuoka	208,7	21	VII	1891	Sapporo	137,8	15	IX	1879
Itsugahara	230,3	3	VII	1891	Kamikawa	75.2	21	VII	1890
Akamagaseki	158,1	18	VI	1889	Soya	51.4	21	X	1889
Sakai	200,8	24	IX	1886	Abashiri	48.2	12	IX	1889
Tsu	171.2	30	IX	1891	Nemuro	93.7	19	IX	1885
Nagoya	112.9	5	VIII	1891	Kushiro	48.4	21	VII	1890
Gifu	150.2	1	VII	1885	Erimo	119.2	21	VIII	1889
Hamamatsu	137.2	11	IX	1889					

The greatest quantities fallen in the course of 4 hours are 177 mm at Naha on the 11th of April 1891, 165 mm at Kochi and 164 mm at Tokushima. The maximum precipitations, during four hours for every station, are:—

MAXIMUM OF PRECIPITATION IN 4 HOURS.

Locality.	in 4 h.	Day	Month	Year	Locality.	in 4 h.	Day	Month	Year
Naha	176.9	11	IV	1891	Numazu	110.4	27	VIII	1886
Kagoshima	66.1	20	VII	1889	Tokio	56.6	15	V	1889
Miyazaki	155.0	21	IX	1886	Utsunomiya	43.4	16	VII	1891
Kochi	165.6	11	IX	1890	Choshi	79.9	29	VII	1888
Tokushima	164.2	2	VIII	1891	Kanazawa	45.7	10	VII	1888
Wakayama	80.6	31	VIII	1888	Fushiki	52.3	18	IX	1891
Oita	81.9	30	VII	1888	Nagano	35.0	30	IX	1891
Yamaguchi	66.3	30	VI	1891	Niigata	51.9	13	VIII	1891
Hiroshima	72.2	23	VII	1889	Yamagata	46.5	8	VIII	1890
Matsuyama	37.8	28	XII	1890	Akita	40.3	28	VIII	1887
Okayama	31.5	16	VIII	1891	Fukushima	92.1	30	IX	1891
Ozaka	63.7	26	IV	1889	Ishinomaki	58.4	7	X	1888
Kioto	71.5	21	VIII	1891	Miyako	80.9	11	IX	1889
Kumamoto	83.2	28	XII	1890	Aomori	71.6	1	X	1890
Saga	90.1	11	IX	1891	Hakodate	56.9	11	I	1891
Nagasaki	90.4	25	VI	1887	Suttsu	62.2	28	VI	1886
Fukuoka	112.3	21	VII	1891	Sapporo	43.7	27	VIII	1889
Itsugahara	123.1	17	VII	1889	Kamikawa	33.3	25	VI	1890
Akamagaseki	80.3	30	VI	1891	Soya	41.1	31	VII	1890
Sakai	78.4	21	IX	1886	Abashiri	33.5	29	III	1891
Tsu	106.1	30	IX	1891	Nemuro	50.8	12	IX	1889
Nagoya	74.9	5	VIII	1891	Kushiro	34.2	9	XII	1890
Gifu	90.2	29	VII	1888	Erimo	55.9	27	VIII	1886
Hamamatsu	75.0	28	VII	1886					

Again if we take the amount fallen in one hour, the maximum is 54 mm at Kumamoto and next to it come 43 mm at Nagoya and Hakodate, and next to them comes 40 mm at Wakayama. For the detail, see the following table:—

MAXIMUM OF PRECIPITATION IN 1 HOUR.

Locality.	in 1 h.	Day.	Month.	Year.
Kumamoto	54.0	28	XII	1890
Matsuyama	21.4	9	IX	1890
Hiroshima	33.6	1	IX	1888
Wakayama	40.1	30	VIII	1888
Ozaka	28.1	24	VI	1891
Nagoya	43.0	5	VIII	1891
Nagano	29.4	24	VIII	1891
Tokio	36.1	27	VIII	1886
Hakodate	43.0	14	I	1891
Sapporo	22.1	29	VIII	1891
Nemuro	21.7	11	IX	1889

We have said that the rainfall in Tanabe was 900 mm in twenty four hours. This is equivalent to the rain of 150 mm in four hours and 37 mm in one hour. Now according to the above table, such rain must be said to be very heavy. If we remember that generally the intensity of rain does not continue constant for a long time, it will be easily seen that this rain at Tanabe must have been very intense for some time.

If we compare the frequency of the heavy rain, and the mean amount of precipitation, we see that those places having plentiful precipitation generally can not be said to have frequent heavy rains. The number of days in which we had an amount of rain more than 100 mm are as follows:—

NUMBER OF DAYS WITH PRECIPITATION, IN WHICH THE AMOUNT WAS GREATER THAN 100 MM.

Locality.	Total.	Mean number of days in a year.	Locality.	Total.	Mean number of days in a year.
Naha	2	1.0	Numazu	9	1.0
Kagoshima	11	1.6	Tokio	9	0.6
Miyazaki	13	1.5	Choshi	3	0.6
Kochi	29	3.0	Kanazawa	2	0.2
Wakayama	9	0.7	Fushiki	1	0.2
Oita	6	1.2	Nagano	0	0
Hiroshima	8	0.6	Niigata	1	0.1
Yamaguchi	5	1.2	Yamagata	0	0
Matsuyama	0	0	Akita	0	0
Okayama	0	0	Fukushima	3	1.5
Ozaka	5	0.6	Ishinomaki	2	0.5
Kioto	3	0.3	Miyako	5	0.5
Kumamoto	2	2.0	Aomori	2	0.2
Nagasaki	27	2.0	Hakodate	2	0.2
Itsugahara	8	1.5	Suttsu	0	0
Akamagaseki	9	1.0	Sapporo	1	0.1
Sakai	2	0.2	Kamikawa	0	0
Tsu	3	1.3	Soya	0	0
Gifu	7	0.8	Nemuro	0	0
Hamamatsu	5	0.6	Erimo	1	0.2

This table shows that the frequency of rain and the amount of precipitation are independent of each other. Thus in Kanazawa, where the mean precipitation for a year is 2,500 mm, we had only two heavy rains during ten years, and moreover in both of them, the quantity was only 130 mm. And they have fallen not in winter when the mean precipitation is great, but in summer which has rather scanty precipitation.

The preceding table shows us that the heavy rain is most frequent in our southern coast facing the Pacific, especially in Kiushu, Shikoku, and Kii. In these regions, heavy rains exceeding 100 mm in a day falls almost certainly two or three times in a year. On the contrary, the coast along the Sea of Japan, and the central Nippon have very few heavy rains. There we have one or two rains over 100 mm only in eight or nine years, and mostly they are totally absent. In Hokkaido, there are very scanty precipitations and also very few heavy rains through the year.

In order to examine, whether heavy rain falls more frequently in summer or in winter, we give its number for each month in the following table. Total number of heavy rains in the table is 198.

	Jan.	Feb.	March	April	May	June	July	Aug.	Sept.	Oct.	Nov.	Dec.
No. of days with heavy rain	0	0	2	12	12	30	45	27	35	28	4	3
% of the total	0	0	1	6	6	15	22	14	18	14	2	2

Thus the maximum frequency is in July, being 22 per cent of the whole number, and the next frequency is in September. Our general conclusion is that the heavy rain is a phenomenon peculiar to the warmer season from April to October. During the colder season from November to March, it is almost entirely absent.

E. FREQUENCY OF PRECIPITATION.

We shall now pass on to the considerations of the frequency and the probability of precipitation.

The numbers of days with precipitation, (that is, the days whose amounts of precipitation exceed 0.1 mm) for several stations are:—

MEAN NUMBER OF DAYS WITH PRECIPITATION.

Locality.	Jan.	Feb.	Mar.	Apr.	May	June	July	Aug.	Sept.	Oct.	Nov.	Dec.	Year	Wint.	Spr.	Sum.	Aut.
Naha	15.0	21.0	21.0	17.0	24.0	16.0	15.5	15.0	22.0	10.5	14.5	15.5	207.0	51.5	62.0	46.5	47.0
Kagoshima	13.7	11.2	14.0	17.2	15.2	17.9	15.2	12.7	13.0	11.3	9.6	12.6	163.6	37.5	46.4	45.8	33.9
Miyazaki	7.2	9.3	12.4	15.9	13.2	16.8	14.3	14.0	16.1	12.4	7.9	6.1	145.6	22.6	41.5	45.1	36.4
Kochi	6.2	7.9	11.8	14.5	14.1	15.7	15.8	14.0	15.3	10.5	8.8	6.9	142.5	21.0	40.4	45.5	34.6
Wakayama	9.8	10.4	11.6	14.2	12.7	14.0	12.1	9.7	14.0	11.9	10.6	11.1	142.1	31.3	38.5	35.8	36.5
Oita	8.0	9.4	11.6	15.2	12.2	11.6	13.4	11.2	13.0	9.2	7.2	9.2	131.2	26.6	39.0	36.2	29.4
Yamaguchi	16.5	17.0	17.7	16.5	13.0	13.7	16.7	10.0	13.3	9.7	12.3	17.3	173.7	50.8	47.2	40.4	35.3
Hiroshima	9.9	8.9	11.5	13.7	12.4	13.5	12.0	8.3	11.4	8.5	8.2	8.6	126.9	27.4	37.6	33.8	28.1
Matsuyama	10.0	12.5	14.0	15.0	12.0	10.0	12.5	9.0	16.5	10.5	7.0	14.0	143.0	36.5	41.0	31.5	34.0
Okayama	4.0	10.0	8.0	9.0	7.0	14.0	16.0	11.0	14.0	7.0	7.0	9.0	116.0	23.0	24.0	41.0	28.0
Ozaka	7.8	8.4	12.1	13.8	12.6	13.8	10.6	9.5	12.2	10.3	9.3	9.1	129.5	25.3	38.5	33.9	31.8
Kioto	13.2	12.6	13.6	14.6	13.8	15.1	14.5	11.2	14.2	10.6	11.2	11.7	156.3	37.5	42.0	40.8	36.0
Kumamoto	16.0	12.0	15.0	16.0	12.0	13.0	15.5	11.0	16.0	6.0	7.0	13.0	152.5	41.0	43.0	39.5	29.0
Saga	12.0	14.0	9.0	12.0	6.0	7.0	15.0	10.5	12.5	6.0	7.5	12.0	123.5	38.0	27.0	32.5	26.0
Nagasaki	16.5	13.2	14.1	15.3	13.7	17.1	14.1	12.6	12.9	10.1	12.4	16.8	168.8	46.5	43.1	43.8	35.4
Fukuoka	14.0	16.5	16.5	14.5	12.5	11.0	12.0	12.5	13.0	8.5	9.5	14.5	155.0	45.0	43.5	35.5	31.0
Itsugahara	11.0	10.2	10.8	13.8	10.6	11.4	15.8	10.2	11.6	7.2	8.2	13.8	134.6	35.0	35.2	37.4	27.0
Akamagaseki	16.2	14.6	14.2	14.7	12.1	14.6	10.4	9.8	12.9	10.2	12.4	16.1	158.2	46.9	41.0	34.8	35.5
Sakai	25.4	21.8	19.0	15.1	13.2	13.9	13.1	9.6	15.3	15.9	20.7	24.2	207.2	71.4	47.3	36.6	51.9
Tsu	12.5	13.5	13.0	14.5	12.0	13.5	19.7	12.7	18.0	10.3	9.3	10.7	159.7	36.7	39.5	45.9	37.6
Nagoya	9.0	10.0	10.0	9.0	8.0	12.0	17.5	15.0	16.0	9.0	9.0	10.0	134.5	29.0	27.0	44.5	34.0
Gifu	12.7	9.1	12.1	14.7	13.0	14.4	16.8	13.1	13.0	10.0	10.0	12.2	151.1	34.0	39.8	44.3	33.0
Hamamatsu	7.2	7.8	11.8	14.8	13.0	13.8	12.4	10.3	13.3	11.3	9.6	7.3	132.6	22.3	39.6	36.5	34.2
Numazu	7.0	9.4	13.4	15.4	13.1	13.7	15.0	11.1	14.7	11.7	9.4	7.8	141.7	24.2	41.9	39.8	35.8
Tokio	7.2	9.4	11.6	14.9	13.2	14.2	14.1	11.9	15.3	12.5	8.8	6.1	139.2	22.7	39.7	40.2	36.6
Utsunomiya	5.0	9.0	11.0	14.0	9.0	17.0	29.0	11.0	15.5	12.5	11.0	6.5	150.5	20.5	34.0	57.0	39.0
Cheshi	8.8	10.6	14.4	14.2	12.8	12.4	12.6	9.6	13.6	13.8	10.2	10.2	143.2	29.6	41.4	34.6	37.6
Kanazawa	26.2	22.2	19.8	15.5	12.7	14.8	14.0	10.4	16.4	15.7	19.9	25.7	213.3	74.1	48.0	39.2	52.0
Fushiki	26.8	21.7	19.0	16.0	12.9	14.6	13.9	9.6	15.9	15.3	19.6	25.1	210.1	73.6	47.9	38.1	50.8
Nagano	19.7	14.7	12.0	14.3	11.0	14.3	17.0	12.3	11.3	7.0	9.3	14.7	157.6	49.1	37.3	43.6	27.6
Niigata	26.6	23.7	20.8	16.1	13.9	14.8	14.1	9.4	16.3	17.7	23.1	27.5	224.0	77.8	50.8	38.3	57.1
Yamagata	23.0	18.5	16.0	13.0	9.0	12.5	17.0	13.3	13.3	15.0	15.7	19.3	185.6	60.8	38.0	42.8	44.0
Akita	27.8	22.7	20.6	14.2	13.4	13.8	15.1	12.1	18.0	17.6	19.8	25.8	221.1	76.8	48.2	41.0	55.4
Fukushima	11.5	11.5	13.5	9.5	10.3	13.3	19.3	16.0	17.0	10.0	11.3	10.7	153.9	33.7	33.3	48.6	38.3
Ishinomaki	9.0	8.5	11.2	11.5	11.2	13.5	15.5	14.5	14.4	9.6	10.6	11.4	140.9	28.9	33.9	43.5	34.6
Miyako	8.0	7.5	11.6	11.0	12.2	14.1	14.2	13.4	15.7	12.7	11.9	10.6	142.9	26.1	34.8	41.7	40.3
Aomori	27.7	21.9	19.8	13.6	11.2	11.9	14.0	10.3	16.8	15.9	21.9	26.6	211.6	76.2	44.6	36.2	54.6
Hakodate	17.2	14.3	14.2	11.1	11.2	13.2	11.7	11.6	17.0	13.7	16.5	19.3	171.0	50.8	36.5	36.5	47.2
Suttsu	26.7	21.7	16.2	10.2	12.2	13.2	11.7	11.5	17.8	16.2	21.8	25.3	204.5	73.7	38.6	36.4	55.8
Sapporo	17.9	16.2	16.0	10.7	12.3	11.7	11.0	11.9	16.7	17.1	19.0	20.9	181.4	55.0	39.0	34.6	52.8
Kamikawa	20.7	17.0	15.0	15.3	11.7	12.0	15.0	11.2	16.7	21.2	19.5	24.5	199.8	62.2	42.0	38.2	57.4
Soya	18.5	12.5	10.7	7.8	11.0	10.5	9.7	8.3	14.7	15.5	17.7	16.5	153.4	47.5	29.5	28.5	47.9
Abashiri	12.5	12.5	8.0	11.0	11.5	11.5	9.0	8.7	11.3	16.3	12.7	13.0	138.0	38.0	30.5	29.2	40.3
Nemuro	10.8	10.7	12.3	12.2	13.0	14.3	14.3	12.3	16.2	13.2	13.3	13.2	155.8	34.7	37.5	40.9	42.7
Kushiro	6.5	9.0	10.5	12.5	13.5	13.0	14.0	21.5	16.5	14.0	10.0	12.5	153.5	28.0	36.5	48.5	40.5
Erimo	10.8	7.8	9.6	9.4	11.8	13.0	12.0	10.8	13.6	13.4	14.2	16.6	143.0	35.2	30.8	35.8	41.2

Thus for the whole year, we have the most frequent rainfalls along the Sea of Japan, the number of days with precipitation being more than 200 in a year. The district with the least precipitation is Setouchi. In this district, the rainy days are only 120 in a year.

The probability of the precipitation differs considerably on the two sides of Nippon, as the amount of precipitation does. The peculiar characteristics of the back Nippon is that the probability of precipitation is greatest in winter and least in summer. And it exceeds $\frac{1}{2}$ throughout the year. Contrary is the case with the front Nippon, where it is greatest in summer and least in winter, and is always less than $\frac{4}{10}$ throughout the year.

The variation in the back Nippon is this. The annual mean of the probability is $\frac{6}{10}$. During the three winter months, it keeps a great value of $\frac{85}{100}$, and its maximum is in January. It gets less and less toward summer and reaches to its minimum in August when it is $\frac{1}{3}$. In September, it suddenly increases, and again decreases in October. From November onward, it again increases greatly.

The front Nippon may be divided into the following four districts.

(1) The southern coast from the southeastern part of Kiushu to Choshi.—Here the mean probability of the year is $\frac{38}{100}$. The minimum value is in winter, and is about $\frac{1}{4}$. It suddenly increases to $\frac{4}{10}$ in April, and in June and July it reaches to its maximum, greater than $\frac{1}{2}$. There is sudden decrease to $\frac{38}{100}$ in August and again in September it increases up to $\frac{1}{2}$. After that month, it decreases on till it reaches the minimum in winter.

(2) Eastern coast of Nippon.—Here it resembles to the preceding on the whole, but that the probability in winter is rather greater.

(3) Setouchi or Inland Sea.—Here the mean probability of the year is $\frac{35}{100}$, which is the least value in this country. There is no remarkable difference between summer and winter. In April, June, and September, it is somewhat greater than it is in other months, being over $\frac{4}{10}$.

(4) Hokkaido, (excepting those parts along the Sea of Japan).—In this division, the mean probability of the year is about $\frac{1}{2}$. Its general feature is very much similar to that of the back Nippon, though the annual mean probability is far smaller than in the last division. The maximum value $\frac{6}{10}$ is in winter. After April, it remains at about $\frac{4}{10}$, but at August it increases quickly to $\frac{54}{100}$, and keeps on increasing till it reaches the maximum in winter.

In the northwestern coast of Kiushu, and the central Nippon, the annual mean probability is more than $\frac{4}{10}$ which lies between the probabilities on the two sides of Nippon, above mentioned. The annual variation, also, is not very marked,—it is somewhat greater in winter and spring, but smaller in summer and autumn. The minimum is in October, which is about $\frac{3}{10}$.

Such are the great differences in the probabilities of precipitation in different parts. And it is the most remarkable peculiarity that the probability in August is everywhere in this country, very much smaller than the probabilities of the preceding and the succeeding months. The annexed table shows this fact clearly.

PROBABILITY OF PRECIPITATION.

	Jan.	Feb.	Mar.	April	May	June	July	Aug.	Sept.	Oct.	Nov.	Dec.	Year
Along Sea of Japan (From Yuasa to Soya)	86	80	62	48	41	46	41	33	55	52	70	81	59
Hokkaido (Excepting the coast along Sea of Japan)	60	54	45	37	37	39	38	31	54	54	61	65	48
Southern coast of Main Island	24	26	40	46	39	48	51	38	50	38	32	26	38
Eastern coast of Main Island	32	26	37	36	38	45	49	43	51	38	41	40	40
Inland sea	27	37	37	45	39	44	38	30	40	30	28	29	35
NW^{rn} coast of Kiushu	49	50	48	52	42	49	43	38	45	30	34	47	44
Central part of Main Island	43	41	43	47	43	49	51	39	45	32	35	40	42

F. ABSOLUTE PROBABILITY OF PRECIPITATION.

What we have said in the last article relates chiefly to the probability of precipitation in 24 hours. But this does not teach us completely our climatological conditions. Hence for the minuter informations, we must know the absolute probability of precipitation, or the probability for every hour of the day. Now, rigorously speaking, for the determination of this absolute probability, we must have at least hourly observations; for, the phenomena of precipitation are subject to great disturbances. But a moment's reflection shows us that a few number of observations in a day is sufficient for this purpose. The reason is that though precipitation is subject to disturbances, yet it is not momentary phenomena, continuing for some time and varying very slowly. Thus if we had precipitations in two consecutive observations, we may safely assume that there was continuous precipitation during the interval.

Indeed, the method, which Prof. Köppen adopted in his calculation of the absolute probability of precipitation, depends on this assumption. Let us now calculate after him the absolute probability from six observations in a day.

The next table gives the result, which is deduced by the six observations in a day taken at thirty stations for five years, 1886—1890. The numbers given in the table are the percentages, which the observed number of the precipitation bears to the total number of observations.

ABSOLUTE PROBABILITY OF PRECIPITATION.

Locality.	Jan.	Feb.	Mar.	April	May	June	July	Aug.	Sept.	Oct.	Nov.	Dec.	Year
Kagoshima	16	15	18	22	23	31	18	12	14	14	7	14	17
Miyazaki	8	10	13	16	18	21	11	7	14	12	5	5	12
Kochi	7	8	16	17	18	15	16	7	15	11	7	7	12
Wakayama	10	10	15	18	16	17	13	7	11	14	11	11	13
Oita	12	10	11	21	21	20	20	9	15	15	10	12	15
Hiroshima	12	14	15	19	17	17	15	6	13	12	10	10	13
Ozaka	7	8	14	17	15	14	11	6	11	12	10	8	11
Kioto	11	11	14	18	16	15	14	6	11	11	12	10	13
Nagasaki	14	13	14	19	19	20	14	7	10	10	8	13	13
Itsugahara	13	6	10	18	13	17	22	6	10	8	7	7	11
Akamagaseki	19	15	17	20	20	18	16	7	14	11	10	16	15
Sakai	31	25	22	17	14	14	14	8	15	15	17	25	18
Gifu	15	13	15	22	20	16	17	10	16	11	12	14	15
Hamamatsu	8	8	14	18	18	15	13	8	11	14	9	8	12
Numazu	10	9	16	20	18	14	16	10	16	15	11	7	13
Tokio	10	9	15	19	20	15	17	9	16	18	10	6	14
Choshi	15	13	18	18	21	15	16	8	15	21	12	9	15
Kanazawa	37	31	20	22	17	17	13	9	11	15	19	30	20
Fushiki	14	35	20	21	15	16	14	7	14	15	19	35	21
Niigata	36	31	17	20	14	16	11	6	12	14	23	33	19
Akita	33	26	16	16	11	18	13	8	12	12	15	31	18
Ishinomaki	12	9	9	13	13	16	11	10	16	11	10	10	12
Miyako	12	7	11	14	12	19	12	13	16	11	12	8	12
Aomori	53	41	26	15	12	15	12	11	15	15	26	43	24
Hakodate	38	30	20	15	13	15	14	12	14	11	24	31	24
Suttsu	57	40	24	13	14	19	10	11	17	15	30	49	25
Sapporo	22	22	20	15	13	15	9	9	16	12	24	27	17
Nemuro	17	15	16	15	15	18	17	12	18	13	13	14	15
Erimo	14	11	11	14	13	16	9	9	12	12	14	[illegible]	13
Soya	35	27	22	10	12	17	17	10	15	17	28	33	20

Thus there are marked difference of the absolute probabilities between the two sides of Nippon. On the back Nippon, it is rather great, the annual mean being $\frac{21}{100}$, but on the front Nippon it is only $\frac{13}{100}$. Also, on the back Nippon it is greatest in winter, especially in January when it is $\frac{42}{100}$; but on the front Nippon it is small in winter and great in summer. It must be remarked, however, that in summer the two sides have nearly equal probabilities. It is a noteworthy fact common to both sides, that in August, the absolute probability is very small, being always less than $\frac{9}{100}$.

The number of hours with precipitation for every month is as follows:—

NUMBER OF HOURS OF PRECIPITATION.

Locality.	Jan.	Feb.	March	April	May	June	July	Aug.	Sept.	Oct.	Nov.	Dec.	Year.
Kagoshima	117,6	111,2	137,6	159,2	168,8	221,0	132,8	89,6	101,1	105,6	58,6	102,4	1501,0
Miyazaki	56,0	67,2	99,2	112,8	131,2	149,6	81,6	52,0	101,6	92,8	36,0	40,8	1020,8
Kochi	58,6	53,6	116,0	124,0	132,8	111,2	118,4	53,6	104,8	82,4	52,8	58,6	1056,8
Wakayama	71,1	65,6	108,8	127,2	120,8	121,6	91,1	52,8	97,6	100,8	78,4	78,1	1121,6
Oita	82,8	68,8	106,8	148,8	158,0	146,8	146,8	70,0	110,0	114,8	68,8	90,0	1312,4
Hiroshima	85,6	64,8	108,8	131,4	128,8	124,8	113,6	17,2	93,6	87,2	71,2	75,2	1135,2
Ozaka	51,2	53,6	103,2	124,0	115,2	98,4	81,6	42,4	80,0	85,6	75,2	57,6	968,0
Kioto	84,8	93,6	106,4	129,6	118,4	108,0	101,6	47,2	80,0	82,4	87,2	75,2	1114,4
Nagasaki	101,6	89,6	101,6	133,6	144,0	145,6	100,8	54,4	69,6	76,0	58,4	100,0	1175,2
Itsegahara	96,0	40,0	73,2	132,0	100,0	122,8	162,8	46,8	72,0	57,2	53,2	52,0	1008,0
Akamaga-seki	140,8	103,2	123,2	141,6	150,4	128,8	120,8	53,6	98,4	82,4	75,2	121,6	1340,0
Sakai	252,8	172,0	164,0	124,0	103,2	101,6	103,2	56,0	106,4	110,4	122,4	188,0	1604,0
Gifu	112,0	71,2	112,0	159,2	117,2	115,2	125,6	71,2	117,6	84,8	89,6	106,4	1312,0
Hamamatsu	60,0	52,0	105,6	128,8	136,8	110,4	100,0	61,6	99,2	102,4	67,2	56,0	1080,0
Numazu	74,4	60,0	117,6	141,6	134,4	97,6	121,6	73,6	111,4	110,4	78,4	53,6	1177,6
Tokio	72,8	58,4	108,0	135,2	151,2	110,4	124,0	64,0	112,8	131,2	72,0	46,4	1186,4
Choshi	110,8	84,8	134,0	132,8	156,8	110,8	118,8	56,8	110,8	158,8	86,0	63,0	1327,2
Kanazawa	276,0	212,0	149,6	155,8	124,8	123,2	98,4	64,0	104,0	115,2	140,0	226,4	1790,4
Fushiki	325,6	233,6	151,2	153,6	108,0	112,0	107,2	51,2	102,4	111,2	139,2	261,6	1856,8
Niigata	271,2	212,0	125,6	140,8	100,8	112,0	78,4	42,4	88,8	102,4	162,4	214,8	1681,6
Akita	243,2	176,8	122,4	112,8	100,8	131,2	97,6	58,4	87,2	92,0	109,6	228,8	1560,8
Ishinomaki	87,2	60,8	68,8	91,2	96,0	117,6	82,4	73,6	116,0	80,8	74,4	71,2	1020,0
Miyako	85,6	49,6	83,2	97,6	85,6	139,2	87,2	96,0	112,8	83,2	88,0	56,8	1064,8
Aomori	394,4	280,0	190,4	108,0	88,8	106,4	90,4	79,2	110,4	109,6	187,2	319,2	2064,0
Hakodate	284,8	203,2	147,2	111,2	97,6	108,8	103,2	92,0	97,6	82,4	113,2	252,8	1724,0
Suttsu	123,2	270,4	195,2	96,8	102,4	139,2	72,0	81,6	123,2	113,6	219,2	361,6	2198,4
Sapporo	161,6	147,2	147,2	109,6	98,4	110,4	67,2	68,8	116,0	90,4	154,4	200,0	1471,2
Nemuro	124,0	100,8	116,8	106,4	110,4	128,8	125,6	91,2	126,4	94,4	92,0	101,0	1320,8
Erimo	106,0	74,8	82,8	102,0	100,0	112,0	65,8	68,0	88,8	92,0	98,0	154,8	1146,0
Soya	264,0	180,8	160,0	68,8	92,0	122,0	126,0	70,8	110,8	128,8	204,0	246,0	1774,0

The numbers in the above table are obtained by multiplying the number of observation, into the numbers given in the preceding table, of absolute probability.

If we calculate now the means for back side and front side, we obtain the following table:—

	Jan.	Feb.	March	April	May	June	July	Aug.	Sept.	Oct.	Nov.	Dec.	Year
Back side	304	216	156	118	102	117	97	66	103	107	158	259	1806
Front side	78	64	106	128	130	119	107	62	104	100	73	66	1135

Thus on the back Nippon, there are precipitations for about 1,800 hours in a year, while on the front Nippon for about 1,100 hours. Thus the difference between them is about 700 hours.

Dividing the above numbers by the respective number of days with precipitation we shall obtain the average numbers of hours with precipitation in a day with precipitation. They are:—

MEAN DURATION OF PRECIPITATION IN A DAY WITH PRECIPITATION, IN HOURS.

Locality.	Jan.	Feb.	March	April	May	June	July	Aug.	Sept.	Oct.	Nov.	Dec.	Year
Kagoshima	8,65	9,36	10,12	9,26	10,68	12,73	8,62	7,72	7,16	9,10	5,96	8,00	9,23
Miyazaki	7,00	6,72	7,75	7,32	9,37	9,47	5,23	4,19	6,54	7,25	4,29	5,83	6,94
Kochi	7,66	6,70	8,53	7,65	8,97	7,83	8,11	3,94	7,18	7,03	6,11	6,09	7,30
Wakayama	7,91	6,19	9,07	8,37	8,39	9,35	8,14	5,50	7,18	8,54	8,17	6,76	7,88
Oita	10,10	8,09	9,13	9,79	11,29	12,03	11,29	6,67	9,17	11,25	8,91	9,28	9,88
Hiroshima	8,73	7,36	9,22	9,08	9,91	9,60	10,23	6,38	7,80	9,28	8,68	7,67	8,80
Ozaka	7,11	6,54	8,32	8,61	8,35	7,81	7,16	4,24	6,45	9,11	8,17	5,65	7,58
Kioto	8,00	7,43	7,29	8,88	8,00	7,71	7,26	4,29	5,26	8,24	7,52	5,87	7,15
Nagasaki	6,68	6,89	7,36	8,15	10,91	9,33	7,30	5,43	6,44	6,55	5,41	5,88	7,26
Itsugahara	8,97	5,48	7,87	9,63	9,71	10,50	10,64	5,85	7,42	8,54	5,91	4,09	8,10
Akamagaseki	9,03	7,37	9,06	9,08	11,06	9,76	10,60	5,70	7,57	8,58	6,90	8,22	8,63
Sakai	9,87	8,19	8,91	8,27	7,82	7,94	8,60	5,49	7,19	6,84	6,37	8,25	7,97
Gifu	8,36	7,57	8,24	10,31	10,67	8,86	8,49	5,24	8,10	8,00	8,00	7,82	8,39
Hamamatsu	7,30	6,84	8,95	8,70	9,91	8,62	7,69	5,50	6,53	8,26	7,15	7,37	7,85
Numazu	9,79	6,38	8,52	9,08	8,73	7,75	7,79	5,94	6,73	8,90	8,52	6,38	7,88
Tokio	10,11	7,89	9,31	8,11	10,08	8,12	8,05	5,08	7,23	9,65	8,78	5,80	8,19
Choshi	11,42	7,57	9,57	9,16	11,45	9,47	9,00	5,57	7,54	9,92	8,43	5,89	8,83
Kanazawa	10,78	9,55	7,87	10,18	9,31	8,56	7,93	6,81	6,34	7,38	7,78	8,98	8,65
Fushiki	12,33	10,72	8,22	9,60	8,06	7,89	8,65	5,45	6,24	7,22	7,10	10,38	8,90
Niigata	10,12	8,69	6,61	8,19	7,20	7,47	6,53	4,24	5,48	5,95	7,38	9,00	7,61
Akita	9,14	7,43	6,00	7,94	8,13	8,63	7,07	4,87	4,79	5,68	5,77	8,80	7,17
Ishinomaki	7,52	7,79	7,02	7,73	8,28	8,40	6,87	6,03	8,06	7,77	6,76	5,74	7,34
Miyako	9,30	7,75	7,56	8,71	6,58	9,53	6,92	6,96	7,73	6,82	7,86	5,36	7,58
Aomori	14,39	12,39	10,02	8,57	9,06	7,71	6,65	8,08	6,07	7,12	9,09	11,82	9,84
Hakodate	14,99	12,10	10,22	9,93	8,87	7,25	9,21	7,42	5,61	5,67	7,62	12,27	9,44
Suttsu	16,28	12,40	12,35	9,98	8,98	10,74	6,79	8,00	6,77	7,38	9,96	11,12	11,01
Sapporo	9,85	8,46	10,22	10,15	8,95	8,62	6,72	6,62	6,82	5,45	8,21	9,52	8,33
Nemuro	10,69	8,26	8,98	8,87	8,62	8,82	9,10	8,44	7,80	6,74	6,57	7,76	8,32
Erimo	11,16	9,74	9,00	10,52	8,70	8,02	6,68	6,48	6,99	6,48	6,53	9,00	8,17
Soya	12,89	13,94	13,11	9,83	9,02	11,02	11,45	7,30	7,15	8,59	12,00	14,73	11,31

And the averages on the two sides of Nippon are:—

	Jan.	Feb.	March	April	May	June	July	Aug.	Sept.	Oct.	Nov.	Dec.	Year
Back side	12,1	10,6	8,2	9,2	8,5	8,6	7,4	6,4	6,2	6,9	7,1	10,9	9,1
Front side	8,6	7,4	8,5	8,8	9,6	9,3	8,3	5,5	7,3	8,6	7,3	6,5	8,0

Thus on the average of the year, we have precipitations for more than 9 hours in a day with precipitation on the back Nippon and for 8 hours on the front Nippon. Again on the back Nippon, the long period of precipitation over 10 hours in a day occurs in winter, but on the front Nippon the long period is 9 hours and is in the late spring and the early summer.

The mean quantity of precipitation in a day with precipitation, and that in an hour with precipitation, are tabulated in the following:—

MEAN AMOUNT OF PRECIPITATION IN A DAY WITH PRECIPITATION, IN MILLIMETRES.

Locality.	Jan.	Feb.	March	April	May	June	July	Aug.	Sep.	Oct.	Nov.	Dec.	Year
Kagoshima	6,7	6,5	10,1	14,5	14,5	22,4	18,3	15,1	14,7	12,9	9,2	6,8	13,2
Miyazaki	13,0	8,6	12,8	17,2	22,9	28,4	18,5	18,3	31,9	24,0	15,4	13,5	19,8
Kochi	11,0	10,7	14,9	22,8	20,3	21,3	27,1	18,6	34,4	28,3	22,0	15,0	21,5
Wakayama	4,7	5,5	8,8	11,8	10,4	11,1	13,4	9,9	14,2	17,1	13,5	6,8	10,8
Oita	6,1	7,0	9,3	11,3	13,9	18,6	26,1	9,8	16,7	19,5	13,9	10,0	14,0
Hiroshima	4,7	5,6	9,1	14,7	13,7	16,2	21,2	10,6	16,8	15,4	12,8	6,7	13,0
Ozaka	6,0	5,8	9,6	13,5	11,2	11,7	9,9	8,6	11,3	17,4	10,8	6,4	10,5
Kioto	5,0	4,0	8,8	13,5	11,2	12,8	14,0	12,2	10,7	16,0	9,5	5,9	10,3
Nagasaki	4,5	5,5	9,6	14,6	16,1	21,6	21,5	12,9	14,8	13,8	10,3	6,7	12,6
Itsugahara	8,9	3,4	11,2	15,2	13,5	23,0	21,3	15,1	25,8	18,9	11,3	4,3	14,6
Akamagaseki	4,1	8,8	7,4	11,1	13,0	16,4	30,0	8,1	12,6	12,2	7,2	5,8	10,7
Sakai	7,3	4,9	6,7	10,1	7,6	11,3	14,9	9,5	11,6	11,8	8,3	7,0	9,0
Gifu	5,2	6,5	9,8	16,9	16,4	15,0	16,6	11,0	16,9	17,0	9,8	8,0	12,6
Hamamatsu	5,7	9,0	13,7	18,2	17,0	14,0	14,7	13,0	15,1	13,6	11,8	11,9	13,8
Numazu	9,1	5,9	10,0	12,8	11,9	12,8	13,1	16,3	13,9	15,3	10,8	9,9	12,5
Tokio	6,5	5,2	9,6	6,9	12,6	9,6	8,3	11,0	10,6	15,3	15,4	5,2	9,9
Choshi	8,0	6,1	8,6	9,0	13,2	10,2	10,4	10,7	12,3	16,4	11,3	5,9	10,6
Kanazawa	10,9	8,2	8,6	12,8	10,1	12,7	11,5	12,8	17,0	11,6	14,1	15,6	11,6
Fushiki	9,1	6,5	6,7	10,1	7,9	10,2	10,6	10,9	10,0	8,3	9,8	12,8	9,4
Niigata	6,3	4,6	4,5	7,6	5,1	9,1	10,6	9,6	8,6	9,7	8,8	9,0	7,6
Akita	4,7	4,5	5,3	9,8	8,5	11,0	10,1	13,4	9,8	9,3	9,2	7,3	8,0
Ishinomaki	3,1	4,0	5,1	8,8	7,6	9,2	6,4	11,0	10,2	10,2	5,1	3,0	7,2
Miyako	6,9	5,1	6,8	10,2	6,9	10,9	11,0	10,5	13,2	10,9	9,1	5,2	9,3
Aomori	4,6	5,0	3,3	5,7	6,0	7,9	7,5	9,8	9,0	8,1	5,9	4,6	6,1
Hakodate	2,4	2,6	4,4	7,9	5,0	6,5	10,0	7,8	11,1	6,9	6,8	4,1	6,1
Suttsu	4,2	5,1	3,7	7,9	5,2	6,7	7,4	9,7	11,2	10,2	5,9	5,4	6,5
Sapporo	4,4	3,9	4,5	7,2	4,0	5,7	6,1	9,9	10,6	5,2	4,9	5,1	5,8
Nemuro	2,1	1,3	3,5	7,1	8,1	6,9	5,0	7,0	10,5	7,1	7,0	6,4	6,1
Erimo	2,4	2,8	4,7	13,5	7,3	8,0	7,0	11,3	10,1	7,9	6,1	5,3	7,3
Soya	2,9	3,3	8,4	6,2	6,7	8,3	8,6	10,3	9,4	8,5	5,9	2,8	6,1

MEAN AMOUNT OF PRECIPITATION IN A HOUR WITH PRECIPITATION, IN MILLIMETRES.

Kagoshima	0,77	0,65	1,00	1,56	1,36	1,76	2,13	1,96	1,96	1,41	1,54	0,85	1,43
Miyazaki	1,85	1,28	1,66	2,35	2,44	3,00	3,54	4,36	4,91	3,31	3,60	2,31	2,87
Kochi	1,43	1,61	1,75	2,98	2,27	2,72	3,38	4,73	4,80	3,72	3,58	2,46	2,95
Wakayama	0,59	0,89	0,98	1,41	1,24	1,19	1,65	1,81	1,99	2,00	1,66	1,01	1,37
Oita	0,61	0,86	1,02	1,15	1,23	1,55	2,31	1,46	1,82	1,73	1,55	1,08	1,41
Hiroshima	0,53	0,76	0,99	1,62	1,38	1,69	2,34	1,66	2,16	1,66	1,48	0,87	1,47
Ozaka	0,85	0,89	1,16	1,57	1,34	1,50	1,38	2,03	1,75	1,91	1,32	1,13	1,42
Kioto	0,62	0,54	1,21	1,53	1,39	1,67	1,90	2,01	2,03	1,95	1,26	1,01	1,45
Nagasaki	0,68	0,80	1,31	1,80	1,48	2,32	2,05	2,51	2,31	2,11	1,90	1,13	1,74
Itsugahara	1,00	0,62	1,43	1,58	1,39	2,19	2,00	2,58	3,47	2,22	1,91	1,05	1,81
Akamagaseki	0,48	0,52	0,82	1,22	1,18	1,68	2,83	1,41	1,67	1,42	1,08	0,71	1,24
Sakai	0,74	0,60	0,76	1,22	0,97	1,42	1,73	1,73	2,03	1,74	1,30	0,85	1,13
Gifu	0,62	0,87	1,19	1,64	1,54	1,69	1,96	2,11	2,01	2,13	1,23	1,03	1,51
Hamamatsu	0,76	1,31	1,58	2,09	1,72	1,63	1,92	2,36	2,32	1,65	1,65	1,62	1,76
Numazu	0,93	0,98	1,18	1,41	1,71	1,65	1,68	2,74	2,07	1,72	1,27	1,55	1,59
Tokio	0,65	0,66	1,03	0,85	1,25	1,19	1,03	2,16	1,47	1,58	1,76	0,89	1,21
Choshi	0,70	0,81	0,89	0,98	1,15	1,07	1,16	1,92	1,63	1,66	1,70	1,00	1,20
Kanazawa	1,01	0,86	1,09	1,26	1,09	1,48	1,46	1,88	1,74	1,57	1,81	1,73	1,35
Fushiki	0,74	0,60	0,81	1,05	0,99	1,30	1,23	2,00	1,60	1,15	1,38	1,23	1,05
Niigata	0,62	0,53	0,68	0,93	0,71	1,26	1,63	2,28	1,57	1,63	1,20	1,00	1,00
Akita	0,52	0,61	0,88	1,24	1,05	1,27	1,43	2,75	2,04	1,63	1,59	0,83	1,12
Ishinomaki	0,41	0,51	0,73	1,13	0,92	1,10	0,93	1,83	1,27	1,31	0,76	0,52	0,98
Miyako	0,75	0,66	0,90	1,17	1,05	1,14	1,59	1,50	1,71	1,59	1,16	0,97	1,22
Aomori	0,32	0,41	0,33	0,66	0,66	1,03	1,12	1,21	1,48	1,14	0,64	0,39	0,62
Hakodate	0,16	0,21	0,43	0,80	0,57	0,90	1,09	1,05	1,99	1,24	0,90	0,34	0,65
Suttsu	0,26	0,41	0,30	0,79	0,58	0,63	1,09	1,21	1,66	1,38	0,59	0,38	0,59
Sapporo	0,45	0,46	0,44	0,71	0,45	0,66	0,91	1,50	1,55	0,95	0,59	0,54	0,70
Nemuro	0,20	0,16	0,39	0,80	0,94	0,78	0,55	0,86	1,35	1,05	1,06	0,83	0,74
Erimo	0,21	0,29	0,52	1,29	0,84	0,93	1,05	1,75	1,44	1,21	0,94	0,59	0,80
Soya	0,21	0,24	0,26	0,63	0,74	0,71	0,75	1,42	1,32	0,98	0,49	0,19	0,54

The localities where the amount of precipitation both in a day and in an hour is greatest are Kochi and Miyazaki. In them, the mean amount in a day is about 20 mm and in an hour 3 mm. Generally speaking, the mean amount in a day or in an hour with precipitation decreases as we go from the southern part to the northern.

If we consider different seasons, the mean amount in a day with precipitation is maximum in September, throughout the country. In this month, it is generally beyond 10 mm even in the northern provinces, in the southern parts it frequently exceeds 30 mm.

In the southern coast, we have plentiful precipitation in June and July. It is generally 15 mm in a day and very often exceeds the quantity in September.

Everywhere in this country, the amount in an hour with precipitation is greatest in the warmer season, particularly in July, August and September. This confirms our previous conclusion that the heavy rain is a phenomenon peculiar to summer.

We must add here some remarks upon the application of Köppen's formula; namely how far the accuracy is attained by applying his formula to the six observations in a day for the determination of the absolute probability of precipitation.

For this purpose, we have compared the absolute probability obtained by the hourly observations in Tokio with that deduced from six observations in a day by Köppen's formula.

Absolute probability of precipitation.

	Jan.	Feb.	March	April	May	June	July	Aug.	Sept.	Oct.	Nov.	Dec.	Year
According to 6 Obs. in a day.	10	9	15	19	20	15	17	9	16	18	10	6	11
According to the hourly observation.	10	9	15	20	20	16	17	8	17	17	10	6	11

Number of hours of precipitation.

	Jan.	Feb.	March	April	May	June	July	Aug.	Sept.	Oct.	Nov.	Dec.	Year
According to 6 Obs. in a day.	73	58	108	145	151	110	121	61	113	131	72	15	1186
According to the hourly observation.	71	62	109	112	152	113	121	62	122	128	75	11	1205

Thus there is no appreciable difference between the two, showing the legitimacy of Köppen's formula.

6. MAXIMUM DURATIONS OF WET AND DRY DAYS.

Though we have frequently precipitations lasting for great many days in the back Nippon, yet on our country as a whole, the period with precipitation is always shorter than that without it.

The longest period with precipitation ever observed in this country is that at Akita which continued for 81 days from the 2nd of December 1883, to the 20th of February of the next year. The second long period of 69 days happened at Niigata, which begun from the 26th of January 1883, to the 4th of April of the same year. The longest period without precipitation is 53 days from the 26th of November 1880, to the 18th of January 1881. The second long one is 47 days from the 4th of July, 1883, to the 19th of August 1883. Both of these happened at Tokio. The longest periods for several stations are given in the following table.

Locality.	Number of occasions, in a year, on which the precipitation has occurred continually over 5 days.							Number of occasions, in a year, on which no precipitation has occurred continually over 10 days.					
	5-10	10-15	15-20	over 20	Max. duration	Date	Year	10-15	15-20	over 20	Max. duration	Date	Year
Naha	10.0	2.0	2.0	..	17	29 VIII 14 IX	1891	2.0	..	..	12	1-12 X	1891
Kagoshima	6.2	0.8	0.2	..	17	5-21 VI	1883	1.6	0.4	..	16	13-28 IX, 8-23 I	1883;85
Miyazaki	6.5	0.8	..		14	20 VI-3 VII	1887	2.5	1.0	0.3	22	2-23 XII, 5-26 I	1883,91
Kochi	5.4	0.5	0.1		17	16 VI-2 VII	1883	2.0	0.7	0.3	29	25 I-22 II	1885
Wakayama	5.4	0.1	..	..	12	3-14 IV	1885	2.8	0.6	0.2	31	19 VII-18 VIII	1883
Oita	6.4	1.2	0.8	0.2	28	12 VI-9 VII	1889	2.4	1.0	..	16	1-16 I, 18 I-2 II	1888
Yamaguchi	7.2	0.2	..		12	27 IV-8 V	1890	0.2	0.7	..	17	7-23 VII	1890
Hiroshima	3.4	0.2	..		14	12-25 VII	1889	2.5	0.6	0.2	44	8 VII-20 VIII	1886
Matsuyama	2.5	0.5	..		12	27 IV-8 V	1891	0.5	0.5	..	18	8-25 VII	1890
Okayama	2.0	..			6	12-17 VII	1891	4.0	1.0	..	15	22 III-5 IV	1891
Ozaka	4.0	0.1	..	..	11	28 IV-8 V	1890	2.1	0.6	0.4	47	4 VII-19 VIII	1883
Kioto	5.4	0.1	0.1	..	18	12-29 VII	1891	1.5	0.2	..	17	3-19 VIII	1886
Kumamoto	6.0	0.5			11	27 IV-7 V	1890	3.0	0.5	..	17	7-23 VII	1890
Saga	3.0	..	..		5	11-15 IV, 12-16 VII, 19-23 VII	1891	3.0	1.0	..	18	9-26 X	1891
Nagasaki	7.3	0.6	0.1	..	19	13 VI-1 VII	1885	1.7	0.3	..	19	7-25 VII	1890
Fukuoka	12.5	2.5	..	..	14	30 I-12 II	1891	0.5	1.0	..	19	7-25 VII	1890
Itsugahara	4.0	0.6	0.2	..	17	5-25 VII	1891	2.6	0.4	..	48	17 I-3 II; 7-25 VII	1889;90
Akamagaseki	5.2	0.8	..	..	14	26 I-8 II, 12-25 VII	1886;89	0.5	0.8	0.2	25	8 VII-1 VIII	1886
Sakai	10.0	1.0	0.3	0.5	39	31 XII-7 II	1888—89	0.7	0.5	0.2	29	21 VII-18 VIII	1883
Tsu	2.5	0.5		..	15	11-25 IX	1891	0.5	..	..	11	2-12 X	1891
Nagoya	3.0	..	..	..	9	14-22 IX	1890	3.0	..	..	13	1-13 X	1891
Gifu	5.2	0.7	0.1	..	15	8-12 VII	1884	1.5	..	0.2	25	11 II-25 III	1887
Hamamatsu	4.7	0.1	..	..	11	12-22 IX	1890	2.7	0.3	0.2	22	2-23 XII	1883
Numazu	6.2	1.4	0.2	..	17	29 III-14 IV	1885	2.4	1.4	0.7	25	20 XI-14 XII; 6-30 I	1884;91
Tokio	5.1	0.7	..	..	13	27 IV-9 V	1890	3.6	1.0	0.2	58	25 XI-18 I	1880—81
Utsunomiya	3.0	2.0	..	..	14	12-25 IX; 19 VII-1 VIII	1890;91	1.0	2.0	1.0	21	10-30 XII	1891
Choshi	11.6	0.2	..	..	12	30 IX-11 X	1890	1.4	0.4	..	18	26 VIII-12 IX	1891
Kanazawa	11.1	1.8	0.5	0.4	35	29 XI-2 I	1883—84	1.2	0.1	0.1	22	10-31 VII	1883
Fushiki	9.2	3.0	0.5	0.5	37	5 I-12 II	1888	1.0	1.0	0.1	25	23 VII-16 VIII	1885
Nagano	4.0	1.0	..	..	12	27 IV-8 V	1890	1.0	0.1	..	20	1-20 X	1891
Niigata	8.1	1.7	1.0	1.7	69	26 I-4 IV	1883	1.2	0.1	0.2	27	21 VII-17 VIII	1887
Yamagata	5.3	0.7	0.3	..	15	12-26 I	1891	1.0	..	..	13	2-14 X	1891
Akita	9.9	1.9	0.9	1.3	81	2 XII-20 II	1883—84	0.6	..	..	14	21 VII-4 VIII	1887
Fukushima	7.5	1.0	..	..	12	13-24 VII	1891	1.5	..	..	15	30 IX-14 X, 1-15 X	1889,91
Ishinomaki	4.5	0.2	..		12	15-21 IX	1889	2.0	..	..	13	1-13 X	1891
Miyako	3.6	0.4	0.2	..	18	15 IX-2 X	1881	0.7	..	..	12	2-13 VIII	1888
Aomori	9.4	2.0	0.7	0.8	38	24 XII-30 I	1884—85	1.0	0.1	..	15	4-18 IV	1883
Hakodate	9.3	1.8	0.2	0.5	50	12 XI-31 XII	1888	1.3	..	..	12	10-21 V	1890
Suttsu	5.7	1.0	0.8	1.5	46	13 XII-27 I	1890—91	1.0	0.5	..	18	3-20 VIII	1889
Sapporo	2.0	1.0	0.1	..	19	15 XI-3 XII	1887	1.0	0.3	0.2	23	31 V-22 VI	1878
Kamikawa	2.7	4.7	1.2	0.7	27	19 XII-14 I	1889—90	4.7	0.5	..	20	2-21 VIII	1890
Soya	5.5	0.8	0.2	0.2	21	8-28 I	1886	1.2	0.5	0.3	21	1-21 VIII	1890
Nemuro	4.7	0.2	..	..	10	3-12 IX	1886	0.7	0.2	..	16	4-19 VIII	1889
Kushiro	5.5	0.5	..	..	12	1-12 VIII	1890	2.5	..	..	14	16 II-1 III	1890
Erimo	5.0	..	..	..	9	18-26 IX, 12-21 XII	1890	0.7	..	..	13	4-17 III	1888

CHAPTER VII.

CONCLUSIONS.

Though our Empire consists of a group of islands on the Pacific Ocean, yet owing to the fact that it lies near to the Continent of Asia, our climate is considerably influenced by it, and the changes of climate are not so slow as in other islands, but it is as intense as in Continents. Thus the difference of the mean temperatures of summer and winter is, at our northern parts, generally very near to and very frequently exceeds 30° C. Even in the southern parts, this range of temperature still exceeds 20° C. If we take the range between the mean maximum temperature at August and the mean minimum temperature at January, it almost always exceeds 40° C.

In winter, the Continent is very much cooled, and consequently by its influence, our temperature in winter falls very low; so low, indeed, that the mean minimum temperature is below the freezing point everywhere throughout the country, except the vicinity of Okinawa (26° N). Even in the southern extremity of Kiushu, i. e. the places at about 30° N, it is no very remarkable phenomena that the daily mean temperature falls down below the freezing point. For this reason, snow falls in almost all places in this country, and even in the extremely southern part, snow sometimes accumulates on the ground. The falling of temperature below the freezing point is not only the case in midwinter, but there is no place where we have no frost, from the middle decades of November till the last decades of March.

In summer, owing to the great heating of the Continent, our temperature also rises comparatively high, though not so markedly influenced as in winter. For the heating and the cooling of the Continent governs the direction of the prevailing wind of this country, i. e. the wind blows from the Continent in winter, and toward the Continent in summer.

At the time of cooling of the Continent in autumn and winter, the atmosphere on the Continent becomes gradually denser and denser, and consequently its pressure is so increased that at last an area of high pressure is formed there. The pressure increases at great rate from the Pacific toward the Continent, and hence isobars run from SSW toward NNE bending toward the Continent, and moreover very closely packed together. But as the heating of the Continent goes on in spring and summer, the atmosphere there becomes rarer and rarer, and lighter and lighter, till at last an area of low pressure is formed. The pressure decreases gradually from the Pacific toward the Continent, and isobars run from SW toward NE, bending slightly toward east. Hence we have WNW winds of strong forces prevailing in winter, but S or SE winds in summer, whose forces are not, however, so strong as in winter. Thus we see that in our country there is a marked difference in the prevailing winds between summer and winter. These two winds replace each other, in spring at about April, and in autumn at about the end of October.

Since, as we have just mentioned, N W winds prevail in winter and SE winds in summer, and moreover the main mountain systems, running from SW to NE, divide the island into the two sides, front and back, the amounts of precipitation in summer and winter differ very much on these two sides.

In winter, the moisture on the Sea of Japan is carried landward by the prevailing wind, and meeting with the main mountain ranges, it suddenly tends to ascend, and thereby it is cooled and condensed there. By this reason, the weather of the back Nippon in winter is always gloomy; the whole sky being so darkened with dense clouds as the sunlight can not find its way through them, and snow or graupel falling constantly. Thus there is not a day without precipitation through the whole winter. Especially in Echizen, Kaga, Echiu and Echigo, whose southern sides are walled completely by a chain of high mountains, the precipitation is so remarkably great that even on the sea coast, there are generally snowfalls one metre deep; and if we leave the coast, and get to mountain feet it happens very often that snow accumulates more than 3 or 4 metres deep. Thus the moist atmosphere coming from the Sea of Japan liberates its moisture almost wholly on the northern side of the main mountain system, and it gets very dry when it crosses over the mountains and comes to the southern side. Hence in the front Nippon, just contrary to the back Nippon, clear weathers succeed day after day in winter, and it is the clearest season during the whole year. Especially in Musashi Plain (the widest plain in Japan, lying to the north of Tokio), since it has high mountains extending on its northern and western sides, the weather is particularly clear in winter, and it happens frequently that for fifty or sixty days, we have not a single drop of precipitation. If a traveller starting for Tokio from the province of Echigo, through the deep snow in winter, looks back the way from whence he came, from the top of the Mikuni-tôge, he will see that great masses of dense gloomy clouds entirely obscure his view that he can scarcely look into its depth; but if he looks on foreward, he will see a glorious prospect lying before him. No single speck of cloud breaks the azure hue of the sky. The atmosphere itself being bright and transparent, the sparkling rays of the sunshine over the high mountains and the wide plain. He could overlook all the lower country for many a miles of wide plains. He will certainly wonder at the totally different sights, which he has on the two sides.

In summer, contrary to winter, south winds prevail and hence on the front Nippon, we have cloudy weathers and abundant precipitations. Especially for about thirty days from the middle of June to the middle of July, or so called the rainy season, gloomy weathers continue, and there is more or less precipitation every day. But since south winds in summer are not so strong as north winds in winter, and moreover owing to the condition of temperature distribution, there are frequent thunder-storms and strong gusts of wind, the weather of the back Nippon in summer can not be so clear as that of the front Nippon in winter.

The amount of precipitation decreases from the southwestern part of our country toward the north-east. Thus for example, the annual precipitation in Hokkaido is about one-third of that in Kiushu. But this rule is not a fixed one, for there are many exceptions due to the configuration of the locality. The places of the most abundant precipitation in our country are, on the front Nippon, the coast of Kumano in the province of Kii, the southern coast of Shikoku, and the southeastern coast of Kiushu, and on the back Nippon, from Echizen eastward to the middle of Echigo. In almost all of these places, long chains of high mountains stand along the coast and form a barrier for the seasonal winds, so that on the front Nippon, we have enormously heavy rains in summer, and on the back Nippon considerably deep snows in winter, thus contributing great amounts to the annual precipitation.

The places of the least precipitation in this country are Setouchi and the central Nippon. These places are surrounded on all sides by mountains. Hence the prevailing winds both in winter and in summer can not find their way into the inner region without crossing over the mountains, and in so doing, the moistures, which they carry, are condensed on the outer side and left there, so that on entering the interior, they are very dry. Consequently on these districts, the precipitation is very scanty all the year round.

Of the oceanic currents near us, those that have some influences on our climate are Tsushima current and Oyashio, but not Kuroshio, as may be supposed. The warm current Tsushima enters into the Sea of Japan through the Strait of Corea and touches the northwestern coast of Nippon. Now the prevailing wind in winter is also northwesterly; hence this current serves greatly to mitigate the cold of winter in the back Nippon, and moreover increases its amount of precipitation. The cold current Oyashio flowing in the southwestern direction along Chishima touches the eastern and the southern coasts of Hokkaido, and also the eastern Nippon, and serves to lower the temperatures of those districts, especially in summer. The Kuroshio, which touches the southern coast of Nippon, though far superior both in its extent and temperature to the two preceding currents, yet its influence on our temperature is almost entirely absent. This is owing to the facts that in winter, the prevailing northwesterly wind checks the warm atmosphere over the current from coming to the land, and in summer, though south-easterly wind then prevails, the land is already sufficiently heated and thus the atmosphere on

Kuroshio has no use whatever in warming it. Though, as we have just mentioned, Kuroshio has no influence in warming the land, yet it has no small influence on its amount of precipitation. The heavy rains of summer in the front Nippon is mainly due to this current.

In short, though our country is very much influenced by the Continent of Asia in its climate, not being as gentle as on other islands, yet as it is surrounded on all sides by seas it still maintains the character of an island. Moreover, since its extent in the south and north direction is very great, and high mountains tower here and there, there is every possible variety in the configuration. If we wish for cold places, there are very cold places (as Hokkaido) and if we wish for hot places, there are very hot places (as Liukiu). And there are abundant precipitations all the year everywhere. Hence in our country, tropical plants can grow luxuriantly, and also plants in frigid zones can flourish equally well. Indeed, there is no plant in the whole vegetable kingdom that does not flourish in this Empire. With regard to climate, the Empire of Japan is really the Paradise on the whole globe.

FINIS.

PLATE I a — DIURNAL VARIATION OF AIR TEMPERATURE

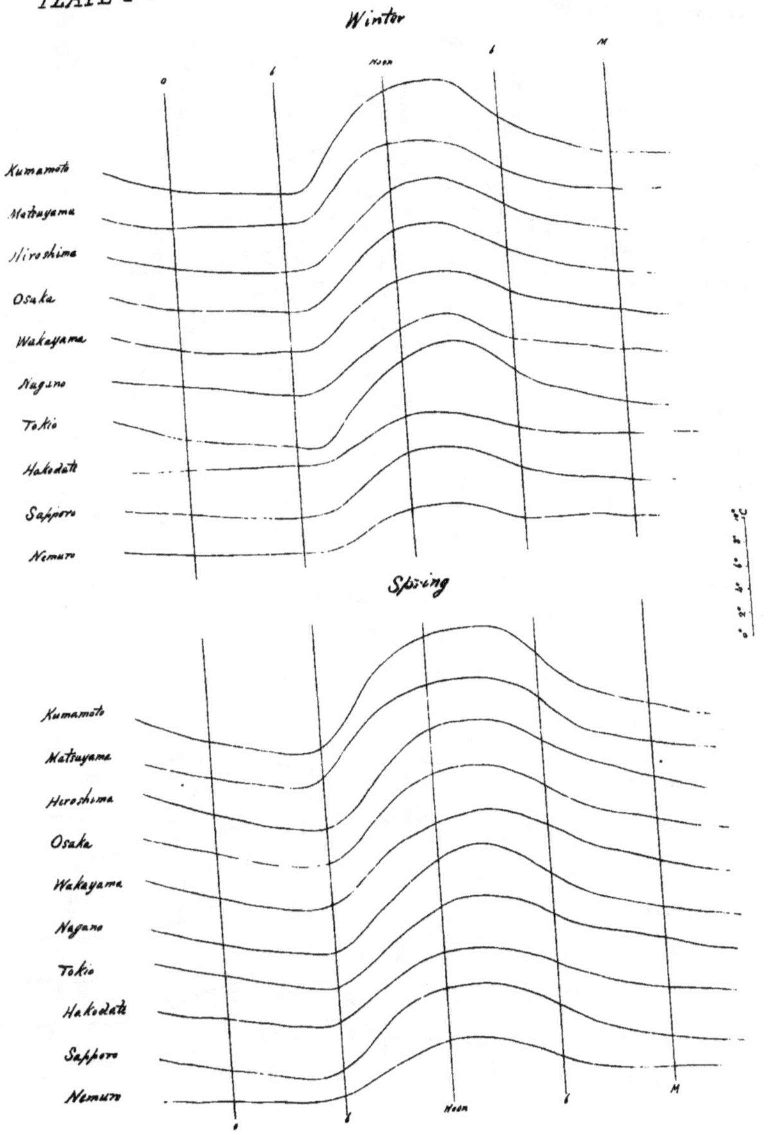

PLATE I b — DIURNAL. VARIATION OF AIR TEMPERATURE

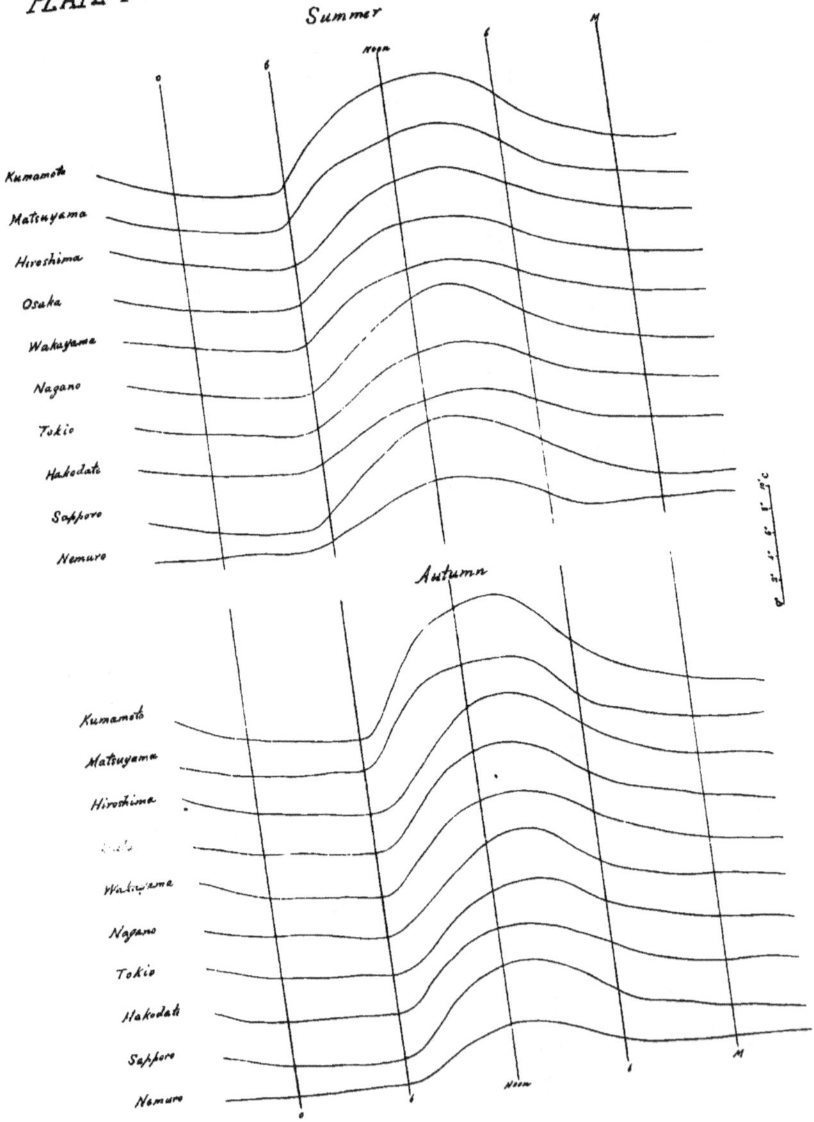

PLATE I c —DIURNAL VARIATION OF AIR TEMPERATURE

Annual

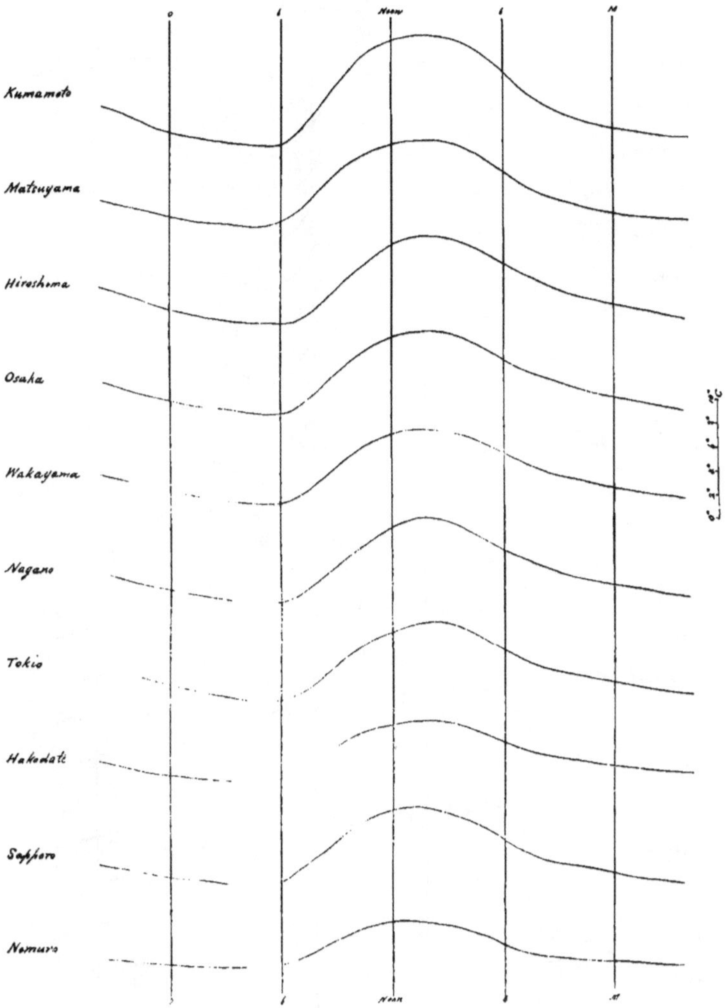

PLATE IIa.—DIURNAL VARIATION OF AIR PRESSURE

Winter

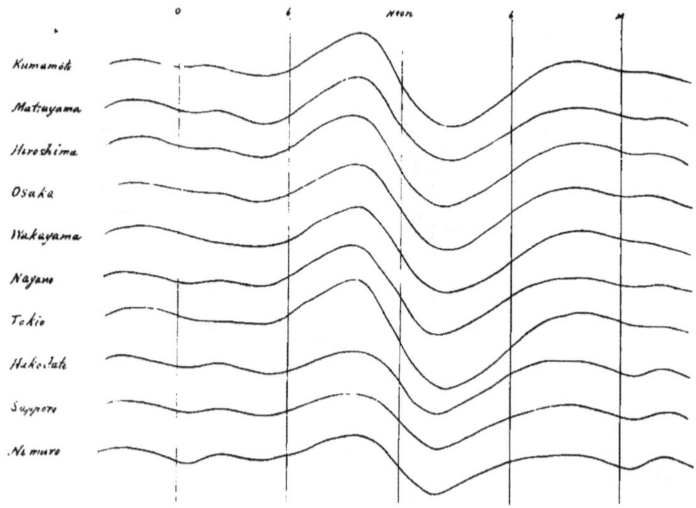

Spring

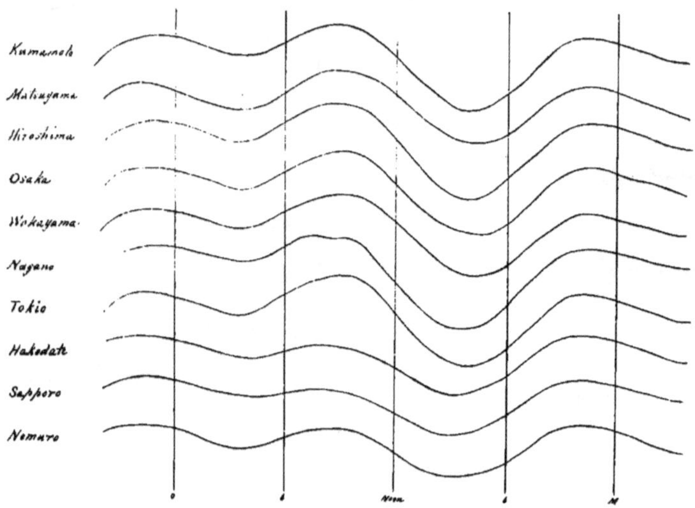

PLATE II b — DIURNAL VARIATION OF AIR PRESSURE

Summer

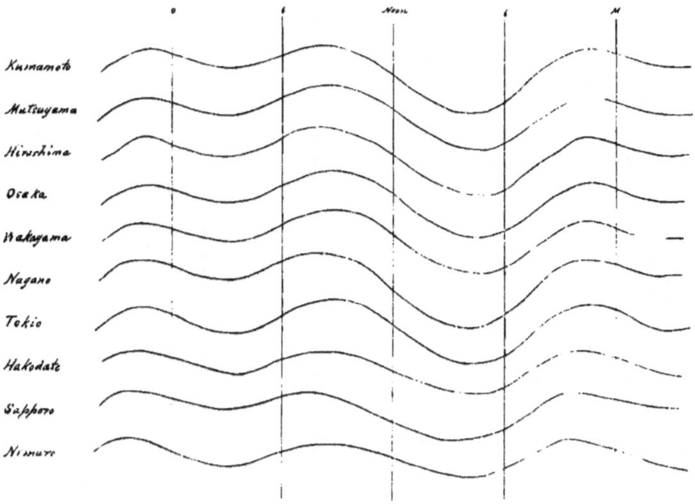

Autumn

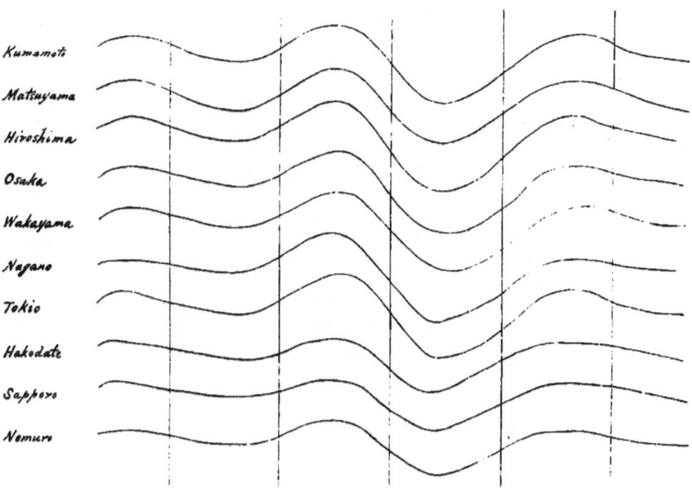

PLATE II c — DIURNAL VARIATION OF AIR PRESSURE

Annual

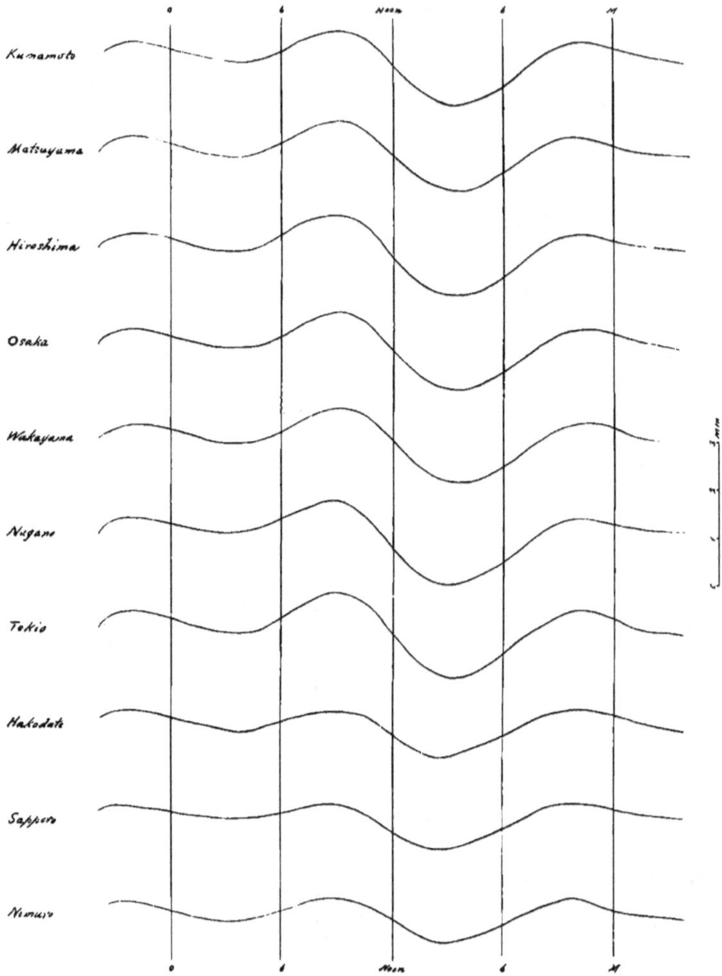

PLATE III — ANNUAL VARIATION OF AIR PRESSURE

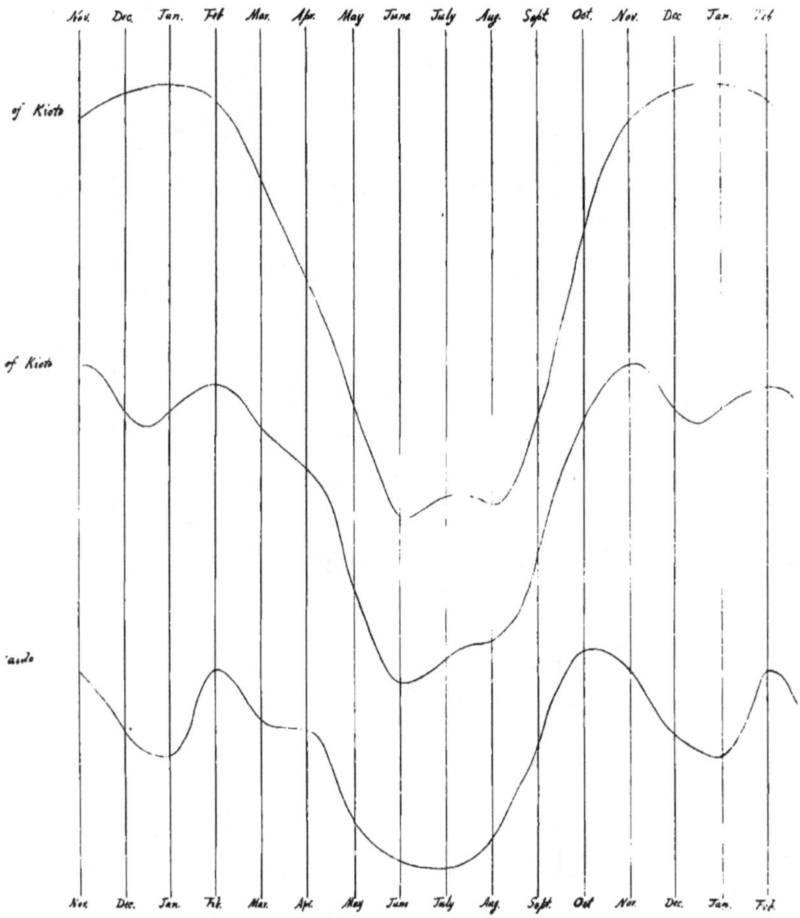

PLATE IV.a — DIURNAL VARIATION OF WIND VELOCITY

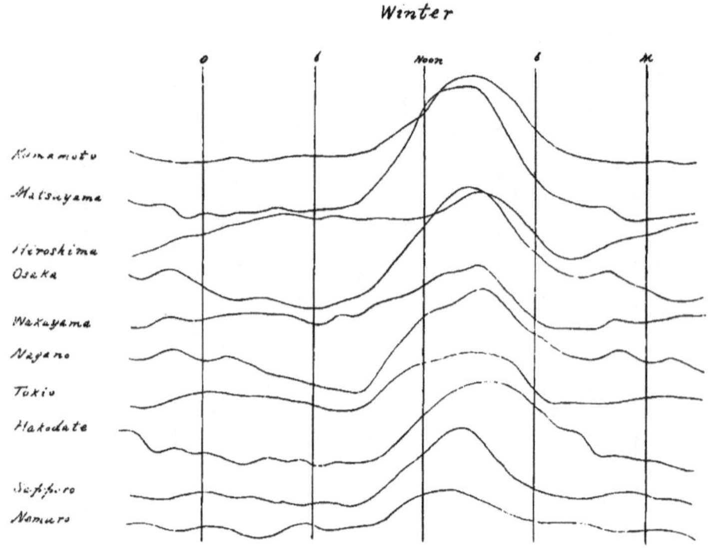

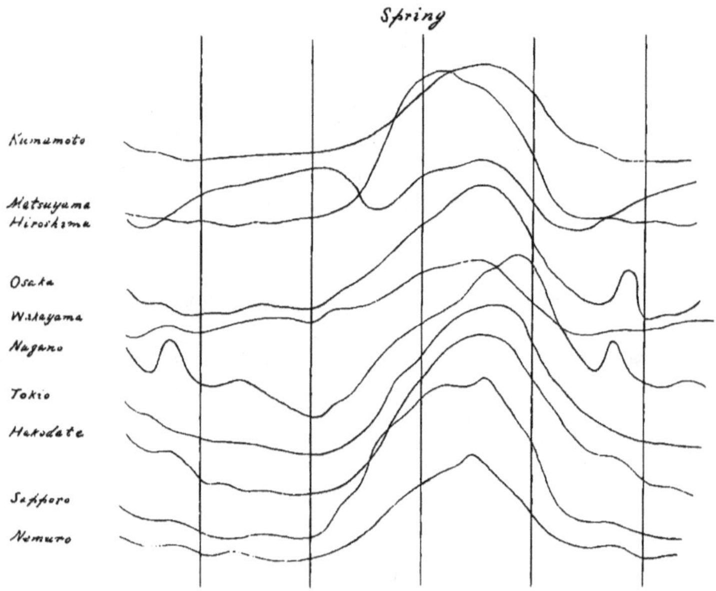

PLATE IV. DIURNAL VARIATION OF WIND VELOCITY

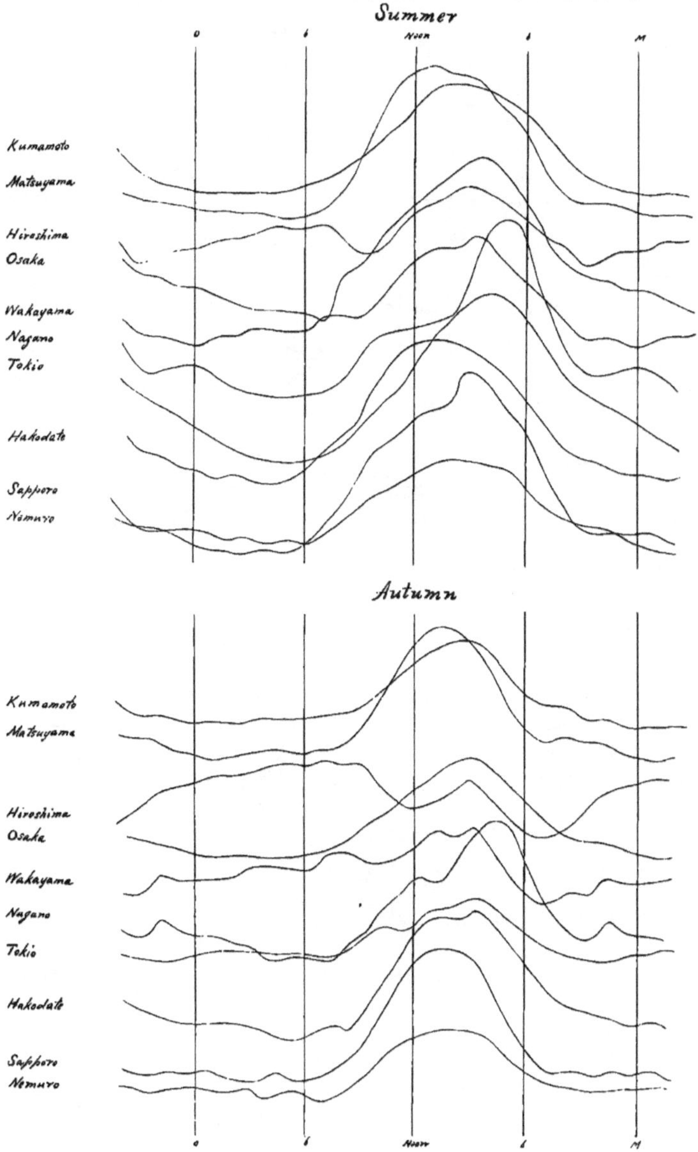

PLATE IVc — DIURNAL VARIATION OF WIND VELOCITY

Annual

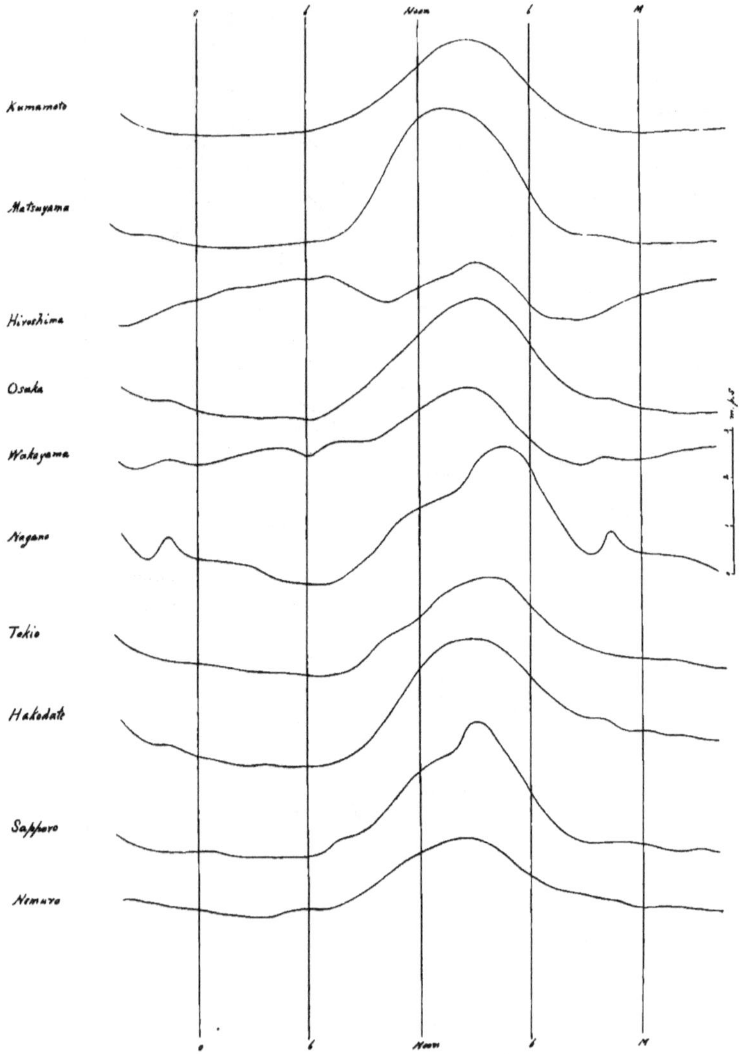

PLATE Va — DIURNAL VARIATION OF RELATIVE HUMIDITY

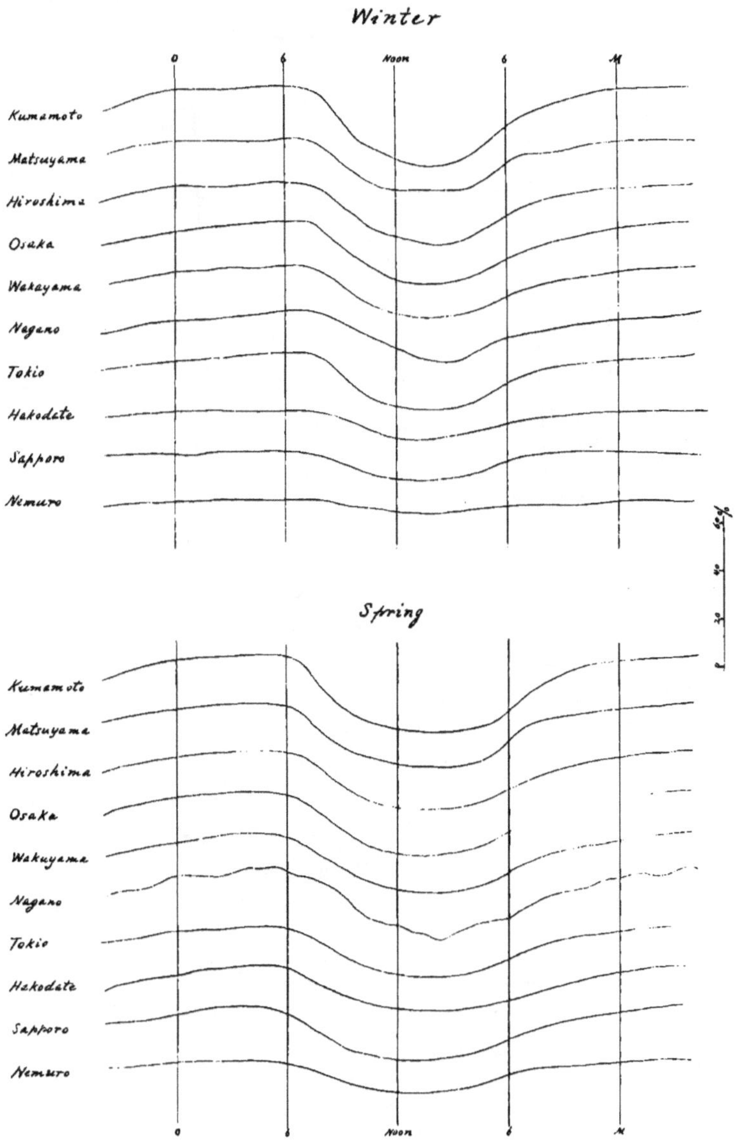

PLATE VI.—DIURNAL VARIATION OF RELATIVE HUMIDITY

Summer

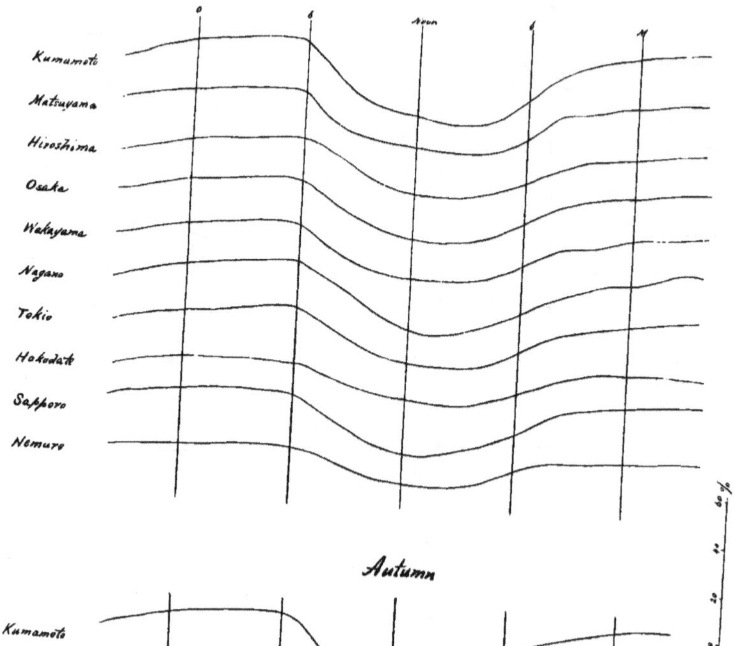

Autumn

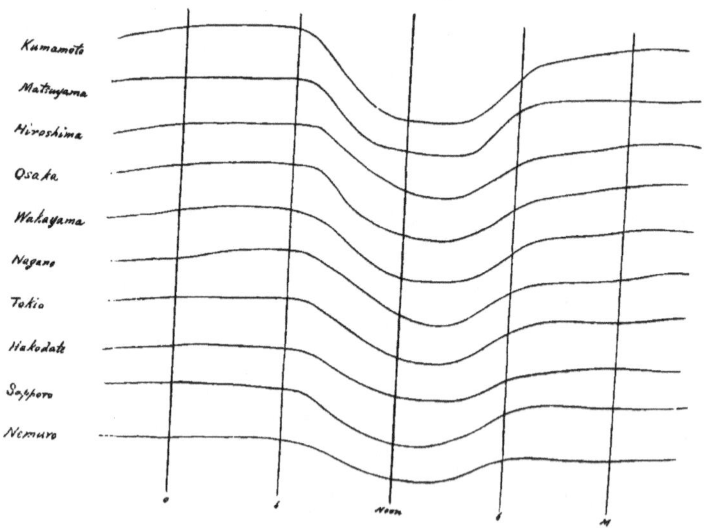

PLATE Vc — DIURNAL VARIATION OF RELATIVE HUMIDITY

Annual

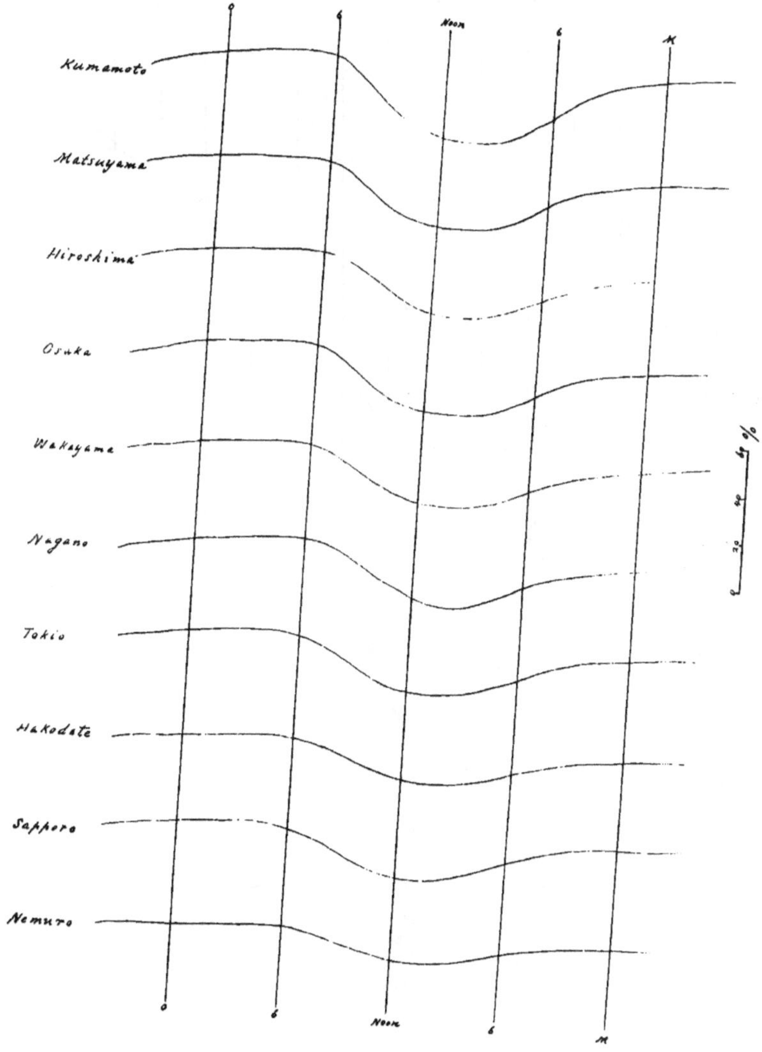

PLATE VI — ANNUAL VARIATION OF RELATIVE HUMIDITY

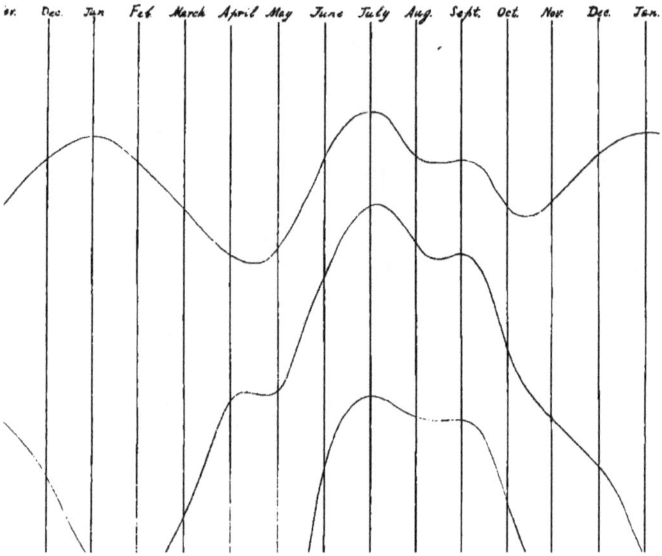

PLATE VIIa — DIURNAL VARIATION OF CLOUD AMOUNT

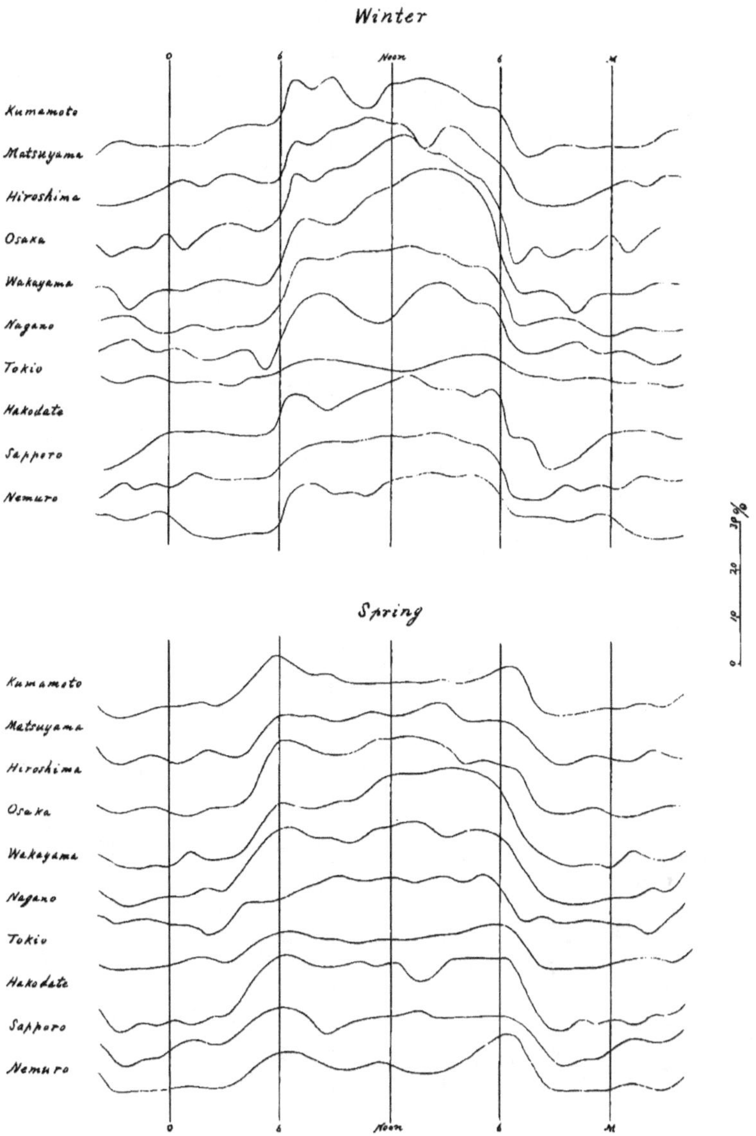

PLATE VIIt — DIURNAL VARIATION OF CLOUD AMOUNT

Summer

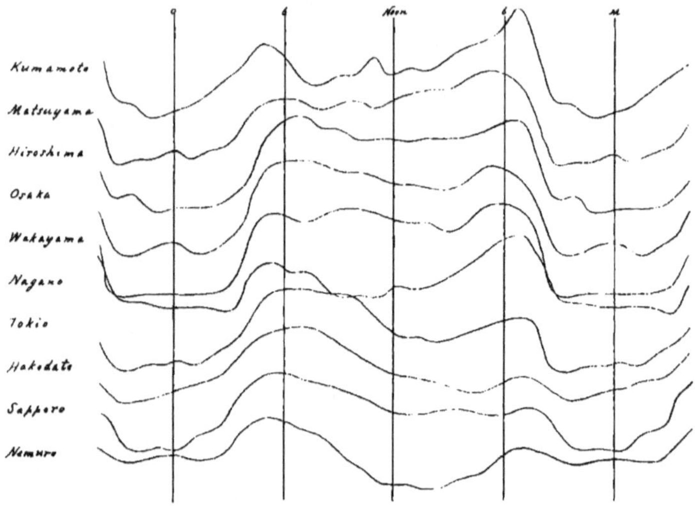

Autumn

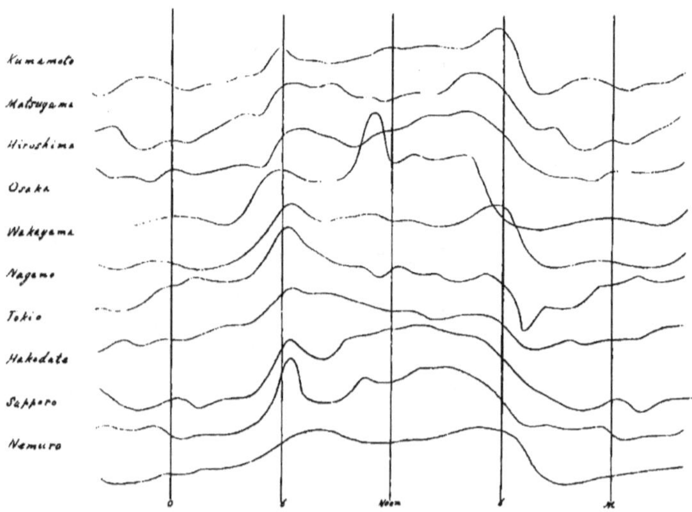

PLATE VIIIc — DIURNAL VARIATION OF CLOUD AMOUNT

Annual

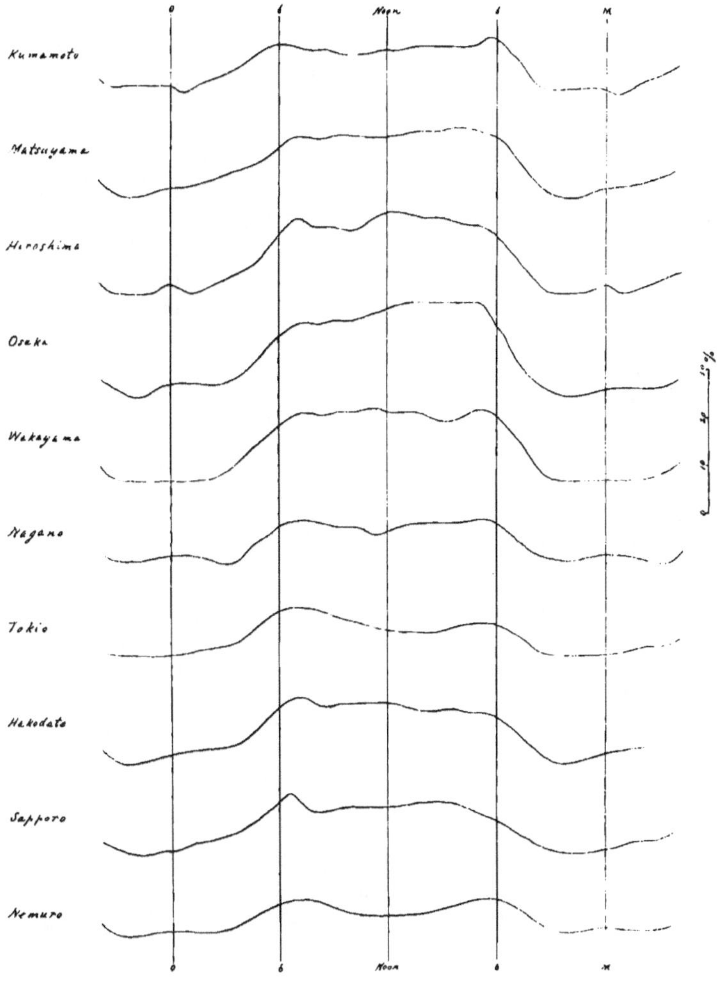

PLATE VIII — ANNUAL VARIATION OF CLOUD AMOUNT

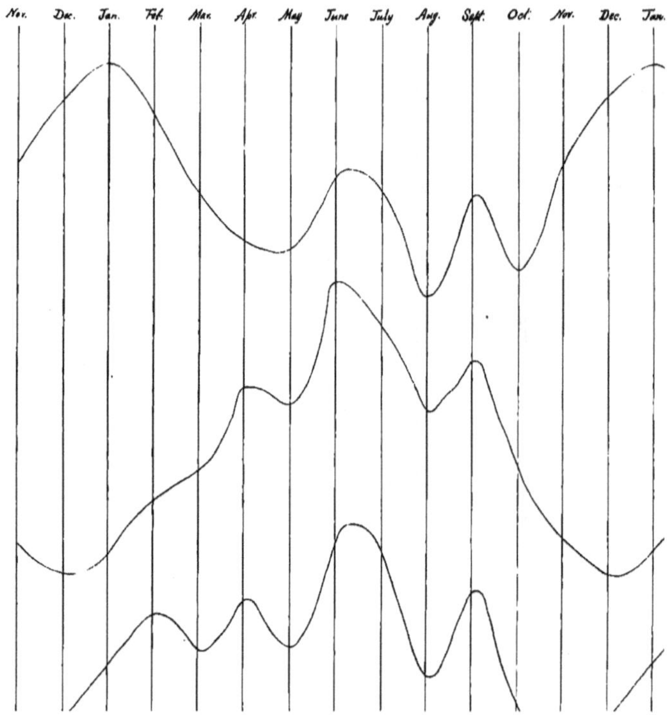

PLATE IX--ANNUAL VARIATION OF THE AMOUNT OF PRECIPITATION

Nov. Dec Jan Feb. Mar. Apr. May. June July Aug. Sept. Oct. Nov. Dec. Jan. Feb.

PLATE Xa

Isotherms.

Winter.

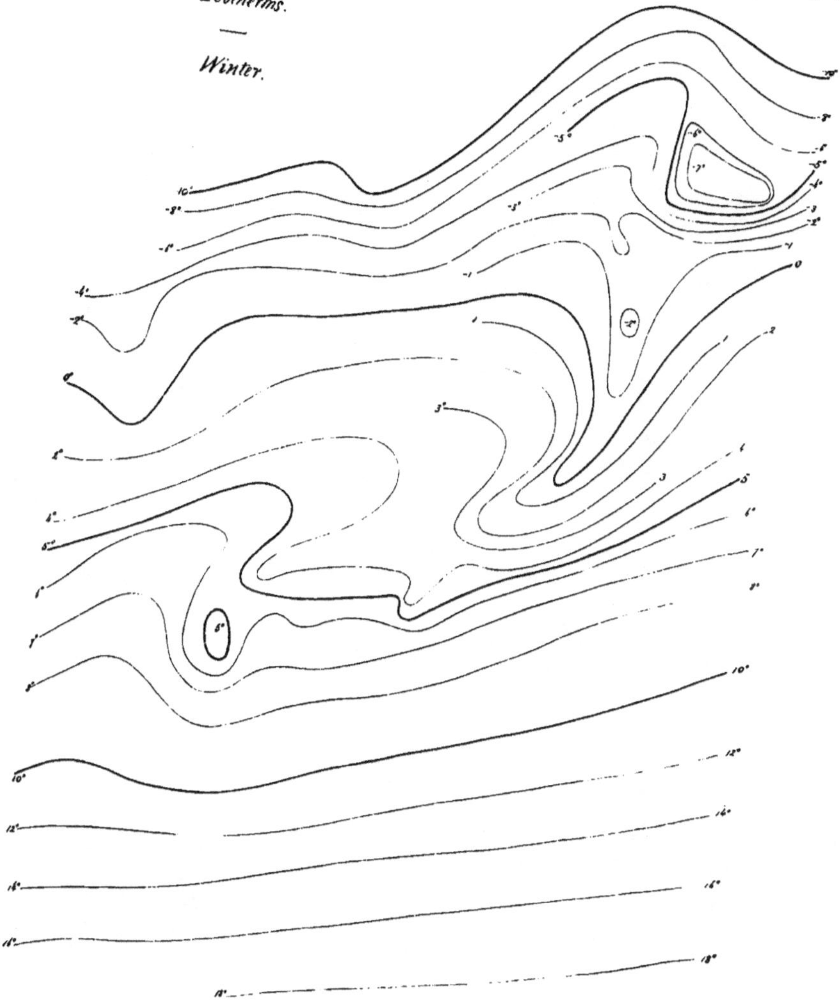

PLATE X b

Isotherms.

Spring.

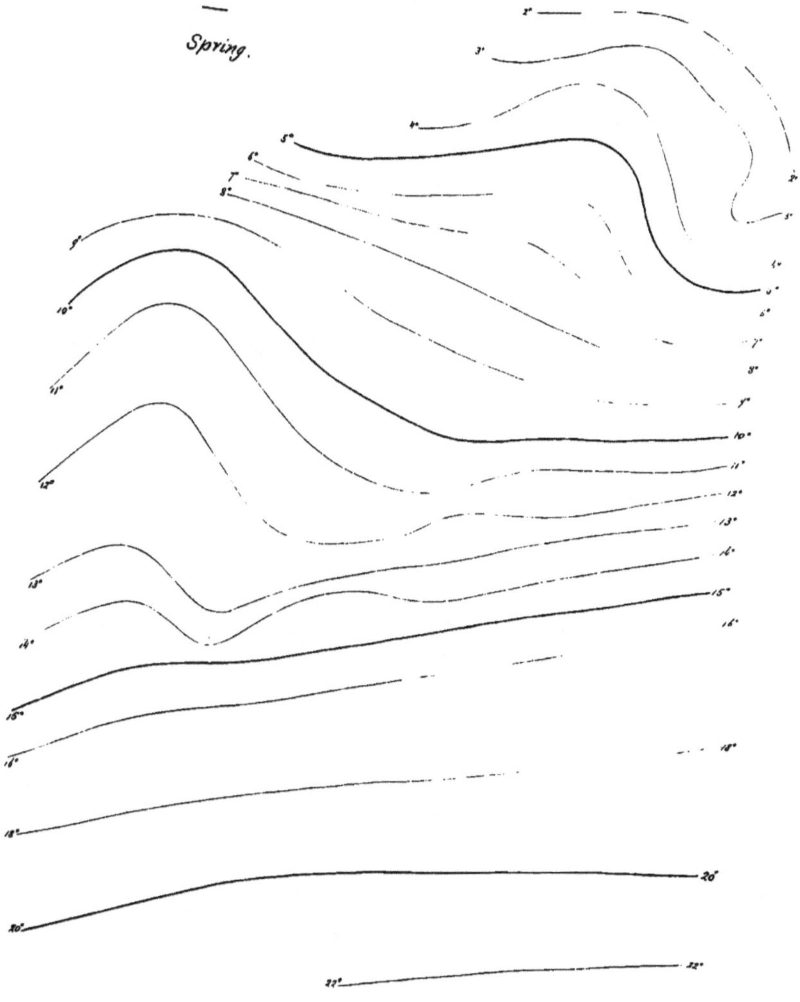

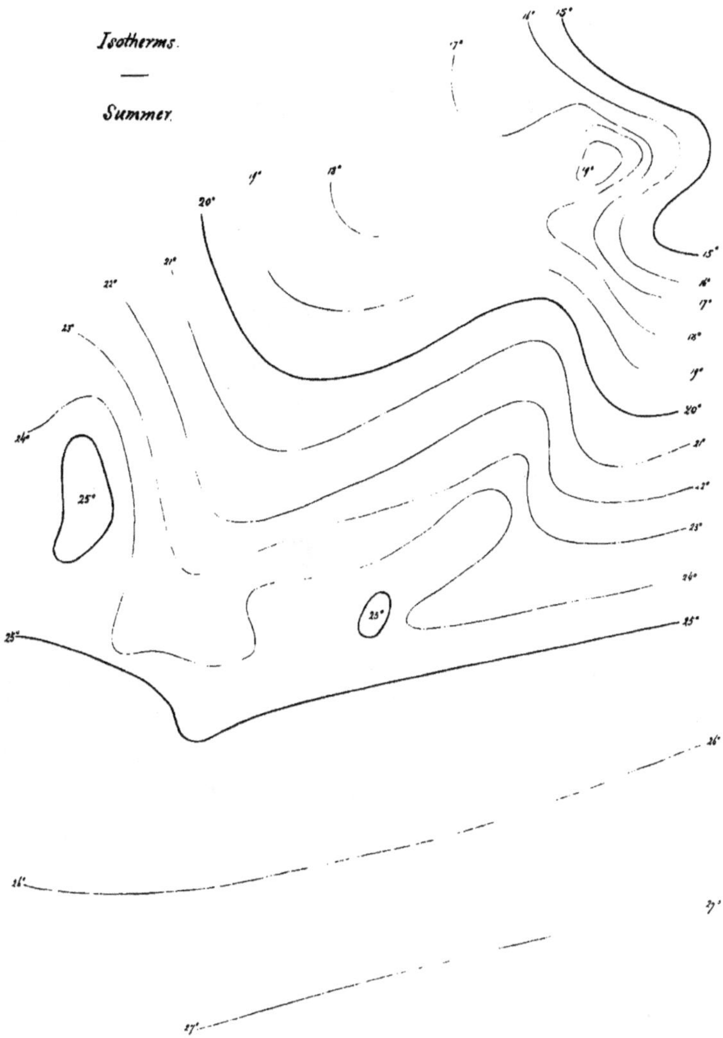
Isotherms.
Summer.
15°
20°
21°
23°
24°
25°
25°
25°
26°
27°
27°

PLATE X d

Isotherms.

Autumn.

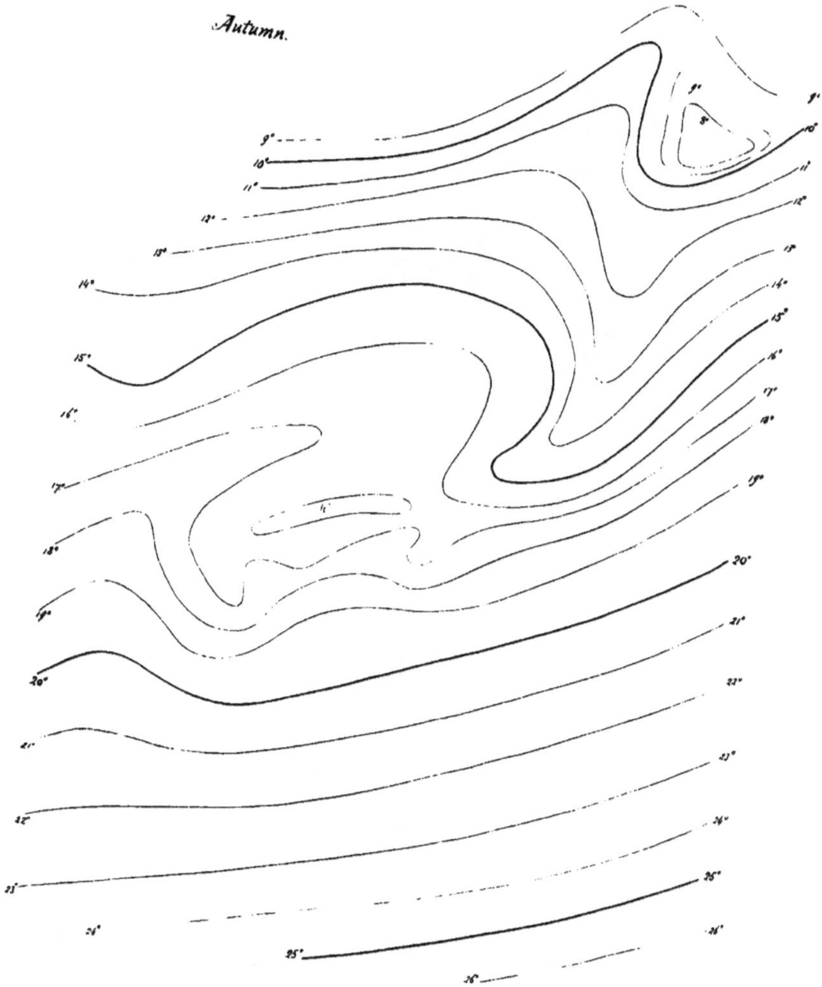

PLATE X e

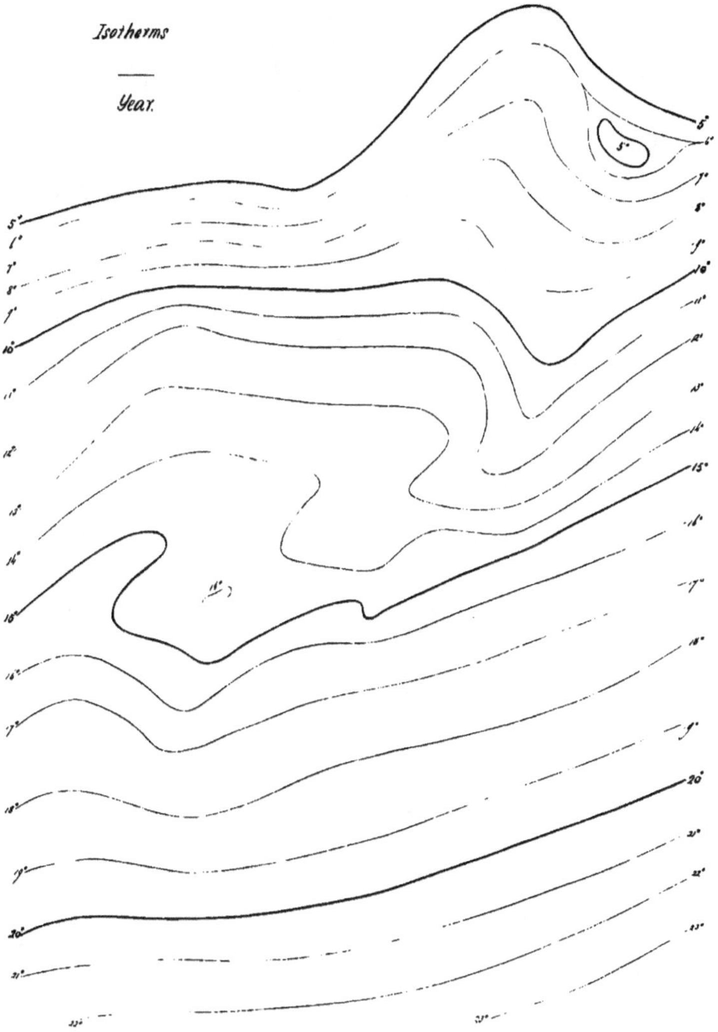

PLATE XI a

Progress of the Isotherm of 0°C toward the South.

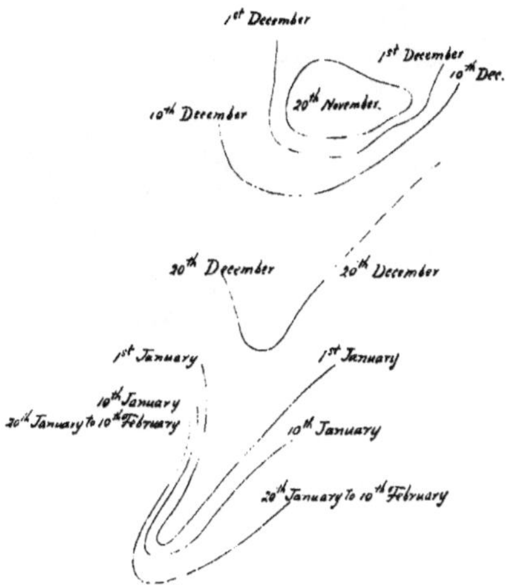

PLATE XI b

Retirement
of the
Isotherm of 0°C
toward the North

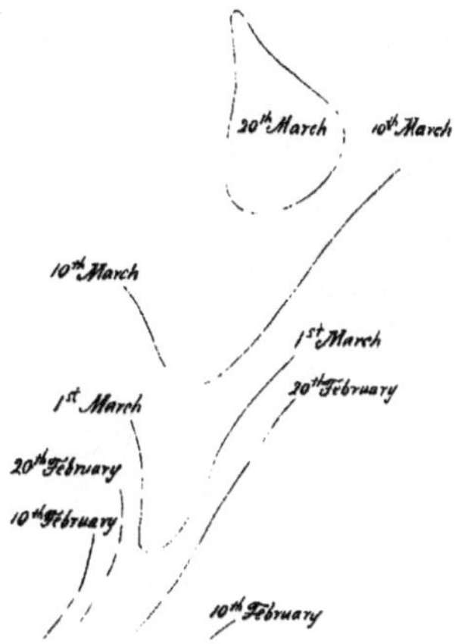

PLATE XII

Annual Amplitude

of

Temperature.

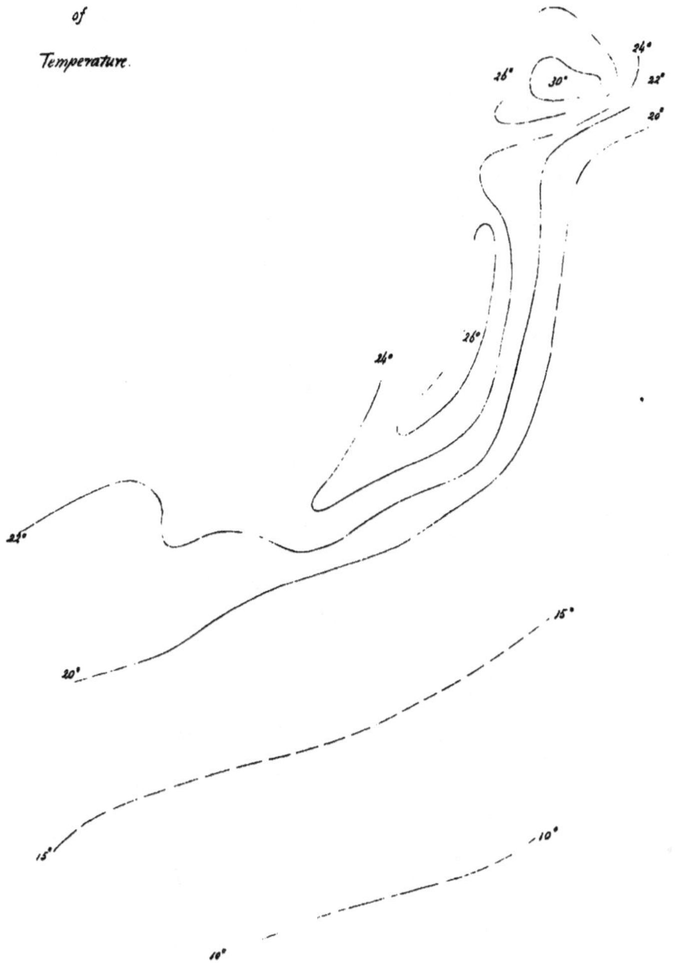

PLATE XIII a

Isanomalies of Temperature.

January.

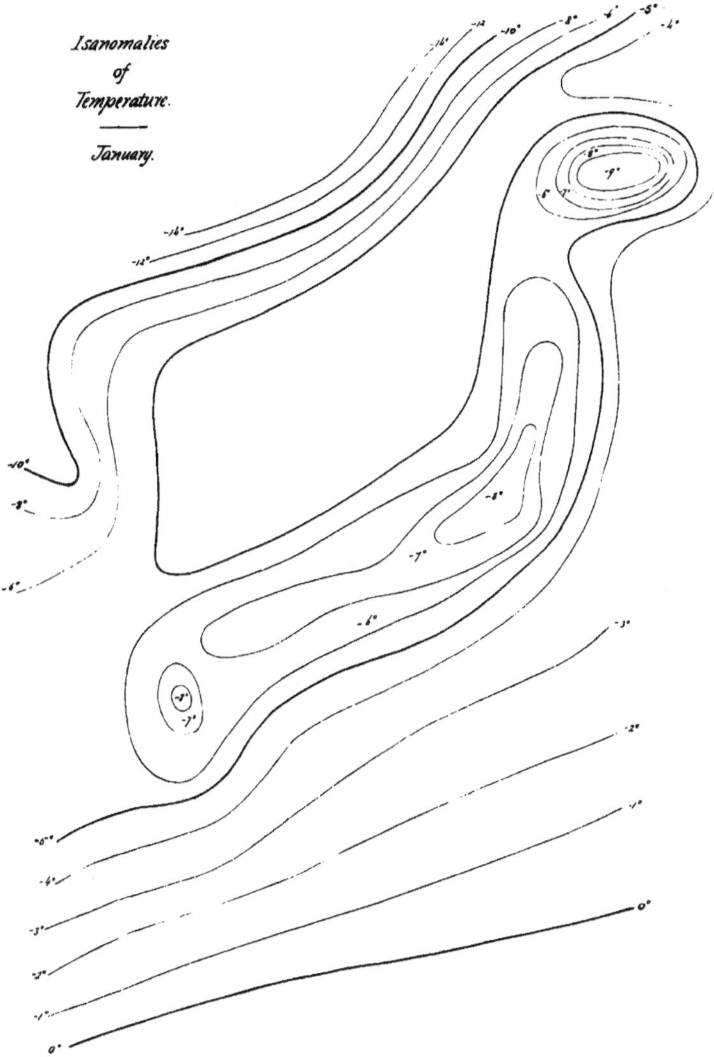

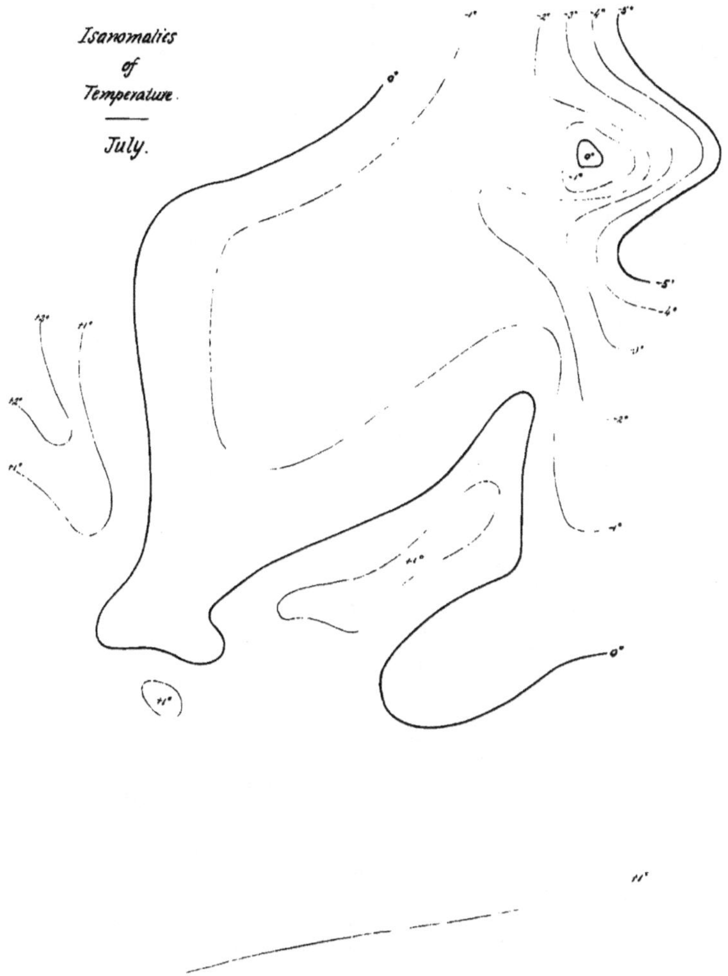
Isanomalies
of
Temperature.
July.
0°
-5°
-4°
0°

PLATE XIII C

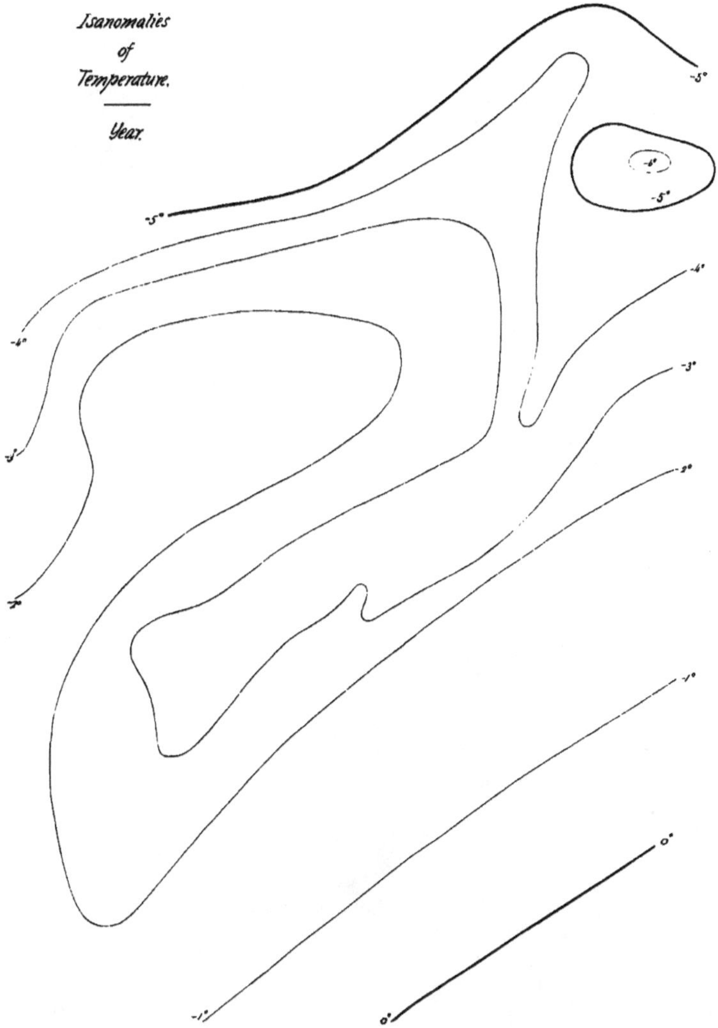
Isanomalies
of
Temperature.
Year.
-5°
-5°
-6°
-5°
-4°
-4°
-3°
-3°
-2°
-2°
-1°
-1°
0°
0°

PLATE XIV a

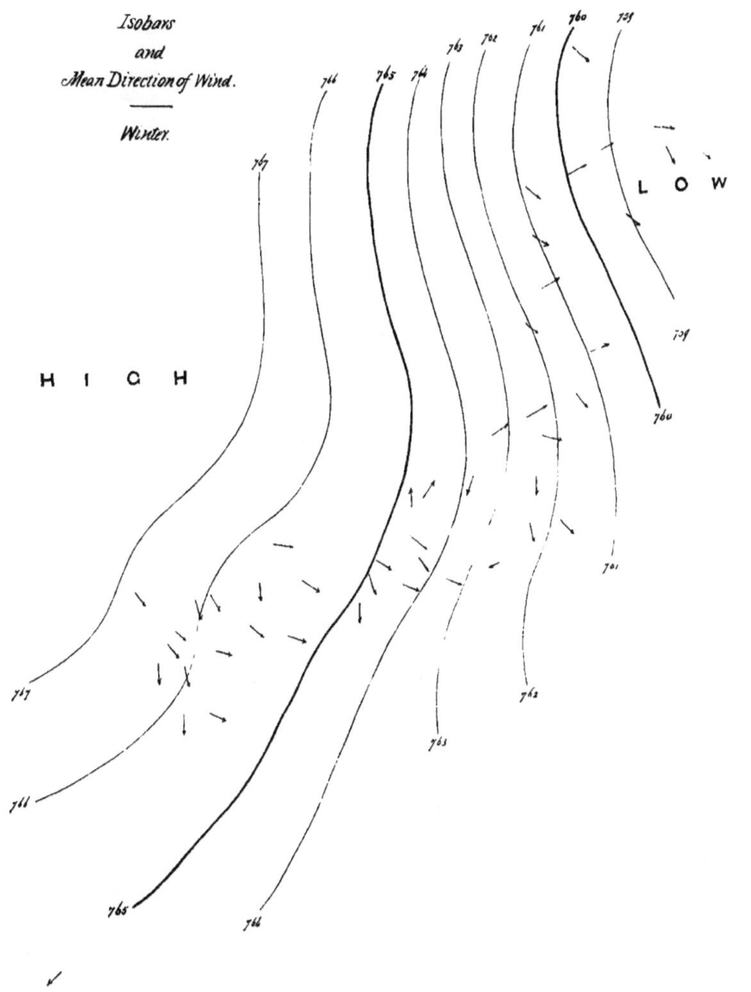

Isobars
and
Mean Direction of Wind.
Winter.
HIGH
LOW
767
766
765
764
763
762
761
760
759

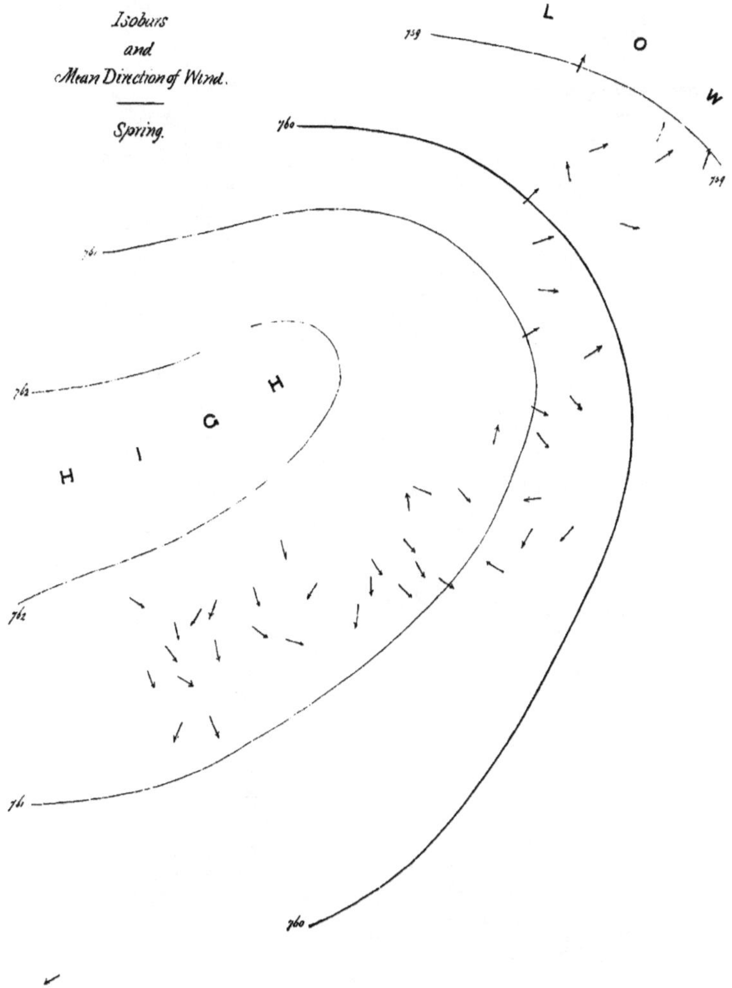
Isobars
and
Mean Direction of Wind.
Spring.
L O W
H I G H
759
760
761
762

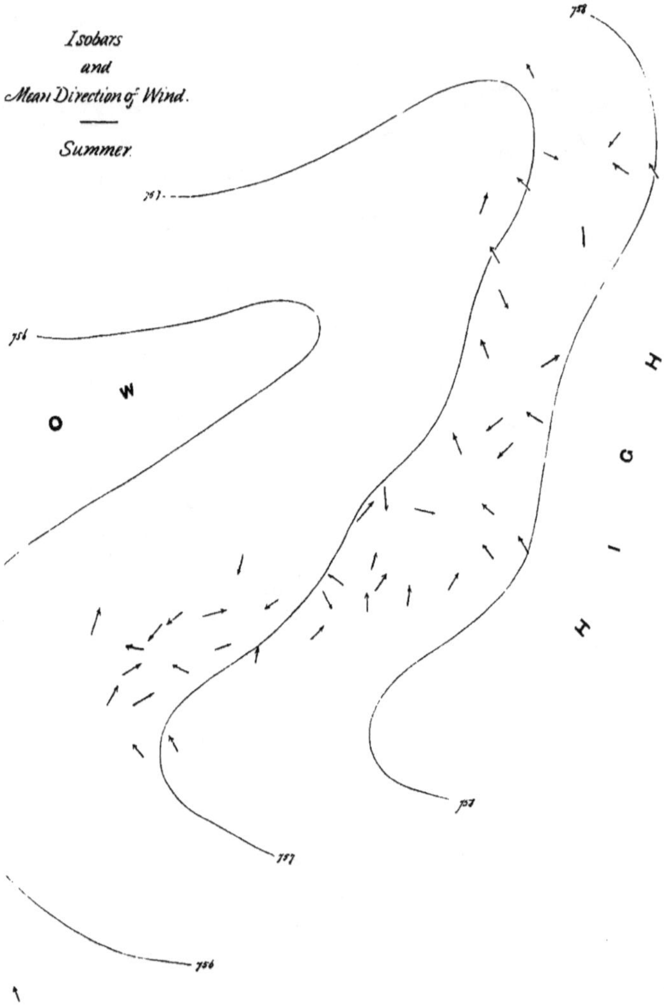
Isobars
and
Mean Direction of Wind.
Summer.
758
757
756
O W
H I G H
757
758
757
756

PLATE XIV. d

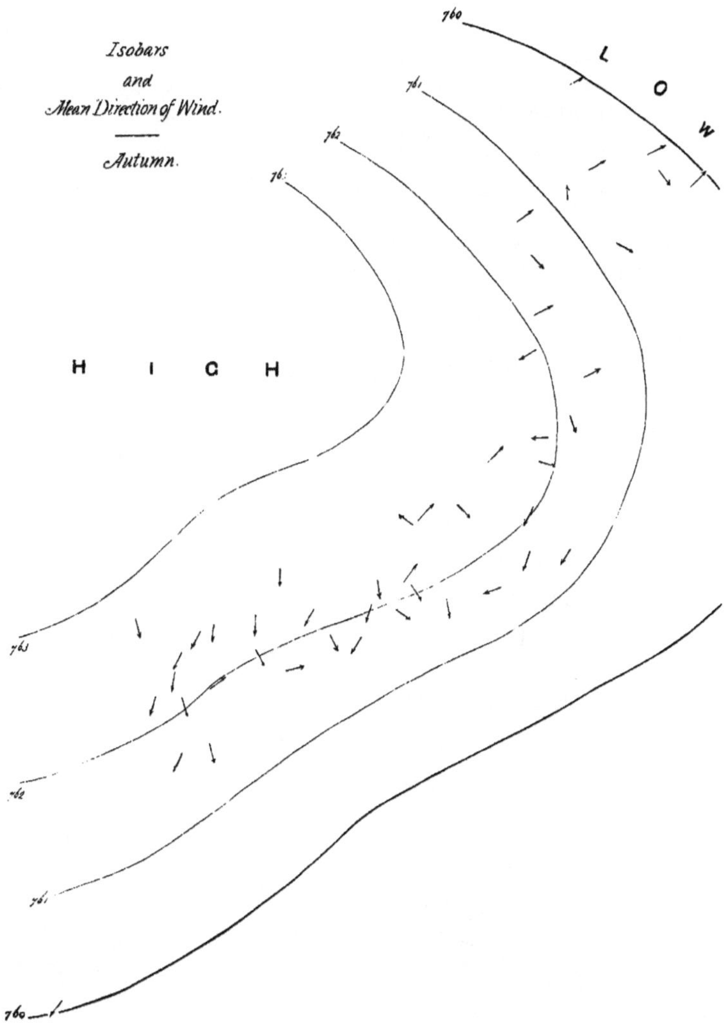
Isobars
and
Mean Direction of Wind.
Autumn.
760
761
762
763
HIGH
LOW
763
762
761
760

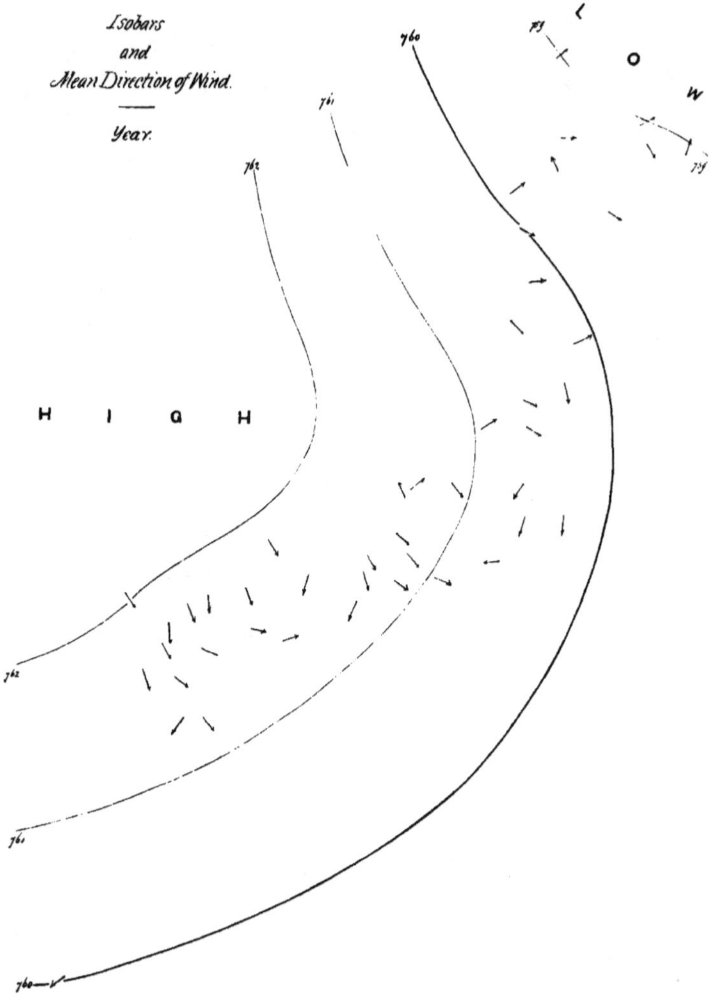
Isobars
and
Mean Direction of Wind.
Year.
HIGH
LOW
760
761
762
759

PLATE XVa

Distribution of Precipitation.

Winter.

PLATE XV.6

Distribution of Precipitation.

Spring.

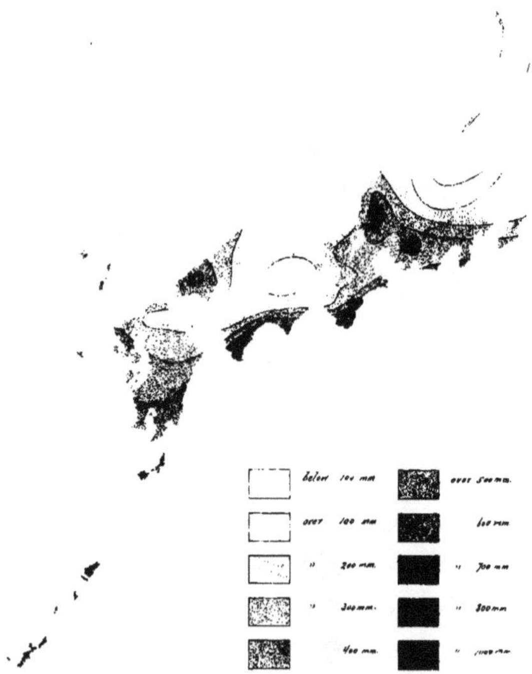

Distribution
of
Precipitation.

Summer.

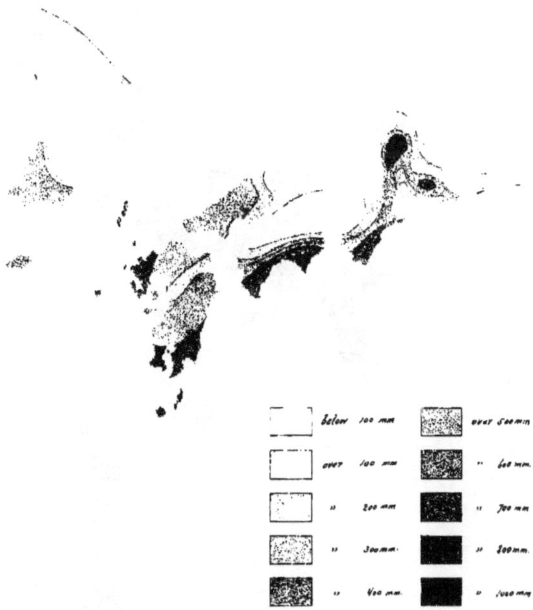

Distribution
of
Precipitation.

Autumn.

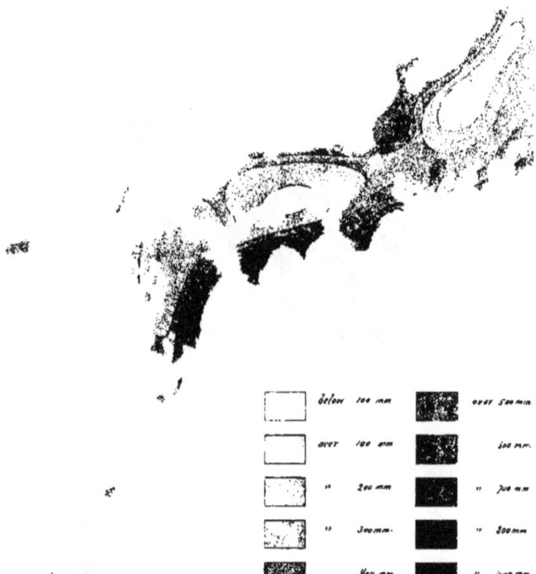

Distribution of Precipitation.

Year.

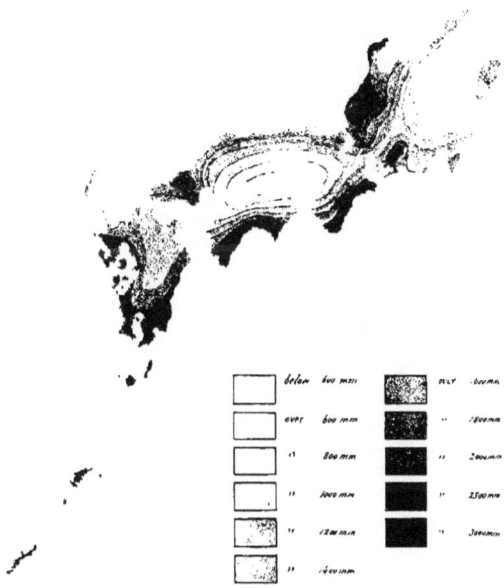